AF695802

COMPUTATIONAL BIOLOGY

COMPUTATIONAL BIOLOGY

B Thiagarajan

Professor and Head
Department of Bioinformatics
Dr G R Damodaran College of Science
Coimbatore

Pa Rajalakshmi

Lecturer
Department of Bioinformatics
Dr G R Damodaran College of Science
Coimbatore

MJP PUBLISHERS

Honour Copyright
&
Exclude Piracy

This book is protected by copyright. Reproduction of any part in any form including photocopying shall not be done except with authorization from the publisher.

Cataloguing-in-Publication Data

Thiagarajan, B. (1975-).

Computational Biology / by B. Thiagarajan. –
and Pa. Rajalakshmi. – Chennai : MJP Publishers, 2009
xxiv, 494 p.; 24 cm.
Includes Glossary, References and Index.
ISBN 978-81-8094-063-7 (pbk.)
1. Biology, Computation I. Rajalakshmi, Pa
II. Title.
570: 511.352 dc22 THI MJP 056

ISBN 978-81-8094-063-7
© Publishers, 2009
All rights reserved
Printed and bound in India

MJP PUBLISHERS
47, Nallathambi Street
Triplicane
Chennai 600 005

Publisher : J.C. Pillai
Managing Editor : C. Sajeesh Kumar
Project Editor : P. Parvath Radha
Acquisitions Editor : C. Janarthanan
Editoral Team : B. Ramalakshmi, N. Pushpa Bharathi, L. Mohanapriya, Lissy John, N. Yamuna Devi, M. Gnanasoundari
CIP Data : Prof. K. Hariharan, Librarian
RKM Vivekananda College, Chennai.

This book has been published in good faith that the work of the author is original. All efforts have been taken to make the material error-free. However, th author and publisher disclaim responsibility for any inadvertent errors.

To

Our Parents,

Teachers,

Students,

Friends and

Relatives

FOREWORD

Bioinformatics has come to stay. It was just a tiny mountain stream flowing down from the lofty peaks of Molecular Biology in the beginning. Rivulets from Structural Biology and Database Management system have conjoined with this and started the swell. Addition of several Omics has made it into a perennial river of unfathomable depth. Spate of computational programs in the field of Systems Biology is making a torrent out of it. Now it has become a sustainable science, needing well-planned academia.

Universities and Colleges have started to offer graduate and postgraduate degrees in Bioinformatics. It has extended its root down into secondary school level Biology also. In this context, the need for standard textbooks has become urgent. Perceptive teachers specialized in Bioinformatics should come forward to write books.

This book brings out the two dimensions of Bioinformatics, namely Computation and Biology, in full. The first part with Computation-related chapters and penned by Prof. B. Thiagarajan has been written elegantly, and is well-sorted with profuse examples. The Biology part of the book written by Ms. Pa Rajalakshmi is informative and comprehensive.

It is indeed a pleasurable experience to read this book. I congratulate both the authors for bringing out such an excellent book. It is undeniably a great gift to the Bioinformatics student community. They will be grateful to the authors forever.

K. Mani, M Sc., Ph D,

Reader in Botany,

PSG College of Arts and Science,

Coimbatore, Tamil Nadu

PREFACE

Computers have become part and parcel of everyone's life and the life scientists are in no way an exception. It is a fact that right from the level of school to that of research, the presence and usage of computers have become inevitable. In less than a decade, computational biology has grown to such a level that even conventional biologists have come to appreciate the application of computers in their disciplines. It has become vital and significant for the biologists to learn computers and acquire the required skills.

There is a common mindset among the people that learning biology or anything related to biosciences is difficult for a computer scientist, and vice versa. This book, for sure, will break the mind block and provide a facilitative opportunity for both the computer scientists and biologists to get into computational biology. We can arguably say that this book would sow the seeds of confidence in the minds of students learning computational biology at the undergraduate, postgraduate and research levels. We can also confidently say that this book provides the rarest combination of biology and computational biology and would definitely be an asset for students and researchers at all levels.

The book is divided into two parts—Part I has the essence of computers and programming in C language, and Part II deals with the 'omics concept in biology, which is the heart of systems biology. The interactive and conversational style of writing followed in this book would, we hope, enable readers to easily grasp the ideas delivered.

When dealing with the essentials of computers and programming in C in Part I, it is assumed that the readers are totally new to the computer science discipline and similarly when dealing with the concepts of systems biology, they are assumed to be beginners in biology. In either way, it would definitely be an easy-go for all categories of readers of this book.

We are sure that the programming examples given in the first part of the book where C programming is dealt with, along with the screenshots, will be well appreciated by the learners, especially those from the biology side. The learners will also be largely benefited once they start applying their logical and reasoning skills to answer the questions at the end of each chapter (self-learning exercises). The programs at the end of chapters 2 through 9 are real food for thought and students should work and come out with handy solutions for them.

Part II of this book focuses on the emerging areas of systems biology. While the precise definition of the term "Systems Biology" is continually undergoing refinement and is open for discussion, researchers agree that the vast quantities of data generated by high-throughput biology, and related techniques used to study living systems, ensure that computation and computers will play an ever-increasing role in biological research. This basic information on computer-based approach gives Systems Biology researchers the opportunity to investigate the ways in which various kinds of data are interrelated in order to go beyond an understanding of individual parts to the functioning of larger biological systems. Using computational tools, the computational biologist analyses, models and simulates biological systems. Biology, once considered primarily to be an observational science, has become quite quantitative.

The iterative and comparative interplay between laboratory experimentation and computer experimentation can increase research productivity, as computation guides the experiment, and the experiment is used to validate and refine computation.

In simple words, this book would certainly meet the needs and demands of both the students and teachers of computational biology.

B Thiagarajan

Pa Rajalakshmi

ACKNOWLEDGEMENTS

First and foremost, we would like to thank God for enabling us to complete this project.

We would like to express our gratitude to our college Management, especially Dr D Padmanaban, Correspondent, GRD Institutions, and Mrs Geetha Padmanaban, Secretary, GRD Institutions, for giving us a platform to prove ourselves in the field of Bioinformatics and constantly motivating us in all our endeavours.

Any work remains absurd and incomplete when not guided properly. In this regard, we are immensely pleased to express our heartfelt gratitude to our Adviser Prof. A N Lakshmanan.

We would be happy to give away a special word of gratitude to our beloved Principal Dr A Ponnusamy, who constantly supported all our initiatives and guided us from the front with his rich experience.

Dr K K Ramachandran, Vice-Principal and Director-GRDSCIB, deserves a big word of thanks from us, since he was a silent motivator for all our academic achievements and always our well-wisher.

We would like to record our sincere gratitude to our mentors in our academic life Dr K Mani and Dr N Vijayaraj.

We are thankful to Dr K Mani for writing a nice foreword to this book.

Our members of staff Ms R Subhashini, Ms K Shobana and Mr R Sridhar deserve a special appreciation for getting in touch with us every now and then regarding the development of this book. Special words of appreciation go to Mr R Sridhar, who was a student and now a colleague, for assisting us in proofreading the manuscript.

Students are always great sources of real inspiration, and we would like to attribute our growth in our career to their constant and constructive criticisms.

We personally thank Mr V Chandrasekar and Mr Imthiaz Meboob Sheriff of Visual Communication Department of our College for helping us out with graphical aspects in this project.

We would like to thank Mr C Sajeesh Kumar and the editorial team of MJP Publishers for taking up a Herculean effort in editing this project and constantly giving their opinions during the course of time.

Mr. Thiagarajan wishes to express his sincere and warm acknowledgements to his parents who gave him a good life to live and provided him the platform for achieving his goals, and his sincere gratitude to his aunt and paternal uncle, both of whom gave their heart in bringing him up as a good human being.

He also wishes to thank all teachers of his school days, with a special reference to Mr K Swaminathan, Mr S Ganapathy Subramanian, Mr K Chandrasekar and Mr S Nagarajan. Dr D Kalidoss of his college had inspired him a lot by his teaching and has been one of those catalysts for his achievements. He expresses his special thanks to Prof. S. T. Rajan, Department of Computer Science and Applications, St. Joseph's College, Trichy, from whom he had acquired most of the skills in C programming.

He also likes to express his heartfelt gratitude to Mr V Sennaraj, Deputy Registrar (Academics), Indian Institute of Space Science and Technology, Trivandrum, and his close friend and associate, for constantly providing him with his opinions and suggestions when the first few chapters of this book were scripted.

He would like to express his sincere gratitude to Dr M N Sundarraman, Former Head, Department of English, St Joseph's College, Trichy, Dr K Palaniappan, Secretary, VLB Janaki Ammal College of Arts and Science, Coimbatore, and Dr R Saravana Selvan, Reader, School of English and Foreign Languages, Bharathiar University, Coimbatore, for their constant support in his academic life.

He would also like to record his sincere thanks to his research supervisor Dr R Bremananth, Professor and Head, Department of Computer Applications, Sri Ramakrishna Engineering College, Coimbatore, for his motivation and guidance.

His wife Ms R Mahalakshmi deserves a greater word of appreciation for constantly helping him out right from the day one of this project. He would love to attribute the achievements in his life to his son Master Akshaye, who has always showed him the lighter part of this world with his smile and mischief.

Both authors would like to thank all those whose names might have missed from their mind without their knowledge, when this acknowledgement was written.

To put in a nutshell, a big THANKS to all.

B Thiagarajan

Pa Rajalakshmi

CONTENTS

PART I

FUNDAMENTALS OF COMPUTING IN BIOSCIENCES

1 ROLE OF COMPUTERS IN BIOSCIENCES

1.1 TRANSFORMATION—A BIOLOGIST TO A COMPUTER SCIENTIST AND VICE-VERSA

Earlier, people who study or do research in a particular discipline restricted themselves to that discipline. This was true with respect to the collegiate level of education or research level. But the situation at present demands everyone (students and researchers as well) to be familiar with the disciplines associated with their area of study or research. Thus the concept of being **interdisciplinary** is stressed at all levels of education. The famous quote **Jack of all trades and master of none** has been rephrased in the past one decade or so and all of us must be **Jack of all trades even if we are not master of any one**.

Knowing to work with computers is not just enough. People must be aware of what is going on at the background. This is similar to the real-life situation, where it would not be enough if one just knows to drive a car. He/she must be aware of the technicality involved in driving a car or any vehicle for that matter. Thus it is essential for a biologist to know to program with the computer science languages and a computer scientist to learn biological concepts.

1.1.1 Biology in Terms of Computers

What do you actually understood by the term **Biology**? It is just the study of things which have life in them. In the earlier biologists analysed the data that they had with microscopes and more such instruments to arrive at a conclusion. In doing so, they spent a lot of time and even though the conclusions and the results that they obtained were worthy, the **time factor** was a cause of concern. But as the technology grew, the biologists could develop their laboratories into sophisticated ones and interpret the data in a much faster way. Now, we have the best technology in our hands due to which the accumulation of data has widened when compared with their interpretation. Every biologist now has the required data at his/her fingertips and so it is easy for him/her to manipulate and interpret the data.

Bioinformatics has really added more feathers to the caps of biologists by providing them with sophisticated applications in the form of software to understand life and its central dogma. It is hardly a decade-old subject, but the contributions to this field by scientists are so much that it has scaled greater heights in less than a decade. When we talk of the contributions, we must know who the contributors are. They are none other than life scientists (biologists, chemists, physicists, biochemists and biotechnologists), mathematicians, statisticians and computer scientists. Bioinformatics can be thought of as a subset of the field of **Computational Biology**, which is an application of quantitative techniques in biological systems. What really does Bioinformatics provide to the life scientists? It provides algorithms, databases, user interfaces, and statistical tools and by providing these, it allows biologists to do wonders such as compare DNA sequences and generate significant results. The results generated are not just significant. They are potential too. The tools offered by Bioinformatics have their own limitations in the sense that they sometimes make the biologists to misinterpret the data. But we cannot deny the fact that the tools and software that are in Bioinformatics would be catalysts to increase the rate of progress of the research works.

1.1.2 How Computing has Changed Life Sciences?

We know very well that any organism's genetic information is stored as DNA, RNA and proteins. We also know that DNA is made up of four nucleotides—adenine, thymine, guanine and cytosine. RNA differs from DNA in its composition, in that thymine is replaced with uracil. Proteins are the building blocks of human life and they are formed out of 20 amino acids. All these are linear macromolecules and hence could be represented as a string of symbols. We can also compare these strings of symbols with one another for obtaining any significant results. Earlier, comparing the sequences was a Herculean task, and the most versatile technology that the biologists could use for the comparison process was just a word processor and it was also time-consuming since the sequences that were compared were aligned manually. Molecular biology has totally changed since the late 1980s due to the introduction of computer programs for comparing sequences. Nowadays, Bioinformatics has provided a variety of tools to the life scientists, such as alignment tools, phylogenetic analysis, ORF (Open Reading Frame) identification, motif identification and homology modelling. Biologists can now find on the web umpteen programs for sequence alignment alone, such as BLAST and FASTA. These tools provide the users with an interface to NCBI (National Centre for Biotechnology Information) and hence the users can access the biological databases.

As mentioned earlier, Bioinformaticians get input from a wide variety of quantitative fields such as statistics, physics, computer science and even engineering. Figure 1.1 shows how these quantitative sciences intersect with biology (one can call it as integrative biology) at different levels. Thus the main purpose of Bioinformatics is not just to develop efficient algorithms.

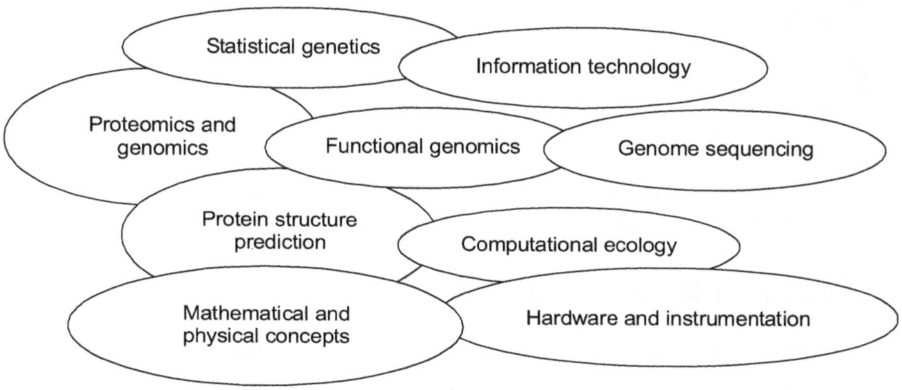

Figure 1.1 Intersection of quantitative sciences with biology at all levels

The ultimate goal is to find out how living things work. Typically all those who learn Bioinformatics land up with the idea that they must be good application developers and their efficiency would be measured based on how good they are in developing algorithms. But in reality it is not so. The fact is that it is very important to understand the biological problems as well as computational solutions in order to produce useful tools and hence the learners of Bioinformatics shall come out of the fantasy of just tool building.

1.1.3 Data Explosion and the Growth of Biological Knowledge

Data explosion was probably the main reason behind the evolution of Bioinformatics. By the end of the 18th century, over 50,000 species of the plant kingdom alone was listed out. Thus biologists were faced with the problem of how to organize, access and sensibly add to the existing list. To memorize the millions of species in the globe would be much complicated for even the most ardent biologist. The evolution of computers has provided a way to maintain and access the taxonomy of species, since this was the very first problem faced by the computer scientists, especially the Bioinformaticians. Fortunately in last 50 years or so, the computers have helped the researchers and student community to see the biological systems in increasingly fine detail. Figure 1.2 shows the rapid growth of biological knowledge in the last 20 years.

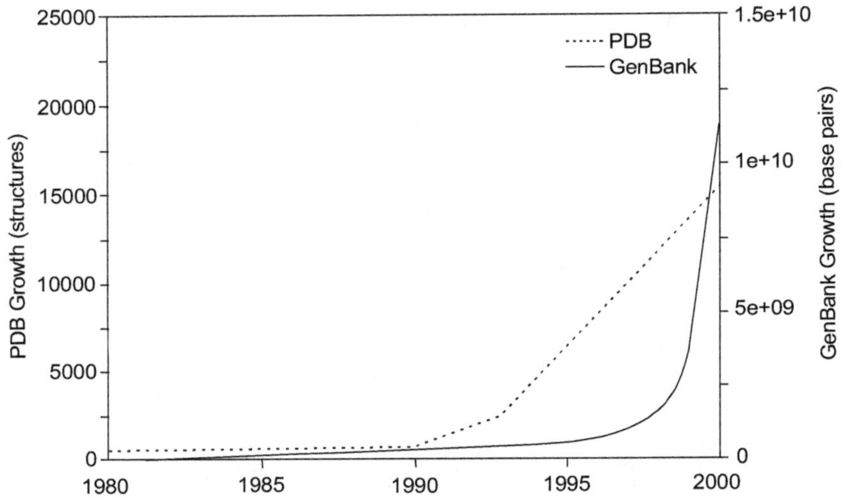

Figure 1.2 Growth of GenBank and Protein Data Bank

Now, it is possible for the biologists to easily catalogue the information gathered by them over a period of time, however voluminous the information may be and the reason for this could be attributed to the development of databases by the computer scientists.

1.1.4 What Informatics Really Means to Biologists?

The primary aspect of Bioinformatics is the representation, storage and distribution of data. The information used in Bioinformatics will be mostly in digital form. The primary job of a Bioinformatician is to concentrate on Bioinformatics infrastructure by designing efficient databases using the formatted data, create tools to query these databases and develop user interfaces that bring together different tools to allow the user to ask complex questions. The above-mentioned aspect is sometimes called as the functional aspect of Bioinformatics.

The scientific aspect of Bioinformatics deals with the development of analytical tools to discover the knowledge in the databases. In doing so, the Bioinformaticians use the biological information at different levels such as comparing sequences to develop a hypothesis about the function of a newly discovered gene, breaking down known three-dimensional structure of a protein into bits to find patterns that can help predict how the protein folds, or modelling how proteins and metabolites in a cell work together to make the cell function. The scientific aspect of Bioinformatics could be achieved only when the scientists and the researchers work together.

1.1.5 Challenges Offered by Biology to Computer Scientists

As we can witness from Figure 1.2, the data in GenBank and the Protein Data Bank are growing at an exponential rate. As such, simply managing, accessing and presenting this to the users in an understandable form is really an uphill task for the computer scientists. Human—computer interaction specialists need to work closely with academic and clinical researchers in the biological sciences to manage such staggering amounts of data.

Computer scientists should understand the fact that biological data is very complex and interlinked. They must be very strong in understanding the biological information of even a small spot on a DNA array that may mean a lot. The profile of a computer scientist must go beyond the ordinary level to understand such information. For example, a small spot on the DNA array might imply the gene location, structure and functions of genes and much more. Only when the computer scientists think from the biologist's point of view, can they easily create information systems that allow the biologists to follow such intricate biological information (quoted above) closely without getting lost in a sea of information.

Computer scientists must understand the fact that each gene in a genome is not an independent entity. Multiple genes interact to form biochemical pathways, which in turn give rise to other pathways. Thus putting the genomic and biochemical data together into quantitative and predictive models will be the work of the computational biologists (Bioinformaticians). The contributions from the computer scientists, mathematicians and statisticians will be very vital towards this effort of a Bioinformatician.

1.1.6 Essential Aspects to Become a Skilled Bioinformatician

There is a wide variety of topics that might be vital for honing yourself to be a skilled Bioinformatician. Obviously it would take enormous time to learn and master all of them. I would like to suggest some of the core skills a Bioinformatician is expected to possess.

Molecular biology is one of those areas one must be well versed in, as it can help the Bioinformatician to learn about all the essentials of life at the molecular level.

Understanding the central dogma of molecular biology is very much important. The central dogma will imply how and why the DNA transcribes to RNA and why RNA is translated to protein.

Molecular modelling and sequence analysis are the two core aspects in molecular biology that stand tall in Bioinformatics and one must be good enough in handling a couple of software related to these two aspects. This will surely facilitate you in learning other similar software.

You must feel comfortable in working in command line environment. You may gain this experience by working in the platforms such as UNIX or Linux.

Learning programming languages such as C, C++ and scripting languages such as Perl or Python is equally essential. This will surely enhance your skills in developing good and efficient algorithms and will give food for your thoughts.

Learning other advanced skills such as molecular evolution, physical chemistry (kinetics, thermodynamics and statistical mechanics), etc. will be an added advantage for the Bioinformaticians. Recall the phrase that I had used in the introductory part of this chapter, **one must be jack of all trades even if master of none** if he/she aspires to become a Bioinformatician.

1.1.7 Why Computers in Biology?

Computers are powerful devices for understanding any system that can be described in a mathematical way. Bioinformatics itself has evolved from classical biology, mathematics and computer science and hence one shall not be surprised on the use of computers in Bioinformatics. The Internet has changed how scientists share data and has made it possible for one central warehouse of information to serve an entire research community.

Consider the live example of human genome project which is going on for close to 10 years. Even though the first draft of the human genome sequence has been completed, still the project is on and we are still fine-tuning the data that had been gathered over the last 10 years. This fine-tuning would not have been possible without the technology and even with this kind of technology, the detailed mapping of the genomic data that is now being produced may take several years.

As another example, consider the techniques that are applied to X-ray crystallographic studies of proteins. The crystallographic reflections of a protein could be obtained in a few minutes instead of spending hours or days. This is possible only because of the refinement of the techniques that are applied in the field of crystallography with the advent of computers. The number of crystallized protein structures will certainly increase in number in 5 to 10 years due to the advances in the methodology applied.

Parallel computing is yet another advancement in computer science that finds application in biology at a greater level. In parallel computing, we break down the given problem into computationally tractable components. Then we solve them by employing multiple processors that work in parallel. This is a case of **divide and rule**. This approach has been fully exploited in microarray technology. This technology allows the researchers to conduct thousands of gene expression experiments simultaneously on a tiny chip.

Experimental molecular biology is fully automated nowadays and the application of information technology in the biological sciences has lead to a fundamental change in the way biological research is done. In addition to locating and studying the details of a gene, the research community has started to catalogue the data, so that it can be used by anyone at any point of time. The trend is towards storage of raw biological data of all types in public databases, with the research community having open access to it. Instead of doing preliminary research in the lab, scientists are going to the databases first to save time and resources.

1.1.8 Principles of Operating Systems

An operating system, abbreviated as OS, is a kind of software in a computer system that is responsible for the management and coordination of activities and the sharing of the resources of the computer. It is a platform on which the users of the computer rely on and interact with various applications installed in it. The operating system acts as a host for application programs that are executed on the machine. One of the main purposes of an operating system is to handle the details of the operation of the hardware. This facilitates the application programs from running smoothly without having to manage the details of the hardware components. Almost all types of computers, including hand-held computers, desktops and supercomputers use an operating system of one type or the other. The earlier models of computer, however, used an embedded OS that was written on a compact disk or a similar kind of a storage device.

Operating systems provide the users and the application programs with a number of services. The applications on the machine could access these services through Application Programming Interfaces (APIs) or commonly termed as system calls. By invoking these interfaces, the application can request a service from the operating system, pass parameters and receive the results of the operation. Operating systems are classified broadly as Command Based Interface (CBI) and Graphical User Interface (GUI). Examples for CBI are DOS, UNIX, etc. Examples for GUI are Windows, Solaris, Linux, etc. Generally GUI type of operating systems are more user-friendly compared to CBI.

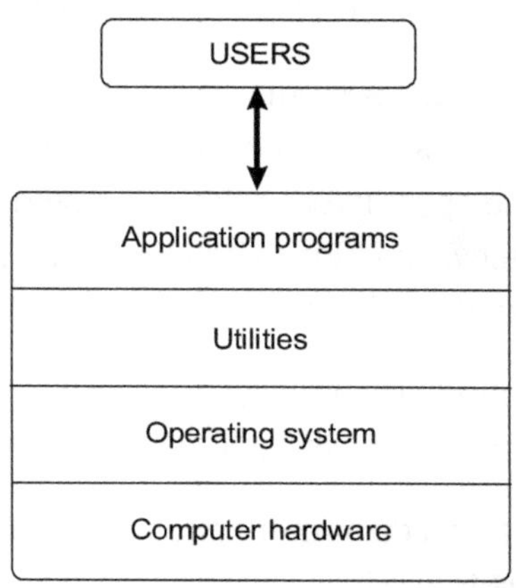

Figure 1.3 Block diagram of operating system

Examples for some of the contemporary operating systems include Microsoft Windows, Mac OS, Linux and Solaris. Microsoft Windows has a significant majority of market share in the desktop and notebook computer markets, while servers generally run on Linux or other UNIX-like systems. Figure 1.3 shows the block diagram of operating system.

Important functions of operating system

The operating system has the following important functions:

Program creation The operating system provides various facilities and services, such as editor and debuggers, to assist the programmer in creating programs. These services are in the form of utility programs that are not actually part of the operating system but are accessible through the operating system.

Program execution An operating system's most basic function is to support the execution of the programs by the users. To execute a program, a number of tasks have to be performed. Instructions and data must be loaded to the main memory. The input and output devices and the files must be made available and so on. These issues are handled by the operating systems.

Access to I/O devices Each input or output device requires its own set of instructions and control for operation. The operating system takes care of the details so that the programmer may think in terms of simple I/O statements.

Interrupts These are nothing but a kind of routines that are written and are the core part of any operating systems since they allow the operating systems to deal with the unexpected activities of the programs under execution as well as the happenings outside the machine. Interrupts allow the most significant aspect namely time-sharing and this is what most of the CPUs follow as a rule of thumb. Interrupt Service Routines (ISRs) are very common to almost all forms of computers.

Controlled access of files The operating system handles the controlled access of files by providing suitable protection mechanism. Access to the files on the disk is considered to be one of the key tasks of any operating system. The specific way files are stored on a disk is called a file system. Normally the operating systems allow the files to be stored in a hierarchy of directories or folders and the arrangement would be like that of a tree.

System access In the case of shared or public system the operating system controls the system and also to system resources. For instance, data will be protected from unauthorized users.

Error detection and response When a computer system is running, a number of errors may occur. For instance, one may get error due to the inability of a particular running application to grant the request of an application. In that case the operating system must respond to clear the error condition.

The response may be terminating the program or simply reporting the error, leaving the decision to the user.

Interpreting the commands The CPU (Central Processing Unit) has not been able to understand the commands given by the user. So the commands are translated into machine-understandable code. The command interpreter is used to do this job and it is an integral part of the operating system.

Memory management This is the key task of any operating system that supports multiprogramming (running more than one application simultaneously). In any operating system we have a kernel which is solely responsible for managing all system memory which is currently in use by programs. This ensures that a program does not interfere with memory already used by another program. Since programs follow the time-sharing concept, each program would definitely have the provision to access the memory independently.

Multitasking This refers to the running of multiple independent programs on the same machine and hence giving an appearance that it is performing the tasks at the same time. This is generally achieved through time-sharing, which means that each program uses a share of the computer's time to execute.

An operating system kernel contains a piece of software called a scheduler and this software would determine the time consumed by each of the program or application to execute. Based on that it schedules the order in which the program or application shall execute and the control has to be passed to them.

Networking Earlier the transactions between the computers was done through the external storage devices like floppy or a compact disk. Nowadays the transactions and the communication between the computers is achieved through the networks. Most operating systems that are used now support a variety of networking protocols, hardware, and applications for using them. This means that computers running dissimilar operating systems can participate in a common network for sharing resources such as computing, files, printers, and scanners using either wired or wireless connections. Networks can essentially allow a computer's operating system to access the resources of a remote computer to support the same functions as it could if those resources were connected directly to the local computer. The idea of networking has laid the basis for the Internet and thanks to its astonishing growth that we are able to share the information available anywhere across the globe.

Processor management The programs given by the users are converted into machine-readable format and are called as process. The devices used to execute the process are called processors. The processor may be hardware or software. The process may be the operating system programs or application programs. The objective of the processor is to allocate the process to the processors and manage the processors. A process can work in three states namely running state, blocked state and ready state.

Information management The management of data and information that is stored in the computing system is also carried out by the operating system. Computers in general are called as information processing systems and the operating systems are the ones doing this job. Each module in the operating system will manage the data. Each of these modules is called a file system. File system plays an important role in managing the information. It allocates the memory for the user and the I/O devices for processing the information.

Examples of operating systems

Microsoft Windows This is probably the most widely used operating system. Windows is not just used as a desktop OS but is also used on servers and supporting applications such as web servers and database servers. The most widely used version of the Microsoft Windows family is Windows XP, released in the year 2001.

In November 2006, after more than five years of development work, Microsoft released Windows Vista, a major new operating system version of the Microsoft Windows family which contains a large number of new features and architectural changes.

UNIX and UNIX-like operating systems The preliminary works for developing the UNIX OS was done by Ken Thompson, Dennis Ritchie and Douglas McIlroy at Bell Labs. They designed and developed the C programming language to build the UNIX operating system. Other operating systems that emerged from Bell Labs were Plan 9 and Inferno, which were engineered for modern distributed environments. Plan 9 was basically a networked operating system, and had built-in graphics, unlike UNIX.

The UNIX-like family is a diverse group of operating systems that consists of Solaris, BSD, Fedora, Ubuntu, Bharath Operating System, etc. and these operating systems are capable of running on a wide variety of machine architectures.

Mac OS Unlike the flavours of UNIX, Mac OS is a proprietary OS and it had the graphical user interface and it is designed, marketed and sold by Apple Inc.

1.1.9 The Best Operating System for Bioinformatics Research

In order to make the most of Bioinformatics, you need to learn UNIX, the classic operating system of powerful computers. Most of the scientific software is developed on UNIX machines and most of the serious researchers will definitely want access to programs that can be run only under UNIX. UNIX comes in a number of flavours, the two most popular being BSD and SunOS. However, a third choice for the learners is there, which is nothing but Linux. Again, there are many flavours of Linux and most of them are open source and easily downloadable. One of the reasons why UNIX or Linux is preferred over other operating systems is that it is more secure. The graphics programs and the visual programming tools that are used in Bioinformatics are designed in such a way that mostly they suit the UNIX or Linux platform. In any case, I would advise the learners to slowly migrate from Windows or any other operating system that you are presently using to UNIX or Linux environment. The main problem with the users of the computers is this migration. But the users must understand the fact that Linux is no more difficult to maintain than a normal PC operating system. You can even go in for a dual boot installation, which will pave the way for learning Linux at a normal pace and still you can have your familiar operating system and continue doing your normal day-to-day work. While learning Linux this way, you will probably start using it in a full-fledged manner and on one fine day you will get rid of your earlier operating system from your computer.

1.1.10 What Bioinformatics Could Offer Us

The questions that trigger Bioinformatics research are very much the same as the researchers have been asking in applied biology for the last few hundred years. Some of those questions are—How can we cure a disease? How can we prevent infection? How to come out with an efficient drug for a disease?

Companies that develop drugs, agricultural chemicals, hybrid plants, plastics and other petroleum derivatives, and biological approaches to environmental remediation, among others, are developing Bioinformatics divisions and looking to Bioinformatics to provide new targets and to help replace scarce natural resources.

Bioinformatics when at its best can let us know how life works at the highest possible resolution and has the ability to manipulate living things with much precision and accuracy.

1.2 EVOLUTION OF PROGRAMMING LANGUAGES

The introduction of the Von Neumann architecture was a crucial step in the development of electronic computers. The basic idea is that instructions can be encoded as data and stored in the memory of the computer. The first consequence of this idea is that changing and modifying the stored program is simple and efficient. In fact, changes can take place at electronic speeds, a very different situation from earlier computers that were programmed by plugging wires into panels. The second, and ultimately more far-reaching, consequence is that computers can process programs themselves, under program control. In particular, a computer can translate a program from one notation to another. Thus the stored-program concept led to the development of **programming languages (PLs)**.

The history of Programming Languages, like the history of any technology, is complex. There are advances and setbacks; ideas that enter the mainstream and ideas that end up in a backwater; even ideas that are submerged for a while and later surface in an unexpected place. With the benefit of hindsight, we can identify several strands in the evolution of PLs. These strands are commonly called **paradigms**.

1.2.1 FORTRAN

FORTRAN was introduced in 1957 at IBM by a team led by John Backus. It is interesting to note that, 20 years later, Backus (1978) criticized FORTRAN and similar languages as "lacking useful mathematical properties". He saw the assignment statement as a source of inefficiency: "The Von Neumann bottleneck". The solution, however, was very similar to the solution he advocated in 1957: programming must become more like mathematics: "we should be focusing on the form and content of the overall result".

Although FORTRAN did not eliminate programming, it was a major step towards the elimination of assembly language coding. The designers focused on efficient implementation rather than elegant language design, knowing that acceptance depended on the high performance of compiled programs. FORTRAN has value semantics. Variable names stand for memory addresses that are determined when the program is loaded.

The major achievement of FORTRAN is its efficient compilation. It has many limitations and a few are listed below:

Flat and uniform structure There is no concept of nesting in FORTRAN. A program consists of a sequence of subroutines and a main program. Variables are either global or local to subroutines. In other words, FORTRAN programs are rather similar to assembly language programs: the main difference is that a typical line of FORTRAN describes evaluating an expression and storing its value in memory whereas a typical line of assembly language specifies a machine instruction.

Limited control structures The control structures of FORTRAN are IF, DO, and GOTO. Since there are no compound statements, labels provide the only indication that a sequence of statements forms a group.

Unsafe memory allocation FORTRAN borrows the concept of COMMON storage from assembly language program. This enables different parts of a program to share regions of memory, but the compiler does not check for consistent usage of these regions. One program component might use a region of memory to store an array of integers, and another might assume that the same region contains real values. To conserve precious memory, FORTRAN also provides the EQUIVALENCE statement, which allows variables with different names and types to share a region of memory.

No recursion FORTRAN allocates all data, including the parameters and local variables of subroutines, statically. Recursion is forbidden because only one instance of a subroutine can be active at one time.

1.2.2 ALGOL 60

ALGOL 60 was designed by an international committee, partly to provide a programming language that was independent of any particular company and its computers. The committee included both John Backus (chief designer of FORTRAN) and John McCarthy (designer of LISP).

The goal was a "universal programming language". In one sense, ALGOL was a failure: a few complete, high-quality compilers were written and the language was not widely used (although it was used more in Europe than in North America). In another sense, ALGOL was a huge success: it became the standard language for describing algorithms. For the better part of 30 years, the Association of Computing Machineries (ACM) required submissions to the algorithm collection to be written in ALGOL.

The major innovations of ALGOL are listed below.

It was **Block Structured** in the sense that the programs written were recursively structured. The recursive structure of programs means that large programs can be constructed from small programs.

It supported **Dynamic Array** usage. The designers of ALGOL realized that it was relatively simple to allow the size of an array to be determined at run-time. The compiler statically allocates space for a pointer and an integer (collectively called a "dope vector") on the stack. At runtime, when the size of the array is known, the appropriate amount of space is allocated on the stack and the components of the "dope vector" are initialized.

The default method of passing parameters in ALGOL was **call by name** and it was described by a rather complicated "copy rule". The essence of the copy rule is that the program behaves as if the text of the formal parameter in the function is replaced by the text of the actual parameter. The complications arise because it may be necessary to rename some of the variables during the textual substitution. The usual implementation strategy was to translate the actual parameter into a procedure with no arguments (called a "thunk"); each occurrence of the formal parameter inside the function was translated into a call to this function. The mechanism seems strange today because few modern languages use it. However, the ALGOL committee had several valid reasons for introducing it, the most important being the fact that it enables procedures to alter their actual parameters.

A variable in an ALGOL procedure can be declared **own**. The effect is that the variable has local scope (it can be accessed only by the statements within the procedure) but global extent (its lifetime is the execution of the entire program).

Some of the important aspects like records were missed out in ALGOL, even though it had above contributions to programming. In many applications, the concept of **own variables** itself was problematic in ALGOL, for, it was difficult to reliably initialize them. The language was quite simple, but not powerful enough.

1.2.3 ALGOL 68

ALGOL 68, which was a successor of ALGOL 60, was much more complex in nature. The main design principle of ALGOL 68 was orthogonality: the language was to be defined using a number of basic concepts that could be combined in arbitrary ways. Although it is true that lack of orthogonality can be a nuisance in programming languages, it does not necessarily follow that orthogonality is always a good thing. The highlighting features of ALGOL 68 are listed below.

The language was described in a formal notation that specified the complete syntax and semantics of the language. The fact that the Report was very hard to understand may have contributed to the slow acceptance of the language.

Operator overloading Programmers can provide new definitions for standard operators such as "+". Even the priority of these operators can be altered. ALGOL 68 has a very uniform notation for declarations and other entities.

Like ALGOL 60, ALGOL 68 was not widely used, although it was popular for a while in various parts of Europe. The ideas that ALGOL 68 introduced, however, have been widely imitated.

1.2.4 COBOL

COBOL introduced structured data and implicit type conversion. When COBOL was introduced, "programming" was more or less synonymous with "numerical computation". COBOL introduced "data processing", where data meant large numbers of characters. The data division of a COBOL program contained descriptions of the data to be processed.

Another important innovation of COBOL was a new approach to data types. The problem of type conversion had not arisen previously because only a small number of types were provided by it. COBOL introduced many new types, in the sense that data could have various degrees of precision, and different representations as text. The choice made by the designers of COBOL was radical: type conversion should be automatic.

During the early 60's, the dominant languages were ALGOL, COBOL and FORTRAN. The continuing desire for a "universal language" that would be applicable to a wide variety of problem domains led IBM (International Business Machines) to propose a new programming language (originally called NPL but changed later due to technical reasons, to PL/I) that would combine the best features of these three languages. Insiders at the time referred to the new language as "CobAlgoltran".

The design principles of PL/I included the following.

- The language should contain the features necessary for all kinds of programming;
- A programmer could learn a subset of the language, suitable for a particular application, without having to learn the entire language.

An important lesson of PL/I is that these design goals are doomed to failure. A programmer who has learned a "subset" of PL/I is likely, like all programmers, to make a mistake. With luck, the compiler will detect the error and provide a diagnostic message that is incomprehensible to the programmer because it refers to a part of the language outside the learned subset. More probably, the compiler will not detect the error and the program will behave in a way that is inexplicable to the programmer, again because it is outside the learned subset. PL/I extends the automatic type conversion facilities of COBOL to an extreme degree.

PL/I did introduce some important new features into programming languages, but they were not well-designed and their existence encouraged others to produce better designs. For example, every variable in PL/I has a storage class, that would be static, automatic, controlled or based and some of these were later incorporated into C. Other issues with PL/I are that it allows programmer-defined types, but they could not be named and it provides a simple and not-very-safe form of exception handling.

1.2.5 Pascal

Pascal was designed by Niclaus Wirth in 1996 as a reaction to the complexity of ALGOL 68, PL/I, and other languages that were becoming popular in the late 60's. The introduction of data structuring was a highlight in Pascal. The following are some of the important features of Pascal.

The language proved the fact that a programming language could be simple, yet powerful.

The type system of Pascal was based on primitives (integer, real, bool, etc.) and mechanisms for building structured types (array, record, file, set, etc.). Thus data types in Pascal form a recursive hierarchy just as blocks do in ALGOL 60.

Pascal provides no implicit type conversions other than sub-range to integer and integer to real. All other type conversions are explicit (even when no action is required) and the compiler checks type correctness.

Pascal was designed to match Wirth's ideas of program development by stepwise refinement. Pascal is a kind of "fill in the blanks" language in which all programs have a similar structure, determined by the relatively strict syntax. Programmers are expected to start with a complete but skeletal "program" and flesh it out in a series of refinement steps, each of which makes certain decisions and adds new details. The monolithic structure that this idea imposes on programs is a drawback of Pascal because it prevents independent compilation of components.

Pascal was a failure because it was too simple. Because of the perceived missing features, supersets were developed and, inevitably, these became incompatible. The first version of "Standard Pascal" was almost useless as a practical programming language and the Revised Standard described a usable language but appeared only after most people had lost interest in Pascal.

Like ALGOL 60, Pascal missed important opportunities. The record type was a useful innovation (although very similar to the ALGOL 68 struct) but allowed data only. Allowing functions in a record declaration would have paved the way to modular and even object-oriented programming.

Nevertheless, Pascal had a strong influence on many later languages. Its most important innovations were probably the combination of simplicity, data type declarations, and static type checking.

1.2.6 Modula-2

Wirth followed Pascal with **Modula-2**, which inherits Pascal's strengths and, to some extent, removes Pascal's weaknesses. The important contribution of Modula-2 was, of course, the introduction of modules. A module in Modula-2 has an interface and an implementation. The interface provides information about the use of the module to both the programmer and the compiler. The implementation contains the "secret" information about the module. This design has the unfortunate consequence that some information that should be secret must be put into the interface. For example, the compiler must know the size of the object in order to declare an instance of it. This implies that the size must be deducible from the interface which implies, in turn, that the interface must contain the representation of the object (a similar problem appears in C++ too).

Modula-2 provides a limited escape from this dilemma: a programmer can define an "opaque" type with a hidden representation. In this case, the interface contains only a pointer to the instance and the representation can be placed in the implementation module. The most striking feature of Modula-2 was the modules with separated interface and implementation descriptions.

1.2.7 C

C is a very pragmatic programming language. Dennis Ritchie designed it for a particular task "systems programming" for which it has been widely used. The enormous success of C is partly accidental. UNIX, after Bell released it to universities, became popular, with good reason. Since UNIX depended heavily on C, the spread of UNIX inevitably led to the spread of C.

C is based on a small number of primitive concepts. For example, arrays are defined in terms of pointers and pointer arithmetic. This is both the strength and weakness of C. The number of concepts is small, but C does not provide real support for arrays, strings, or Boolean operations.

C is a low-level language by comparison with the other programming languages discussed in this section. It is designed to be easy to compile and to produce efficient object code. The compiler is assumed to be rather unsophisticated and in need of hints such as register. C is notable for its concise syntax. Some syntactic features are inherited from ALGOL 68 (for example, += and other such assignment operators) and others are unique to C and C++ (for example, postfix and prefix ++ and --, which are discussed later in the Chapter 4 of this book).

Thus, mushroom of programming languages were in practice and C has emerged a successful language and stands high among them and we will discuss the programming techniques in C from the Chapter 2 of this book.

1.3 PROGRAMMING LANGUAGES USED IN LIFE SCIENCES

There are umpteen programming languages used in life sciences and one can categorize them as scripting languages and general purpose languages. The following paragraphs would give a brief idea on what kind of role these languages play and how vital they are. There is a foundation called Open Bioinformatics Foundation (OBF), which is a working coalition of several of Bioinformatics programming language projects. Specifically, it grew out of the volunteer projects like BioPerl, BioJava and BioPython.

Currently, the most popular use of computer programming languages in the biological sciences is in the field of Bioinformatics. A frequent task needed in Bioinformatics is to find certain patterns in data files. The problem is that these data files often contain information in various formats. A solution is to use regular expressions to specify these patterns so that the information can be mined from the data files. Scripting languages are often used in Bioinformatics due to their ability to work with regular expressions.

We shall discuss some of the scripting languages and general purpose languages that are considered to be utmost essential for the Bioinformatics projects and applications.

1.3.1 Scripting Languages

BioPerl is arguably the most successful project that is part of the OBF. BioPerl code is an extensive library of core modules written in Perl to support the processing, manipulating and managing of biological information in the form of sequences. One of the reasons that the Perl programming language was chosen was that it had already gained popularity in the Bioinformatics community for its support of text processing and pattern matching task. The BioPerl project attempts to emulate the object-oriented programming paradigm through the use of Perl modules and by adhering to three design principles. The first principle is to separate the interface from the implementation. The second principle is to provide a base framework for the respective operation by generalizing common routines into a single module. The third principle is quite a complicated one and called as factory and strategy pattern. One can download the manual and working procedure from the link www.bioperl.org.

In general, high-level scripting languages are popular language choices for researchers in the field of Bioinformatics. In addition to its abilities as a scripting language, **Python** has the added support of advanced numerical capabilities through the Scientific Tools for Python (SciPy) project. The **BioPython** project was created in 1999 and is modelled after the successful Bioperl project. Most of the work of the BioPython project has focused on creating parsers for biological data and designing a useful interface to represent sequences. One of the unique features of the BioPython project is the use of a standard event-oriented parser design. One can visit www.biopython.org for more information and for getting the manual for this language.

Formerly known as **GenePHP**, the **BioPHP** project seeks to extend the PHP language so that it can be used to develop Bioinformatics applications. The main purpose of the BioPHP project is to encourage the use of PHP as a "glue" language to bind web-based Bioinformatics applications and databases. Some of the tasks that BioPHP can currently do is read biological data in the GenBank, SwissProt, FASTA, and Clustal ALN formats and perform simple sequence analysis tasks.

BioRuby is part of the OBF and is primarily supported by the Human Genome Center at the University of Tokyo and the Bioinformatics Center at Kyoto University. One of the advantages of using Ruby for Bioinformatics is that it has native support for object-oriented programming with a simple but powerful syntax. In addition to the BioRuby project, several educators have advocated the desirability of introducing undergraduate students to the Ruby programming language.

1.3.2 General-Purpose Languages

Sometimes, researchers in the biological sciences will choose a programming language based on its general use and popularity. Below is a list of several projects that use traditional high-level general purpose programming languages.

The Laboratory of DNA Information Analysis at the University of Tokyo has produced an open source C library of the most commonly used clustering algorithms. Clustering routines are used to analyse gene expression data. The library, as written, is callable from any program written in C or C++. Extensions have also been developed to allow these programs to be used in Perl and Python programs.

BioJava is an open source Java library and part of the OBF. The BioJava project is primarily concerned with how to represent sequences. The major feature of the BioJava project is that two unique representation schemes for sequences have been defined in the Java programming language.

One of the more disappointing aspects of examining programming languages and how they are being used in the biological sciences is to find a project that seems promising but is ultimately a disappointment. Squeak is an open source implementation of the object-oriented programming language Smalltalk. With it's built-in graphics and truly object-oriented paradigm, Squeak would seem an ideal language for the biological sciences. In fact, the BioSqueak homepage seems to make such promises. However, there is no evidence that any work has actually been completed on the BioSqueak project.

Haskell is a general purpose, purely functional programming language. Haskell has been used to implement dynamic programming algorithms, including those involved in RNA folding grammars. The group claims that their work adds a significant amount of flexibility and versatility in the development of new dynamic programming algorithms.

LISP, formerly known as BioLingua, is a web-based programming environment that enables biologists to analyse biological systems by combining knowledge and data through direct end-user programming. The goal of BioBike (an offspring of Lisp) is to enable biologists to program directly by providing an environment that is more natural to trained biologists. The main BioBike language is BioLisp. BioLisp is simply common Lisp with added biological functionality.

Biomedical Logic Programming (Blip) is a collection of logic programming modules intended primarily for Bioinformatics and biomedical applications. Blip allows users to program in a declarative way and facilitates both query-oriented and application-oriented programming.

While not technically a programming language, no discussion of biological oriented programming would be complete without a brief discussion of XML. When dealing with massive amounts of data in different formats, there must be a way to exchange this information from one application to another. XML is an extensible, universal format for structured data exchange and documents on the web. Two of the most notable attempts to use XML as a framework in biology have been the Bioinformatics Sequence Markup Language (BSML) and the BIO polymer Markup Language (BioML).

1.4 FUNDAMENTALS OF NUMBER SYSTEMS

Numbers are fundamental to everything in computers. A computer scientist shall have a good knowledge in subjects like mathematics, statistics and digital electronics. These three fields have a great influence on Computing and one cannot become a good programmer without having a brief knowledge on these. In this section, we will be discussing the different number systems and their significance and inter-conversions, which will facilitate the programmer to have a clear idea when he/she writes programs.

1.4.1 Categories of Number Systems

Number systems can be broadly categorized into two, namely weighted/positional and non-weighted/ non-positional. This is based on the weight/position assigned for each of the digits in the number. Each digit will have weight/position, which is unique to itself.

Weighted/positional number systems are further classified into decimal, binary, octal and hexadecimal numbers. A good in-use example for non-positional number systems is the Roman numerals (All the number systems use what we informally call in mathematics as Arabic numerals). Our discussion will be only on the weighted/positional number systems.

1.4.2 Representation of Numbers in Computers

Computers store information in the form of numbers and that too it prefers storing in binary or hexadecimal format. No computer would possibly understand any language that is being spoken by us and it understands only the electronic signals, which are 0 and 1. Hence this section of the chapter would deal with the number systems that are important from the point of view of a novice programmer. We would consider the number systems decimal, binary, octal and hexadecimal for our discussion in the forthcoming sections and we would also learn how to convert from one system to another.

1.5 COUNTING NUMBERS IN DIFFERENT NUMBER SYSTEMS

1.5.1 Decimal Numbers (Denary Number System)

Let us begin the discussion with the decimal numbers. In Greek, 'deca' means 'ten' and hence there are ten digits in this system of numbers and the counting would begin with '0' and proceeds as 1, 2, 3, 4, 5, 6, 7, 8 and 9 (this is the first level of the decimal numbers which is very often called as single digit numbers). Once we reach '9', the next number is formed with the help of the second digit of the sequence '1', by combining it with all the other digits and hence we get 11, 12, 13, 14, 15, 16, 17, 18 and 19 (this is a part of the second level of the decimal numbers or two-digit numbers).

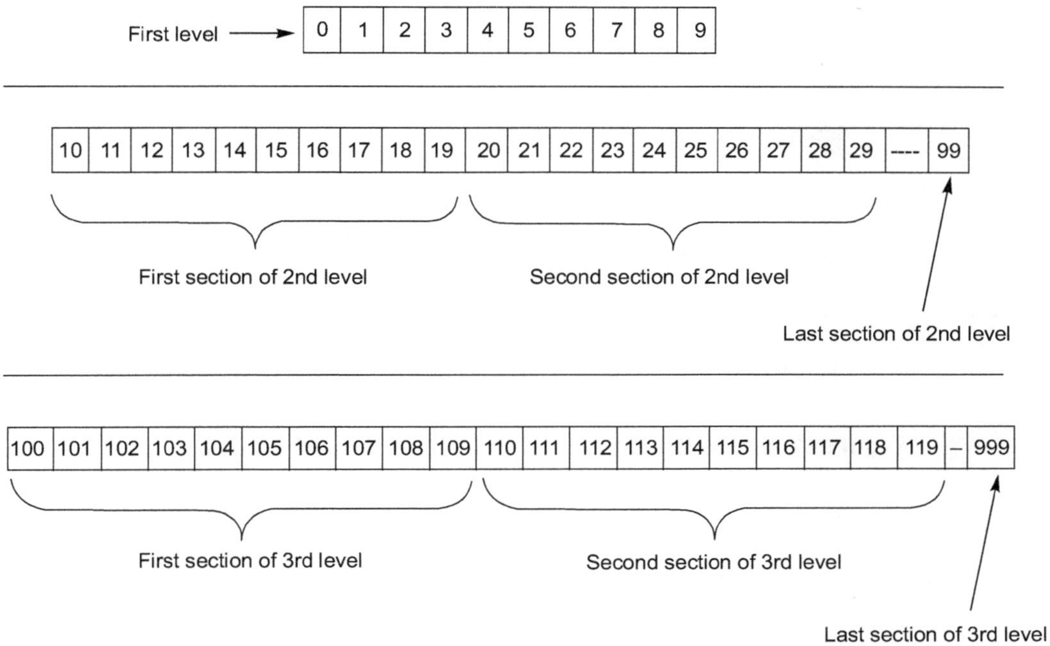

Figure 1.4 Levels of counting in decimal number system

This process continues till all the digits in the first level are used to get the complete set of second level of decimal numbers. Once we reach 99, we are said to have completed the second level of the decimal numbers. With the help of all the numbers in the second level and combining with each of those in the first level, we get the third level of the system (three-digit numbers). Like this, the process continues without any end. Figure 1.4 shows a few of the levels of counting in the decimal number system.

1.5.2 Binary Numbers

In Greek, 'bi' means 'two' and hence we would have only two digits in this number system and they are 0 and 1. We would follow the same procedure for forming the other numbers in this system, given the first level of the system (0 and 1). We start with 0 and 1 in the first level (single digit numbers) and get the second level by suitably combining the digits of the first level and hence we would get the two new numbers 10 and 11 (two-digit numbers). We proceed in the same way and the process would be endless as the case with the decimal number system. Figure 1.5 shows a few of the levels of counting in the binary number system.

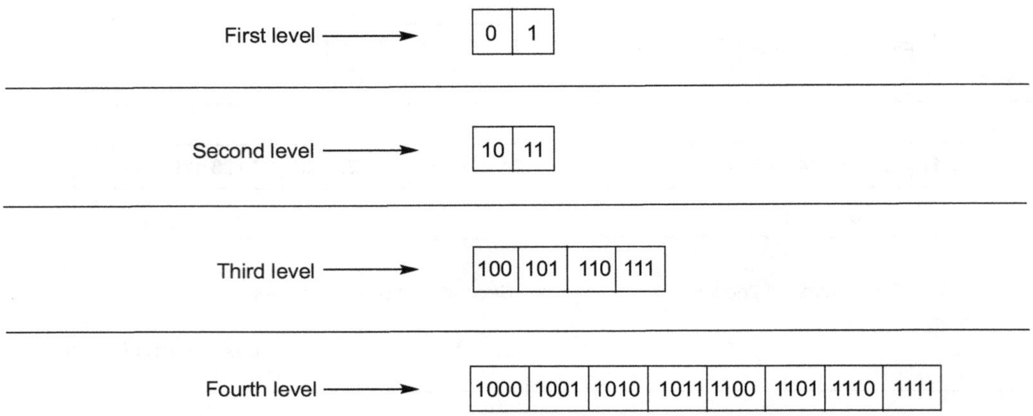

Figure 1.5 Levels of counting in binary number system

1.5.3 Octal Numbers

In Greek, 'octa' means 'eight' and hence there are eight digits in the first level of this number system. Using the combination of these eight digits, we get the second level of the octal numbers and we use the same kind of procedure as we did for the earlier two systems of numbers. The first level would contain 0, 1, 2, 3, 4, 5, 6 and 7. The Figure 1.6 shows a few of the levels of counting in octal number system.

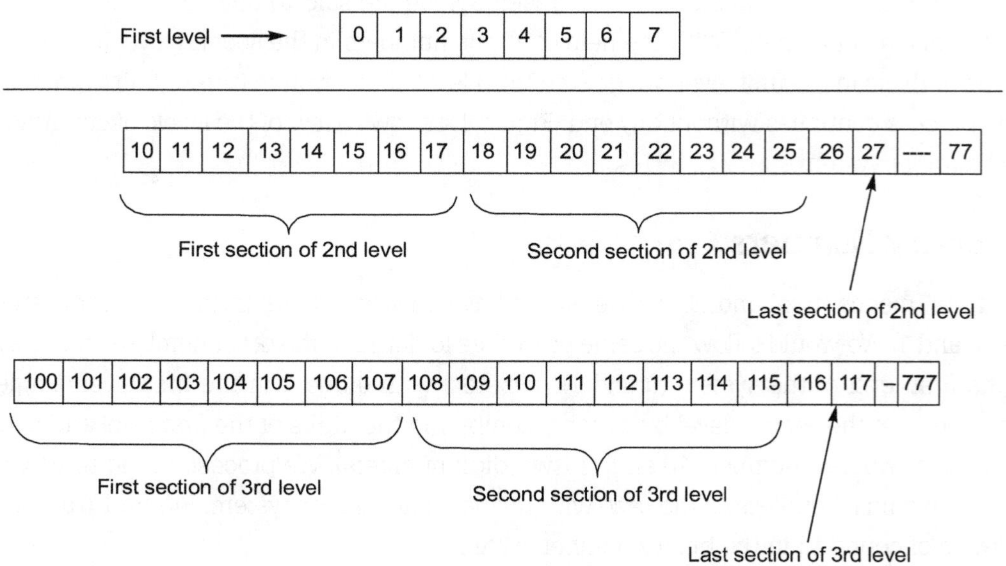

Figure 1.6 Levels of counting in octal number system

1.5.4 Hexadecimal Numbers

In Greek, 'hexa' means 'six' and we already knew that 'deca' means 'ten' and hence this system of numbers would have sixteen numbers in the first level and these would be used to generate the second, third and nth levels of the system. The speciality of this number system is that the counting begins from 0 and goes till '9' as usual. After '9', the number would be counted not as '10', but as 'A', 11 as 'B', 12 as 'C', 13 as 'D', 14 as 'E' and 15 as 'F'. The first level of counting ends with 'F' and the next level begins with the usual procedure of combining the digits of the previous level. This is again an endless process. Figure 1.7 shows a few levels in hexadecimal counting.

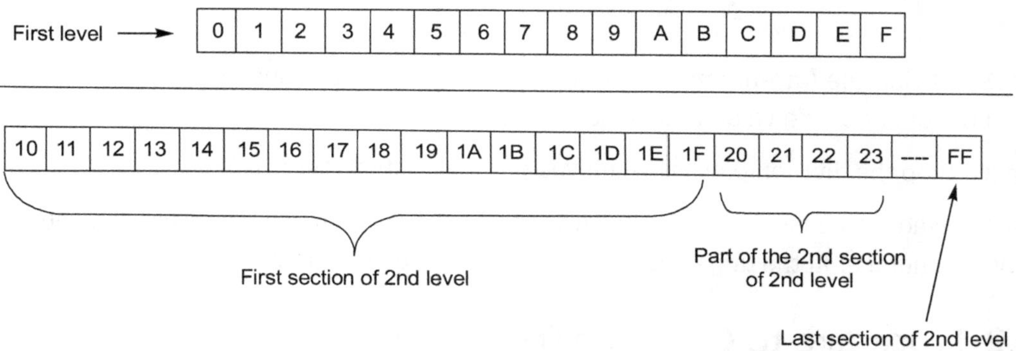

Figure 1.7 Levels of counting in hexadecimal number system

1.6 CONVERSION FROM ONE NUMBER SYSTEM TO ANOTHER

The conversion process is quite simple and we shall deal with the following conversions in this section.

Decimal integer/fraction to binary integer/fraction

Decimal integer/fraction to octal integer/fraction

Decimal integer/fraction to hexadecimal integer/fraction

Binary integer/fraction to decimal integer/fraction

Binary integer/fraction to octal integer/fraction

Binary integer/fraction to hexadecimal integer/fraction

Octal integer/fraction to decimal integer/fraction

Octal integer/fraction to binary integer/fraction

Octal integer/fraction to hexadecimal integer/fraction

Hexadecimal integer/fraction to decimal integer/fraction

Hexadecimal integer/fraction to binary integer/fraction

Hexadecimal integer/fraction to octal integer/fraction

As we do the process, you would understand clearly the conversions and to have a better practice, go through the exercises given at the end of this Chapter.

1.6.1 Procedure to Convert an Integer in One Base to that in Another Base

The following are the steps to convert an integer (of any number system) to the other number system.

Step 1 Divide the (given) number by the desired base (target number system). As you divide, record the quotients down and the remainders beside.

Step 2 Stop the division process when you get the quotient zero.

Step 3 Read the remainders that you have entered from bottom to top. The resulting number is the equivalent number in the target number system for the given number.

1.6.2 Procedure to Convert a Fraction in One Base to that in Another Base

The following are the steps to convert a fraction (of any number system) to the other number system.

Step 1 Multiply the (given) number with the desired base.

Step 2 As you multiply, record the fractional part and the integer part of the product separately. Take the integer part of the result and repeat the above two steps (a kind of recursive process)

Step 3 Once the required precision is reached, stop the process of multiplication. Read the fractional parts of the products obtained in each step from top to bottom and that will be the equivalent number in the target number system.

1.6.3 Decimal Integer/Fraction to Binary Integer/Fraction

Let us consider a number such as 17 which is a decimal integer number and let us convert it into its equivalent binary by following the procedure given in the section 1.6.1. Figure 1.8 shows how the process of division is done. This method is sometimes called as ladder method of division. The remainders recorded beside each division process (shown in small squares) are read from bottom to top to get the equivalent binary number. In this case, the binary for the number 17 is 10001.

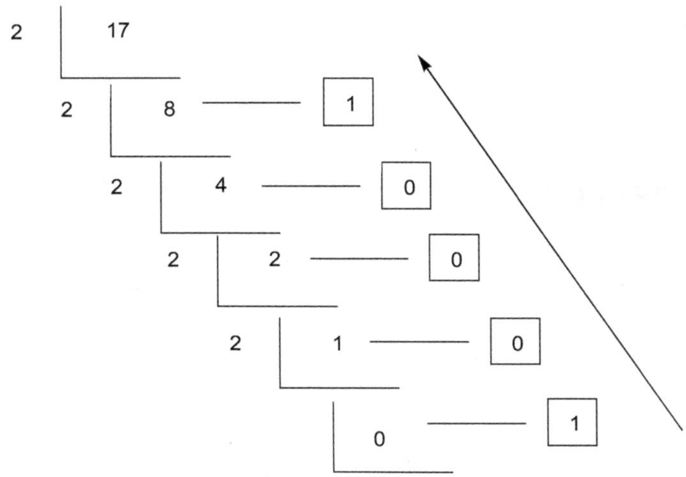

Figure 1.8 Converting a decimal integer to binary integer

This method is called as ladder method of converting the number in one system to another. Let us see what would be the result when we have a decimal fraction. Consider for example the number 0.74. Let us apply the steps that are given in the section 1.6.2 to convert this fraction into its equivalent binary fraction. Table 1.1 shows the conversion process.

Table 1.1 Conversion of decimal fraction to its equivalent binary

Multiplication	Result	Integer part	Fractional part
0.74 × 2	1.48	1	0.48
0.48 × 2	0.96	0	0.96
0.96 × 2	1.92	1	0.92
0.92 × 2	1.84	1	0.84
0.84 × 2	1.68	1	0.68

In this conversion process, we have taken the result to correspond to five places of binary point. That is the reason why we have stopped the conversion process once we get the integer part of the result in the fifth column. Now reading the column **integer part** from top to bottom, we get the equivalent binary fraction of the given decimal. Hence the result in binary is 0.10111 for the given decimal fraction 0.74. This is only an approximate value, since the multiplication of values is stopped abruptly. If the number is a mixed number, that is, say it is 17.74, you may follow the conversions separately for integers and fractions and combine the result. For the number 17.74, the result would be 10001.10111.

1.6.4 Decimal Integer/Fraction to Octal Integer/Fraction

The ladder method could be followed in the conversion of decimal integer to an octal fraction. Let us consider the number 17 in decimal and convert the same to octal. Figure 1.9 shows how this is done.

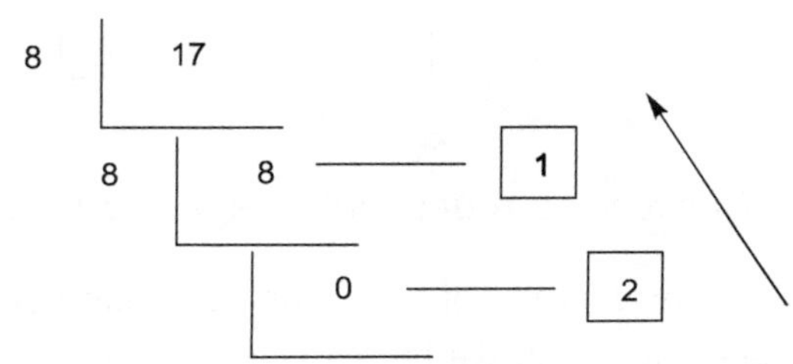

Figure 1.9 Converting a decimal integer to octal integer

From the figure, we can get the octal equivalent of the decimal number 17 as 21. Let us consider a decimal fraction 0.74 and find out the equivalent octal fraction. Table 1.2 shows the conversion process and the result could be read from the column integer part from top to bottom.

Table 1.2 Conversion of decimal fraction to its equivalent octal

Multiplication	Result	Integer part	Fractional part
0.74 × 8	5.92	5	0.92
0.92 × 8	7.36	7	0.36
0.36 × 8	2.88	2	0.88
0.88 × 8	7.04	7	0.04

From the table, we could see the result corresponding to four octal points for the decimal 0.74 is 0.5727. If the given number is 17.74, the result would be 21.5727.

1.6.5 Decimal Integer/Fraction to Hexadecimal Integer/Fraction

Let us consider the number 17 and try to find out its equivalent hexadecimal integer by applying the ladder method. Figure 1.10 shows the procedure and the result is 11 in hexadecimal for the number 17.

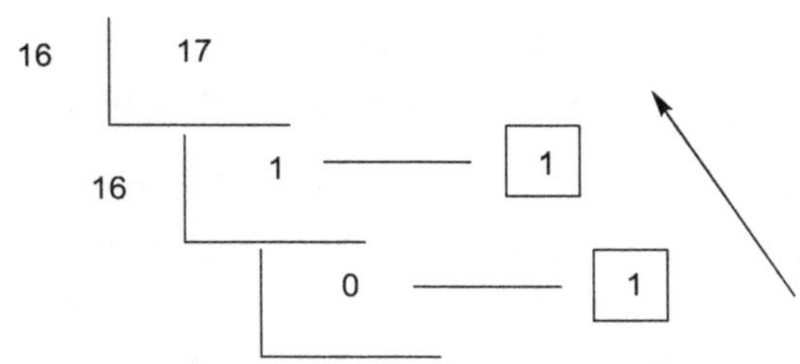

Figure 1.10 Converting a decimal integer to hexadecimal integer

Consider the number 0.74 and let us convert this decimal fraction into its equivalent hexadecimal. Table 1.3 shows the conversion process and the result.

Table 1.3 Conversion of decimal fraction to its equivalent hexadecimal

Multiplication	Result	Integer part	Fractional part
0.74 × 16	11.84	11	0.84
0.84 × 16	13.44	13	0.44
0.44 × 16	7.04	7	0.04
0.04 × 16	0.64	0	0.64
0.64 × 16	10.24	10	0.24

From the table we could see that the equivalent value is 0.(11)(13)70(10) in hexadecimal for the given decimal fraction 0.74. The value could be written also as 0.BD70A, since the numbers 10 to 15 in hexadecimal are coded with the alphabets A to F, as we had discussed earlier. If the given decimal number is 17.74, the corresponding value in hexadecimal would be 11.BD70A.

1.6.6 Binary Integer/Fraction to Decimal Integer/Fraction

This is the reverse conversion process. For example, consider the number 10001, which is a binary number. We knew that the number systems are categorized according to the weights assigned to the digits of the number. For example, consider the decimal number 189. We read this number as one hundred and eighty nine, assigning weight to each of the digits of the number. The digit '9' is assigned with 'units', '8' with 'tens' and '1' with 'hundreds'. Thus we are able to read and understand the number properly. In other words, the digit '9' is said to have 9 units and it is obtained by multiplying 9 with 10^0 and the digit '8' is said to have 8 tens and it is obtained by multiplying it with 10^1 and as we proceed towards left we have to increase the power of 10. Thus each digit is multiplied with 10^x with x taking the values from 0 to n, where n value is one less than the number of digits in the given number. If the given number has four digits, the value of x would be from 0 to 3.

The above discussion is true for any number system. The base of the number system alone changes and not the procedure. For example, consider the binary number system, whose base is 2 and if we take an example number, say, 10001, then each of the binary digit has its own unique weight and it is obtained by multiplying the binary digit with 2^x where x would take values from 0 to n, where n value would be one less than the actual number of digits in the given number. In the number that we have taken, there are five digits and hence x value would be from 0 to 4. Each digit of the binary number has to be multiplied with the corresponding weight to obtain its position in the number and once we add them, we will get the equivalent decimal number. Figure 1.11 shows the conversion process. We see that we get the result as 17.

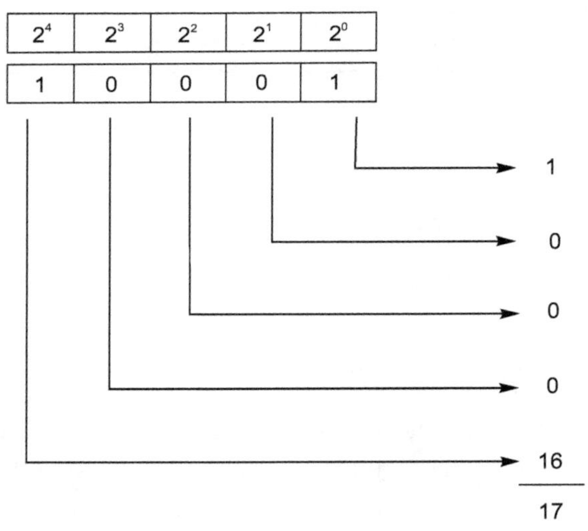

Figure 1.11 Converting a binary integer to a decimal integer

Let us consider a binary fraction 0.10001 and find out the decimal equivalent of this. We have to proceed in a similar way as shown above, except that the base has to be raised to the negative powers of 2 as we proceed towards right of the binary point. Figure 1.12 shows how the number is getting converted. We get the result as 0.53125.

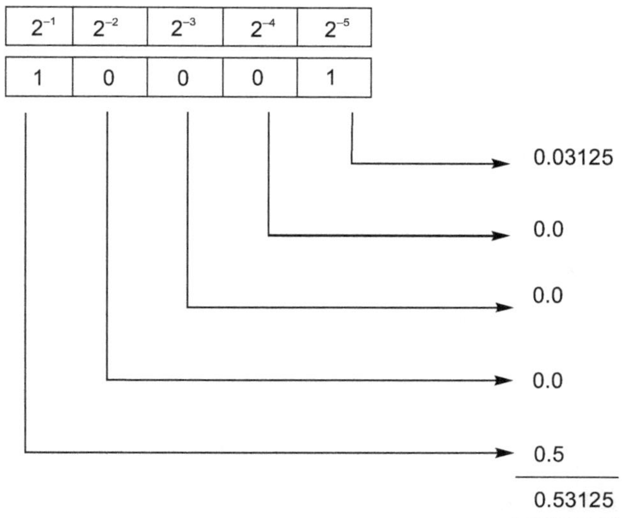

Figure 1.12 Converting a binary fraction to a decimal fraction

1.6.7 Binary Integer/Fraction to Octal Integer/Fraction

Consider the binary number 10001. If we want to convert this to its equivalent octal, we have two options. Convert the number to decimal (follow the procedure given in the section 1.6.6) and then convert the decimal to octal (as explained in the section 1.6.4). But this process will be quite time-consuming and hence I would like to suggest an easy way to do this. Consider the number and start from the right extreme position and as you proceed towards left, group the digits of the given number such that three bits are in each group. If any digit of the number is left alone, include as many leading zeroes to form a group of three digits. After you have grouped them, convert each group to their decimal equivalent and record the results down each group. Combine the results from left to right to get the final result. Figure 1.13 shows the result of converting 10001 to its octal equivalent and we see that the result is 21. (You may check this result by following the other procedure that I had mentioned in the beginning of this section).

The binary fraction is also converted to its equivalent octal in a similar way. In this case, one has to start the grouping from the left side, proceed towards right, align three digits in each group and do the conversion for each group to get the final result. Figure 1.14 shows the way the number 0.1011011 is converted to its octal equivalent. In case of non-availability of digits, you may include as many trailing zeroes. We see that the result of the conversion is 0.554.

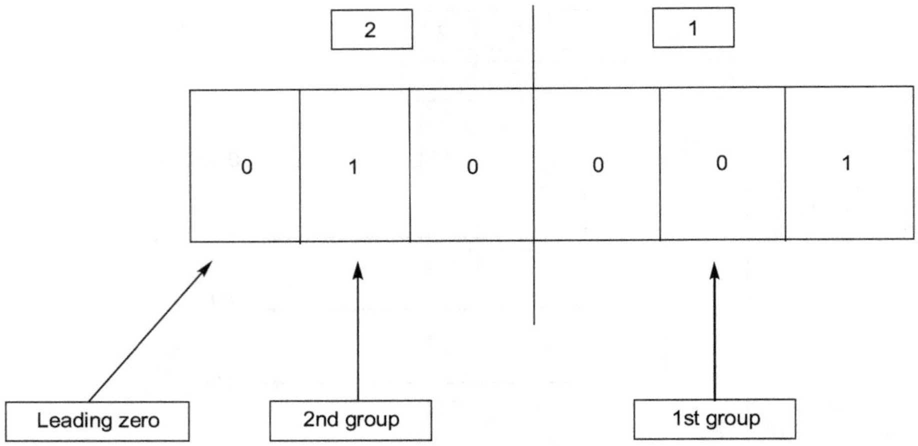

Figure 1.13 Converting a binary integer to an octal integer

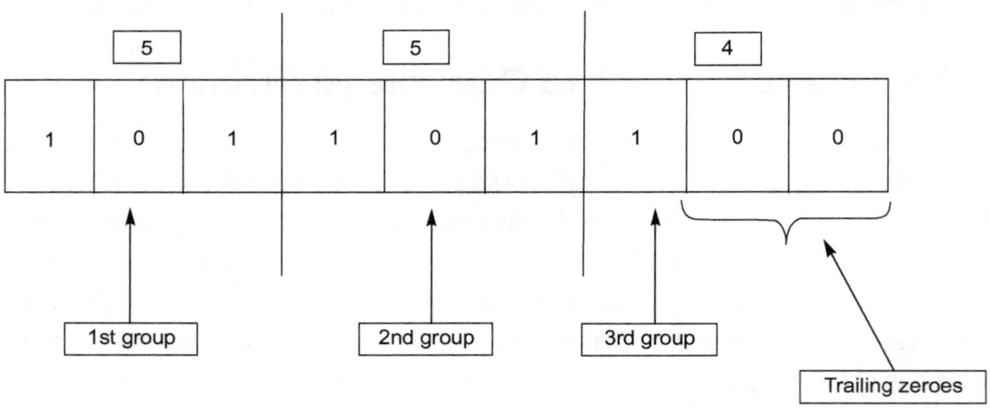

Figure 1.14 Converting a binary fraction to an octal fraction

1.6.8 Binary Integer/Fraction to Hexadecimal Integer/Fraction

We could follow a similar method discussed in the section 1.6.7 for converting a binary integer to its equivalent hexadecimal. Instead of forming a group of three, we would form a group of four and we would convert each group to decimal and combine the results to get the final result. Figure 1.15 shows how we convert the binary 10001 to its hexadecimal equivalent. We see that the result in hexadecimal is 11.

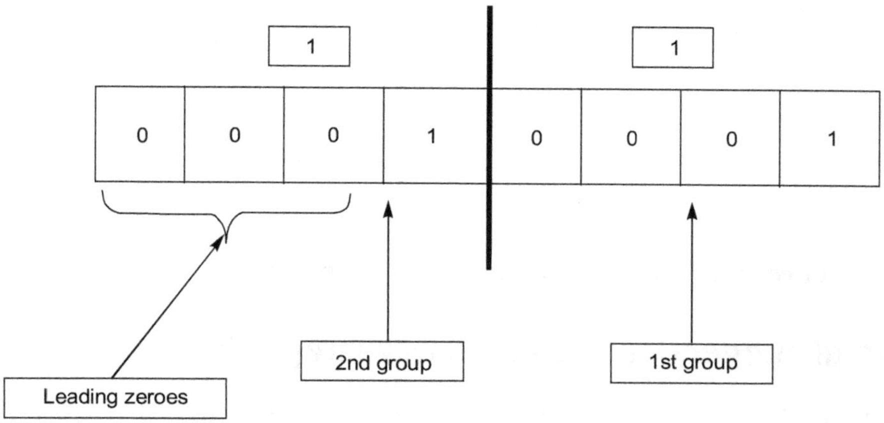

Figure 1.15 Converting a binary integer to hexadecimal integer

Consider the binary fraction 0.10111 and we follow the grouping from the left to right and convert each group to their decimal equivalent and combine the individual results to get the final answer. Figure 1.16 explains how it is done. The final answer would be 0.B8.

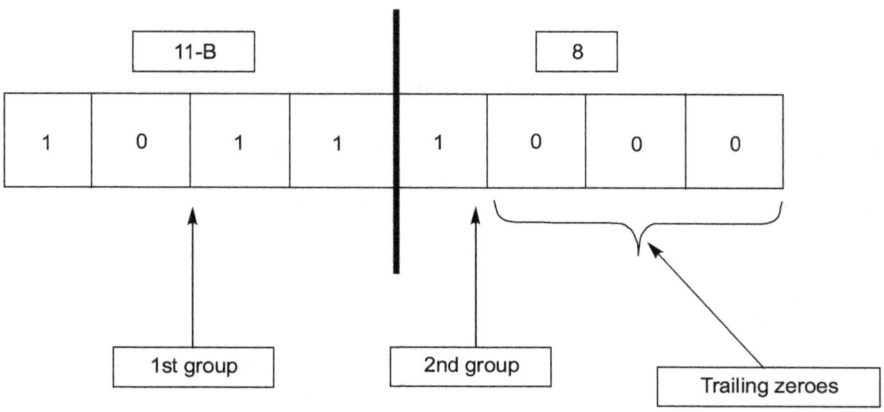

Figure 1.16 Converting a binary fraction to hexadecimal fraction

1.6.9 Octal Integer/Fraction to Decimal Integer/Fraction

Consider the octal integer 73. If we want to convert this to its decimal equivalent, multiply each of the digits with their weights (8^0, 8^1, etc.). Figure 1.17 shows how to do this and the result of the conversion is 59 in decimal.

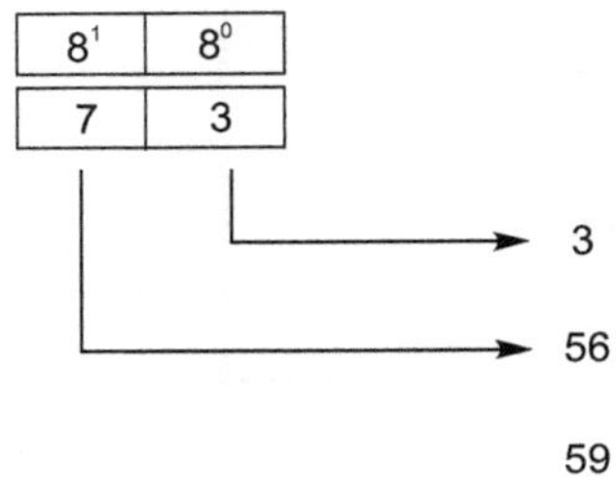

Figure 1.17 Converting an octal integer to decimal integer

1.6.10 Octal Integer/Fraction to Binary Integer/Fraction

Consider an octal number 73. Converting this to its binary is just the reverse of that discussed in the section 1.6.7. There we grouped the given binary digits in threes and then wrote the decimal equivalent of each group and got the final result by combining the individual results. In this case, we would write the binary equivalent of the given octal number such that each digit of the octal is represented by three binary digits. Then we would combine all the binary digits to get the binary equivalent of the given octal. Figure 1.18 shows how it is done for the octal 73. The result is 111011.

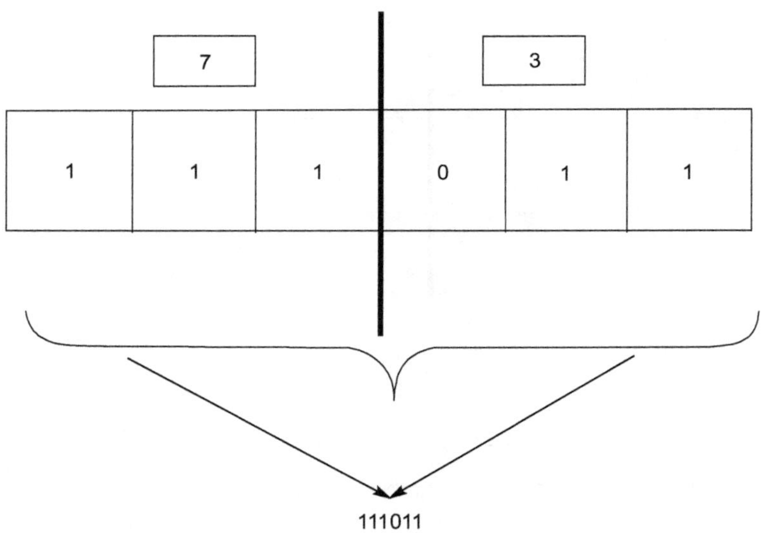

Figure 1.18 Converting an octal integer to binary integer

If we are given an octal fraction, say, 0.72, we would proceed in a similar way, by converting each of the digits towards the right of the octal point to binary digits in threes and then combine the results to get the final result in binary form. Figure 1.19 shows how it is done. The result is 0.11101.

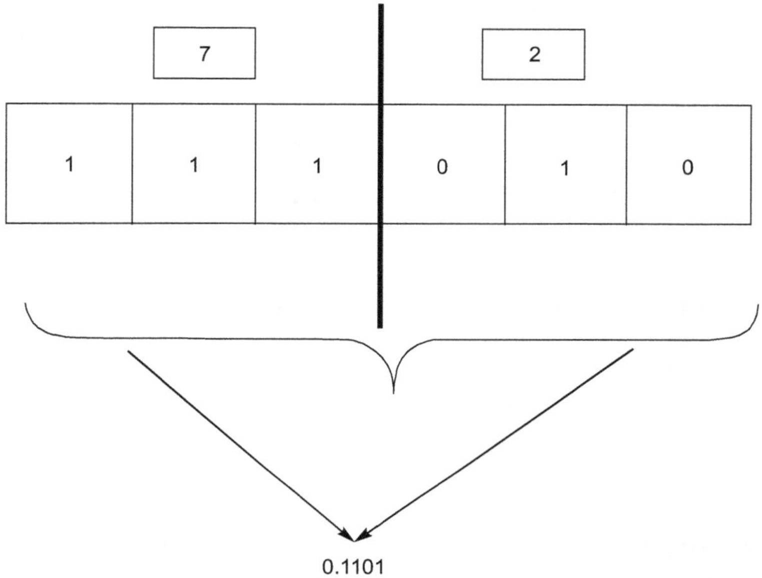

Figure 1.19 Converting an octal fraction to binary fraction

1.6.11 Octal Integer/Fraction to Hexadecimal Integer/Fraction

Consider the octal number 73. For converting this to its hexadecimal equivalent, first we have to convert it to its binary (Refer the section 1.6.10) and then from binary we can convert to hexadecimal (Refer the section 1.6.8). The result would be 3B.

For converting an octal fraction to its hexadecimal equivalent, a similar procedure as said in the sections 1.6.10 and 1.6.8 could be followed (Readers are advised to work out and cross check).

1.6.12 Hexadecimal Integer/Fraction to Decimal Integer/Fraction

Consider the hexadecimal number 1B. For converting this to its decimal equivalent, we have to multiply the digits with their weights and the weights would range from 16^0 to 16^n. The value of n depends on the number of digits in the given hexadecimal number (it will be from 0 to n–1 in practice). Figure 1.20 shows how the number is converted to decimal. Before converting the above number, we must change the alphabet notation with actual number. Thus, in the above number, the alphabet B would be replaced with 11 and hence the number that is rewritten would be 1(11). The result is 27.

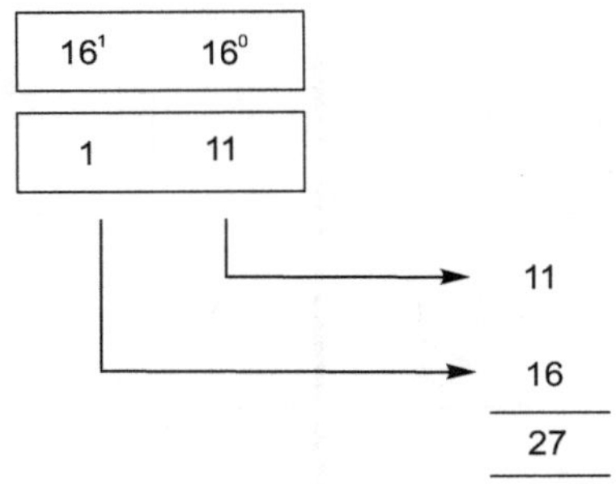

Figure 1.20 Converting a hexadecimal integer to decimal integer

For converting the hexadecimal fraction to its equivalent decimal, we would apply a similar method, except for negative powers for the base 16, since we proceed from left to right starting from the digit after the hexadecimal point of the given number. Consider for example, the hexadecimal number 0.2A.

Figure 1.21 shows the result of converting this to its equivalent decimal number. See to that the alphabet is converted to the corresponding digit of the hexadecimal system, before the conversion algorithm is applied. The equivalent value is found to be 0.4375.

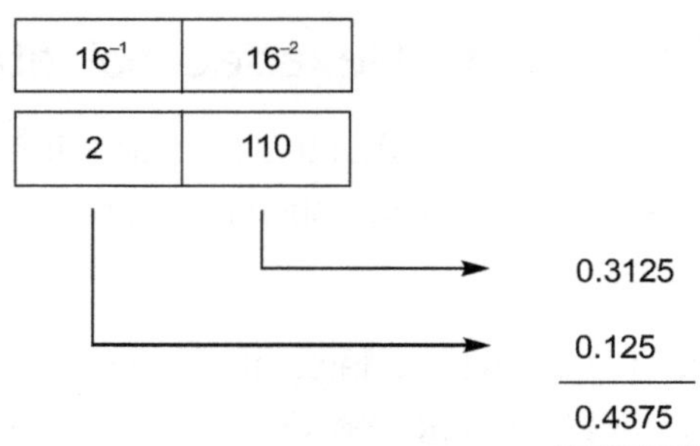

Figure 1.21 Converting a hexadecimal fraction to decimal fraction

1.6.13 Hexadecimal Integer/Fraction to Binary Integer/Fraction

Converting a hexadecimal integer to its binary equivalent is achieved by just reversing the idea discussed in the section 1.6.8. Each digit of the given hexadecimal number is written in their equivalent binary form in a four bit format and the final result is obtained by combining the results of the individual digits. Consider the hexadecimal number 4B and let us write its equivalent binary number by writing the binary equivalent of each of the digits of the number. Figure 1.22 shows how the result is obtained following this procedure. The alphabet B is first replaced with its numerical equivalent before the conversion. The result of the conversion is 1001011.

Similarly, when we are given a hexadecimal fraction, a similar kind of procedure is followed and the result is obtained. Consider for example, the hexadecimal fraction 0.AB. When we want to convert this to its equivalent binary fraction, we would convert each of the digits towards the right of the hexadecimal point to its binary format with four bits. We then combine the results to obtain the final result. The alphabets are converted to their numerical equivalents before the conversion process. Figure 1.23 shows how this conversion takes place and the result obtained is 0.10101011. When a mixed number is given for conversion, you have to carry out the conversion process individually and then combine the results to get the final result.

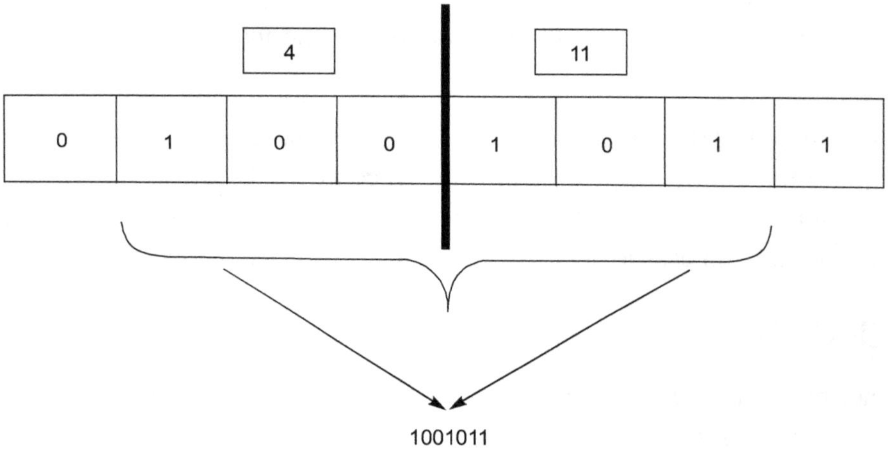

Figure 1.22 Converting a hexadecimal integer to binary integer

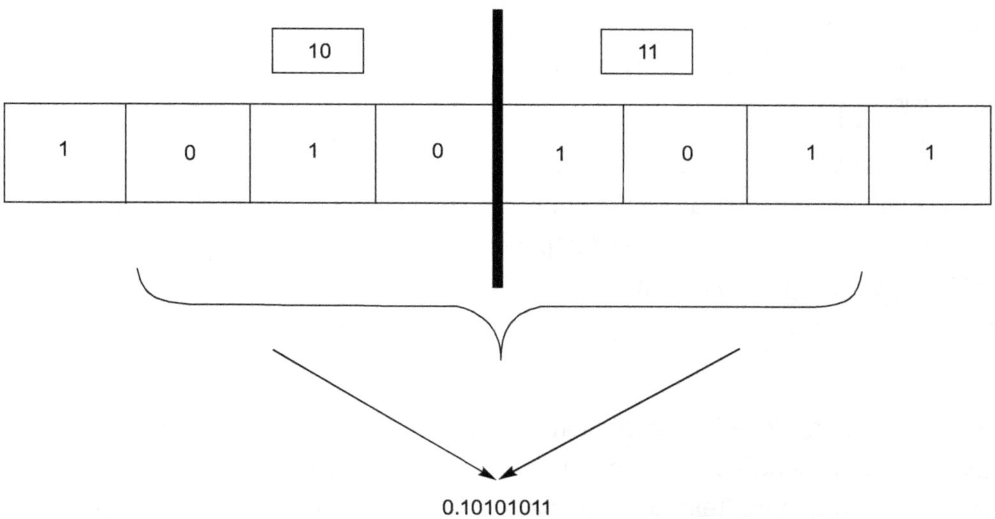

Figure 1.23 Converting a hexadecimal fraction to binary fraction

1.6.14 Hexadecimal Integer/Fraction to Octal Integer/Fraction

Conversion of hexadecimal integer to its octal is a two-step process. First convert the given hexadecimal number to its equivalent binary (Refer the section 1.6.13). Convert the binary number thus obtained to its equivalent octal (Refer the section 1.6.7). Hence the conversion is complete. Take for example, the hexadecimal number F6. The result of converting this to its octal equivalent would be 366.

If a hexadecimal fraction such as 0.F3 is given, the conversion process is simple. First convert the given fraction to its binary equivalent and then the binary fraction to its octal equivalent. (Refer the sections 1.6.13 and 1.6.7). I would request all the readers to check this with a few examples.

We have discussed all the possible inter-conversions between the number systems in use and I am sure all the readers would be thorough with the concepts discussed. I would advice all the readers to go through this section of this chapter to have a clear understanding of the number systems, since they are the base for learning any programming language.

1.7 ASCII CODE

If you are somewhat familiar with computers, then you know that all modern computers are "digital", i.e., internally they represent all data as numbers. In the very early days of computing (during 1940s), it became clear that computers could be used for more than just number crunching. They could be used to store and manipulate text. This could be done by simply representing different alphabetic letters by specific numbers. For example, the number 65 to represent the letter "A", 66 to represent "B", and so on. At first, there was no standard, and different ways of representing text as numbers developed.

Around the early 1960s an 8-bit byte was becoming the standard way that computer hardware was built, and you can store 128 different numbers in a 7-bit number. When all possible alphanumeric characters are counted (A to Z, lower and upper case, numeric digits 0 to 9, special characters like "%, *, /, ?" etc.) we would end up with a value of 90-plus. It was therefore decided to use 7 bits to store the new ASCII code, with the eighth bit being used as a parity bit to detect transmission errors.

American Standard Code for Information Interchange (ASCII) is a character encoding based on the English alphabets. ASCII codes represent text in computers, communications equipment, and other devices that work with text. Most modern character encodings, which support many more characters than did the original, have a historical basis in ASCII.

ASCII includes definitions for 128 characters: 33 are non-printing, mostly obsolete control characters that affect how text is processed; 94 are printable characters (excluding the space). The ASCII character encoding or a compatible extension is used on nearly all common computers, especially personal computers and workstations. Figure 1.24 represents the most commonly used 7-bit ASCII table and it has the ASCII values for the 128 characters supported and defined by ASCII encoding.

Even as these new standards are phased in, the 7-bit ASCII table continues to be the backbone of modern computing and data storage. It is one of the few real standards that all computers understand, and everything from e-mail to web browsing to document editing would not be possible without it. It is so ubiquitous that the terms "text file" and "ASCII file" have come to mean the same thing for most computer users.

ASCII	Hex	Symbol	ASCII	Hex	Symbol	ASCII	Hex	Symbol	ASCII	Hex	Symbol
0	0	NUL	16	10	DLE	32	20	(space)	48	30	0
1	1	SOH	17	11	DC1	33	21	!	49	31	1
2	2	STX	18	12	DC2	34	22	"	50	32	2
3	3	ETX	19	13	DC3	35	23	#	51	33	3
4	4	EOT	20	14	DC4	36	24	$	52	34	4
5	5	ENQ	21	15	NAK	37	25	%	53	35	5
6	6	ACK	22	16	SYN	38	26	&	54	36	6
7	7	BEL	23	17	ETB	39	27	'	55	37	7
8	8	BS	24	18	CAN	40	28	(	56	38	8
9	9	TAB	25	19	BM	41	29	)	57	39	9
10	A	LF	26	1A	SUB	42	2A	*	58	3A	:
11	B	VT	27	1B	ESC	43	2B	+	59	3B	;
12	C	FF	28	1C	FS	44	2C	,	60	3C	<
13	D	CR	29	1D	GS	45	2D	-	61	3D	=
14	E	SO	30	1E	RS	46	2E	.	62	3E	>
15	F	SI	31	1F	US	47	2F	/	63	3F	?

ASCII	Hex	Symbol	ASCII	Hex	Symbol	ASCII	Hex	Symbol	ASCII	Hex	Symbol
64	40	@	80	50	P	96	60	'	112	70	p
65	41	A	81	51	Q	97	61	a	113	71	q
66	42	B	82	52	R	98	62	b	114	72	r
67	43	C	83	53	S	99	63	c	115	73	s
68	44	D	84	54	T	100	64	d	116	74	t
69	45	E	85	55	U	101	65	e	117	75	u
70	46	F	86	56	V	102	66	f	118	76	v
71	47	G	87	57	W	103	67	g	119	77	w
72	48	H	88	58	X	104	68	h	120	78	x
73	49	I	89	59	Y	105	69	i	121	79	y
74	4A	J	90	5A	Z	106	6A	j	122	7A	z
75	4B	K	91	5B	[	107	6B	k	123	7B	{
76	4C	L	92	5C	\	108	6C	l	124	7C	|
77	4D	M	93	5D	]	109	6D	m	125	7D	}
78	4E	N	94	5E	"	110	6E	n	126	7E	~
79	4F	O	95	5F	-	111	6F	o	127	7F	□

Figure 1.24 ASCII Coding

1.8 ALGORITHMS

Algorithms are nothing but a type of effective method in which a definite list of well-defined instructions for completing a task, when given a proper initial state, will proceed through a well-defined series of successive states, eventually terminating in an end-state.

Well, what is the etymology of this term? **Abu Abdullah Muhammad bin Musa al-Khwarizmi**, a Persian astronomer and mathematician, wrote a paper in Arabic, which was later translated to Latin, entitled as *Algoritmi de numero Indorum*, which meant "Book by Algoritmus on the numbers of the Indians". But the word "Algoritmi" in the title was treated as a Latin plural and this led to the word "Algorithm". The real meaning of the word is "Calculation Method".

Every one of us does calculation in our routine life and that may be the one related to the academy or business or personal. For the calculation, we normally adopt a formula and a method and the methods may be different for each of us, even though the result is going to be one and the same.

The concept of algorithm was introduced to define the notion of **logic**. Algorithms are essential to the way computers process information, because a computer program is essentially an algorithm that tells the computer what specific steps to perform and in what specific order, in order to carry out a specified task. Typically, when an algorithm is associated with processing information, data are read from an input device, written to an output device, and/or stored for further processing. Stored data are regarded as part of the internal state of the entity performing the algorithm. In practice, the state is stored in a data structure, but an algorithm requires the internal data only for specific operation sets called abstract data types.

Since an algorithm is a precise list of precise steps and hence the order of computation will almost always be critical to the functioning of the algorithm. Instructions are usually assumed to be listed explicitly, and are described as starting "from the top" and going "down to the bottom", an idea that is described more formally by flow of control.

1.8.1 How to Express an Algorithm?

Algorithms can be expressed in many kinds of notations, including natural languages, pseudocode, flowcharts and programming languages. Natural language expressions of algorithms tend to be verbose and ambiguous, and are rarely used for complex or technical algorithms. Pseudocode and flowcharts are structured ways to express algorithms that avoid many of the ambiguities common in natural language statements, while remaining independent of a particular implementation language. Programming languages are primarily intended for expressing algorithms in a form that can be executed by a computer, but are often used as a way to define or document algorithms.

1.9 FLOWCHARTS

There is a famous saying that goes like **A picture is worth thousand words** and this is exactly implied by the flowcharts, which are structured ways to express algorithms. As already mentioned, the flowcharts resolve the ambiguities that might be a common problem in using a natural language. Moreover, pictures are the easiest way to convey the messages to human beings belonging to any community or group. A flowchart is one of the seven basic tools of quality control (The other tools being histogram, Pareto chart, check sheet, control chart, cause-and-effect diagram and scatter diagram). The first structured method for documenting process flow, the flow process chart, was introduced by **Frank Gilbreth** (An advocate of scientific management and a pioneer of motion study).

Any vector-based drawing program can be used to create flowcharts. Some tools offer special support for flowcharts, e.g. OpenOffice.org Draw, ConceptDraw, SmartDraw, Visio, OmniGraffle and Dia.

1.9.1 Symbols Used in a Flowchart

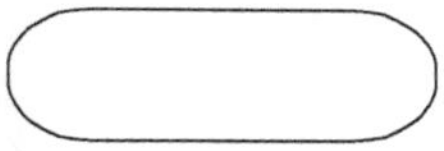

Start and **End** symbols, represented as lozenges, ovals or rounded rectangles, usually containing the word "Start" or "End", or another phrase signalling the start or end of a process, such as "submit enquiry" or "receive product".

Arrows show the "flow of control". An arrow coming from one symbol and ending at another symbol represents that control passes to the symbol the arrow points to.

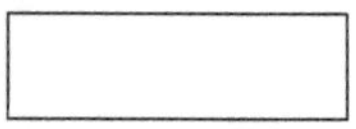

Processing or **Calculating** steps are represented as rectangles.

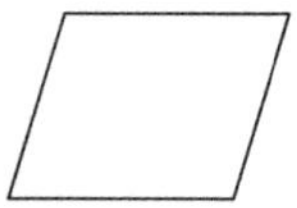

Input/Output is represented as a parallelogram.

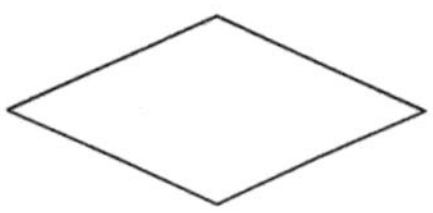

Conditional (or decision) is represented as a diamond (rhombus). These typically contain a Yes/No question or True/False test. This symbol is unique in that it has two arrows coming out of it, usually from the bottom point and right point, one corresponding to Yes or True, and one corresponding to No or False. The arrows should always be labelled. More than two arrows can be used, but this is normally a clear indicator that a complex decision is being taken, in which case it may need to be broken-down further, or replaced with the "pre-defined process" symbol.

A **Document** is represented as a rectangle with a wavy base.

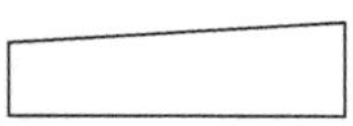

A **Manual input** is represented by rectangle, with the top irregularly sloping up from left to right.

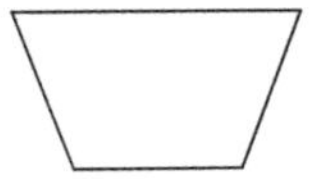

A **Manual operation** is represented by a trapezoid with the longest parallel side at the top, to represent an operation or adjustment to process that can only be made manually.

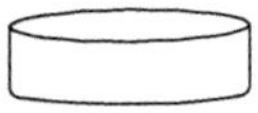

A **Data File** is represented by a cylinder.

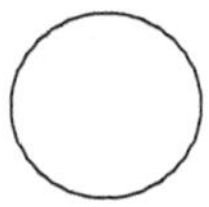

Connectors are usually represented as circles, to represent converging paths in the flowchart. Circles will have more than one arrow coming into them but only one going out. Some flowcharts may just have an arrow point to another arrow instead. These are useful to represent an iterative process. A loop may, for example, consist of a connector where control first enters, processing steps, and conditional with one arrow exiting the loop, and one going back to the connector.

Off-page Connectors are often used to signify a connection to a (part of another) process held on another sheet or screen. It is important to remember to keep these connections logical in order. All processes should flow from top to bottom and left to right.

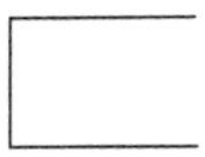

Comment Symbols are used when additional explanation or comment is required. This symbol is usually connected to the symbol it is explaining by a dashed line.

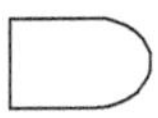

Delay—Wait for elevator, papers waiting, material waiting.

SUMMARY

I am sure that this chapter would have been a good preparatory one for all those who had just started their affinity towards computers and programming languages. Especially those who come from pure life sciences would have surely felt comfortable with computing after learning the essentials.

To summarize, I would say this chapter on its entirety has dealt with various concepts that are essential to become a good and a worthy Bioinformatician. The people from biology were introduced to the computing field and they were made comfortable in that they are not learning anything new. The reasons why the biologists must know and use computers were explained. Also the issues handled by Bioinformatics were briefed. The significance of transforming from a biologist to a computer scientist was also explained.

We understood the importance of an operating system and it was followed by a very explanatory section on the different functions of an operating system and how important they are while we handle the computing system. A few examples for the operating systems that are in use today have been quoted.

We also learnt the various programming languages that evolved during different points of time and also their achievements and shortcomings. The main focus of this section was to introduce the biologists and other life scientists to the fact that programming languages are nothing new to the science and they are in existence for decades and scores together. We also discussed the variety of programming languages that are used in life sciences, both in developing applications and research. We categorized them as scripting languages and general-purpose languages.

We discussed the different number systems and how they are useful in storing the information. We also discussed the inter-conversion of number systems with sufficient examples and figures. To conclude the chapter, we had also learnt the ASCII character coding and its significance while programming and the concept of algorithms and flowcharts are also discussed. I am sure that the figures and illustrations that are given in this chapter would be of profound use for the learners of Bioinformatics.

CRITICAL THINKING QUESTIONS

1. Convert $0100\ 1011_2$ to decimal.
2. Convert 173.34375_{10} to binary.
3. Convert the number 01111101.1001_2 to octal and to hexadecimal.
4. Convert 123.45_8 to binary.
5. Convert $C8.D3_{16}$ to binary.
6. Write down the binary equivalent of 2AC3.F1.
7. Convert 13_{10} to its equivalent octal and hexadecimal number systems.
8. Convert the number 13_8 to its equivalent hexadecimal system.
9. Identify the general equation for representing a number in any number system.
10. Convert the number 9F to the other three number systems.

SELF-LEARNING EXERCISES

1. Try to learn about the different codes followed in computing (To quote a few, we have BCD, Gray, 8221, 2421, etc.).
2. Try to evaluate the performance of the various versions of Windows operating system. Do you think the recent version is the best one when you compare with any of the earlier ones? If so, what are your opinions for the view and if not, what are your opinions against the view?
3. Make a thorough study on different operating systems that are available as open source. Are you of an opinion that these open source operating systems are reliable in terms of security? If so or not, what are the reasons for your answer?
4. Suggest suitable points on how can you contribute to the open source community to enable freedom in using the software and operating systems that are available as open. Have you used any such software or operating system? Share your experience of using them and how advantageous they are when you compare them with the proprietary stuff?
5. Suggest suitable measures that will enable the computer users to share the information more freely and build their own knowledgebase by developing more efficient operating systems
6. Make an elaborate discussion on the different flavours of UNIX and how each one is unique in their functionality and efficiency?
7. Get the list of different programming languages that evolved over a period of time and compare the efficiency of each language and explain their significance.
8. Learn how negative numbers are represented in Computing System (use the knowledge of binary, octal and hexadecimal numbers).
9. Learn the basic arithmetic operations that you can perform with different number systems.
10. Learn about compliment method of storing numbers and explain their significance.
11. Learn about the IEEE method of representing the floating point numbers (real numbers).
12. Learn how octal and hexadecimal numbers are represented in C programming language while they are accessed through the functions scanf() and printf().
13. Give your suggestions and opinions on the applications of programming languages in various disciplines.
14. As a biologist, analyse your aptitude to transform yourself to a computer scientist and then to a Bioinformatician.
15. As a computer scientist, analyse your aptitude to transform yourself to a biologist and then to a Bioinformatician.

2

ESSENTIALS OF C PROGRAMMING

2.1 INTRODUCTORY CONCEPTS

2.1.1 History of C

C, even after 36 years of its discovery, stands tall in the stream of Programming Languages. As you know, learning programming languages is very important, since you have to communicate with the computing system. There are umpteen programming languages in the world of computing and only a few are surviving and being used by the programmers and software developers now. One among them is C language. As mentioned earlier, it has its own place in the stream of programming languages and I can confidently say that it is the foundation for all the languages that were developed after its discovery. If you look at its history, it is quite interesting to note that it emerged from many other languages and surpassed the same to get ascendancy in the computing world. Let us learn the language starting with its history.

The C programming language and its direct descendants are by far the most popular programming languages used in the world today. This structurally tiny language originated at **Bell Labs** in order to write the first version of the Unix Operating System. In spite of its small beginnings, C has scaled up to run on powerful super-computers with gigabytes of memory. The language is extremely portable; it is possible to write a program in C for just about every platform in existence. Yet, if one takes a more in-depth look at C, one realizes that it is rather weak and has a small vocabulary. But, when we look at the efficiency and power of C, the weakness of C could easily be excused.

The history of C and that of the UNIX operating system are intertwined. You might almost say that C was invented and written for the purpose of writing the UNIX operating system. The seeds of UNIX began in 1965 with the completion of a project at MIT (Massachusetts Institute of Technology) called Project MAC, which was one of the first time-sharing computing systems. It allowed simultaneous use of the same computer, for at most thirty people, by using any of the 160 typewriter terminals placed around the campus and in the homes of faculty. It became so overloaded that MIT decided to embark on a more ambitious time-sharing system as a joint venture with General Electric and Bell Labs. This project was given the name MULTICS, which stands for MULTIplexed Information and Computing Service. The trio (Thompson, Ritchie and Kernighan) worked on this project for a few years, but in 1969, Bell Labs decided to discontinue the project MULTICS since it was too expensive and would take longer time to complete. At this point, an informal group of Bell Labs employees, led by Ken Thompson, started to investigate the alternatives.

Dennis Ritchie (Father of C Language)

Thompson and his group started designing an operating system (UNIX) that would support coordinated teams of programmers in the development of products, and would simplify the dialogue between human and machine. Thompson decided that UNIX needed a high-level system language. A high-level language is one which abstracts the hardware away from the programming environment. However, he could not just pick one that already existed. He was restricted by the computing system that he was working, PDP-7, which had only 8 kilobytes of memory. This was such a problem that at the start of the project, he had to compile his program on the more powerful GE-635 machine and then transfer the program to the slower computer using paper tape. Thompson had had some previous experience with the BCPL (Basic Combined Programming Language) system language. But it was too slow and lacked the run-time support that Thompson thought was necessary. So, he took his experience of BCPL and created his own language which he called B. The language was not really a new language. More correctly, it was BCPL squeezed into 8 Kilobytes and filtered through Thompson's brain. B had some problems though, and small UNIX community felt limited by the restrictions in the language. It seemed that B needed some enhancements to satisfy the developers. The main enhancement that it needed was data types to properly handle the characters and floating point numbers. A data type is defined by two properties: a set of values and a set of operations. This means, a language with strong typing would not allow the programmer to do meaningless operations, for example, taking the square root of a string of characters.

One of Thompson's co-workers, Dennis Ritchie, in 1971, began extending B by first adding data types and continued to expand and rewrite the parts of language. In 1973, Ritchie thus completed the essentials of C and then rewrote the entire UNIX kernel using the language. This time he wrote it for the PDP-11, with its huge 24K memory. In 1978, Brian Kernighan and Ritchie published The C Programming Language, which has since then become somewhat of a Bible to C programmers, and is often called the "White Book" or "K&R".

Now that C was the main system language for the UNIX environment, its use quickly spread. As the popularity of UNIX grew, so did that of C. During the 1980s, the use of C language spread widely, and the compilers became available on nearly every machine architecture and operating system. As the language spread, it began to change slightly for each platform that it was written for, and at this point the first edition of K&R no longer fully described the language and it became clear that a standard version of C was needed. Therefore, in the summer of 1983, the American National Standards Institute (ANSI) established the X3J11 committee to define the C language. They took 6 years and finally defined ANSI C, in 1989.

C is now such a popular programming language that there have been several descendants from it that is in wide use. One such descendant Objective C was adopted by the NeXT Computer Company for their OpenStep operating system. Apple Computer recently bought out NeXT, and Objective C is now the main development language at Apple for its next generation operating system, codenamed Rhapsody. Another descendant from C is Concurrent C, designed at Bell Labs by Dr. Narain Gehani and Dr. William D. Roome. This language is particularly well-tailored for parallel programming to write concurrent programs that can run efficiently on single computers, loosely coupled distributed computer networks, or on tightly coupled distributed multiprocessors. And of course there is the extremely popular C++, which is an Object Oriented extension of C.

How did C become so popular despite its deficiencies? C has many unique characteristic features that made it quite popular among the programmers.

2.1.2 Discovery of C—Is that Serendipity?

One can easily argue that C was written accidentally rather than being written intentionally. The reason is that, the intention of Ken Thompson was to enhance the performance of UNIX OS and he wanted to rewrite the entire kernel of UNIX. This persuaded Thompson to write a language and he started exploring various languages that existed and finally came out with B, which was later developed as C with the help of Dennis Ritchie and Brian Kernighan. So, the discovery of C is nothing but serendipity.

2.1.3 People Behind C

C was the brain child of many people, like Ken Thompson, Brian Kernighan and Dennis Ritchie. The contribution of Dennis Ritchie was so much that he is fondly called by the programmers as Father of C Language.

In 1978, Brian Kernighan and Dennis Ritchie published the first edition of *The C Programming Language*. This book, known to C programmers as K&R, served for many years as an informal specification of the language. The version of C that it describes is commonly referred to as K&R C. The second edition of the book covers the later ANSI C standard.

K&R introduced several language features:

- Standard I/O library
- Long integer data type
- Unsigned integer data type

Compound assignment operators =op (shorthand operators as we call it in our book) were changed to op= to remove the semantic ambiguity created by the construct i=–10, which had been interpreted as i =– 10 instead of the possibly intended i= –10.

Even after the publication of the 1989 C standard, for many years K&R C was still considered the "lowest common denominator" to which C programmers restricted themselves when maximum portability was desired, since many older compilers were still in use, and because carefully written K&R C code can be legal Standard C as well.

2.1.4 Importance and Role of C in Bioinformatics

There is data explosion happening now in almost all the fields, exclusively in the fields like Biotechnology, Biochemistry and Microbiology. Maintaining the data, and sequencing and analysing them have become Herculean tasks for the biologists or, even more aptly, life scientists. Database management systems have come to rescue the life scientists and they could manage the data explosion. But, the analysis of the data and sequencing are still an uphill task before them. Programming languages come in handy for these things and once a life scientist learns to program by learning a language like C or PERL, he/she becomes a fully qualified Bioinformatician. So is the case with the computer scientist. A computer scientist becomes a fully qualified Bioinformatician when he/she has a thorough knowledge of Biology (when I say biology it means everything that comprise of the aspects related to the life science like biochemistry, biotechnology and microbiology).

Bioinformaticians generally have either a biology or programming background, and then receive additional training in the other field. The common wisdom is that it's easier for biologists to pick up programming than the other way around; but, of course, it depends on the individual. How does one take the skills learned while programming in, say, the telecommunications industry, and bring them to a job programming for biology? Especially now that Bioinformatics is seen as an important field, many biology researchers are adding Bioinformatics to their grant proposals and research programs.

Anyone who has experience with designing and carrying out an experiment to answer a question has the basic skills needed to program a computer. A laboratory experiment begins with a question, which evolves into a testable hypothesis, that is, a statement that can be tested for truth based on the results of an experiment or experiments. The processes developed to test the hypotheses are analogous to computer programs. The essence of an experiment is: if you take system X, and do something to it, what happens? The experiment that is done must be designed to have results that can be clearly interpreted. Computer programs must also be carefully designed so that the values that are passed from one part of a program to the next can be clearly interpreted. The human programmer must set up unambiguous instructions to the computer and must think through, in advance, what different types of results mean and what the computer should do with them. A large part of practical computer programming is the ability to think critically, to design a process to answer a question, and to understand what is required to answer the question unambiguously.

Even if you have these skills, learning a computer language isn't a trivial undertaking, but it has been made a lot easier in recent years by languages like C. The key to learning C is to use it and to use it right away. Just as no amount of reading the textbook can make you speak a language like Spanish fluently, no amount of reading any book on C is going to be as helpful as getting out there and trying to program in it. Taking into consideration the striking features of C, one can justify its role in Bioinformatics. Not only that it is easy to program in C, it is easy to adopt the program to the level of a biologist, which is not possible with other programming languages. I hope the examples and exercises in this book would inspire you to try a little programming of your own.

2.1.5 C as a Parent Language for Languages Emerged After It

Almost all the languages that were written after C, inherited many of its features. Take for example, C++, which is the so-called superset of C, has all the procedure-oriented features inherited from C. Pascal, another pretty language has most of the aspects taken from C, but it has its own style of programming. PERL (Practitioner Extract Report Language), which is a frequently used language in developing tools and algorithms in Bioinformatics, has many of its features directly inherited from C.

Even though C does not support object-oriented features or database ideas, the languages which are designed with object-oriented features and database aspects, take their fundamentals from C only.

2.1.6 Significant Features of C

- *It is powerful, flexible and imposes few constraints on a programmer.* This means, it is easy to learn and handle this language. A user need not struggle with the syntax and semantics of the language.
- *It provides faster program execution.* C's power and fast program execution comes from its ability to access low-level commands, similar to assembly level language, but with high level syntax. Its flexibility comes from the many ways the programmer has to accomplish the same tasks. C includes bitwise operators along with powerful pointer manipulation capabilities. This can be an advantage to an experienced programmer but disadvantage to a novice.
- *It provides modularity.* If you can divide your work into pieces and complete it, it won't be a burden to you. This is very often called as **Divide and Rule** concept. Sections of code can be stored in libraries for re-use in future programs. This structure of modularity also helps with C's portability and execution speed. The core C language leaves out many features included in the core of other languages. These functions are instead stored in the C Standard Library where they can be called on when needed. An example of this concept would be C's lack of built-in I/O capabilities. I/O functions tend to slow down program execution and also be machine-independent when running optimally. For these reasons, they are stored in a library separately from the C language and only included when necessary.
- *C is robust and transportable.* This means, it is quite easy for us to learn and easy to carry, install and work. It is not specific to any particular operating system and compatible with all the operating systems such as UNIX, DOS, etc. In simpler words, it is machine-dependent.

- *It is a programmer-friendly and free-form language.* C language is very flexible for the programmers as it does not impose any regulations for typing a program. The programmer is just required to open the editor and start typing the program. He/she can start at any part of the editor (an editor is also called as form and the editor normally used for typing a program in C is called as the Turbo Editor) and end anywhere. Editors of some of the other languages have a few restrictions. For instance, in the case of FORTRAN-77, the programmer has to start the coding of the program only from the 7th column of the editor. Similarly in COBOL, the coding has to be started from the 8th column of the editor. Regulations like these are not there in C. Hence, it is called as a free-form language.
- *It is a case-sensitive language.* C is a case-sensitive language. This means the uppercase and lowercase alphabets are treated differently when a program is written in C. So, be very careful when you start writing programs in C.
- *It is a middle-level language.* There are three categories of languages in all: High Level, Middle Level and Low Level. Those languages that use English like statements for writing the source code are put in High Level category, those which use binary digits (0 and 1) or mnemonics are put in the Low Level category and those that use both the high level and low level syntax are included in the category of Middle Level Languages. C fits in the middle level just because of its efficiency in handling both application-level programming and machine-level programming. You will probably be surprised if I say that you can make your computer monitor to burst by executing a program written in C. You can program your keyboard, mouse, etc. using C and make them work the way you like. That is the reason why C is regarded as the most powerful language among the stream of programming languages we have.

2.1.7 Importance of Learning a Language

Why do we learn languages? The fundamental difference between a human being and an animal is that the former knows to communicate and the latter does not know either to speak or to write. So, the reason behind learning a language is mainly to communicate with fellow human beings. Just as languages help to communicate with others, programming languages help to communicate with the Computing System. Not only that, we are able to come out with applications and interactive software, which make our task simple and easy.

2.2 BUILDING BLOCKS OF C

2.2.1 Methodology of Learning C

The next question in our minds will be How to learn a language? Different people adopt different methods to learn a language. Some languages could be learnt just by speaking, some could be by writing, some just by listening. But, almost all of the languages in this world fall under one domain, where we have to adopt a systematic procedure to learn and master it. Take for example; we would like to learn English. What is the procedure we normally follow? This is depicted in Figure 2.1.

Figure 2.1 Methodology of learning English

In Figure 2.2, we can see the methodology of learning C, which is very much similar to that of learning English.

Figure 2.2 Methodology of learning C

Grammar is a very important part of learning any language that human beings speak. Without that, the learning process would be incomplete. In learning English, we would probably learn the grammar after we learn how to form words with the help of the alphabets. It is not explicitly shown in the diagram since it is understood. There is a counterpart for grammar in C, which is nothing but syntax of the language. The characters, identifiers and keywords must be properly connected up in building a meaningful program. This stringing together of words to make meaningful statements is called the syntax. Also, we have to learn the data types and their qualifiers and operators that are available in the language, before we learn to make up the instructions. Let us start from the first block in the Figure 2.2.

2.2.2 The C Character Set

The character set of C language contains more characters as compared to English, which has just 26 alphabets. The following list gives the character set of the C language:

1. Uppercase alphabets A to Z (ASCII—65 to 90)
2. Lowercase alphabets a to z (ASCII—97 to 122)

3. Digits — 0 to 9 (ASCII—48 to 57)
4. Special symbols — @, $, #, $, ^, &, *,~, !, etc.
5. Mathematical symbols — +, –, <, >, =, /, etc.
6. Punctuation symbols — ?, ., ,, :, ;, ", etc.
7. Other symbols
   a. Symbols which we normally do not have on the keyboard, but still may use in programming (like playing card symbols—spade, clubs, diamond, etc.)
   b. Symbols that have no proper notations, but still on the keyboard (like space bar, backspace, enter, arrow keys)

We see that the character set is quite big and totally there are 256 characters in all, that could be used in C. We have to keep in mind that the language is case-sensitive and hence uppercase and lowercase alphabets are treated differently. Besides the first three categories of the characters listed above, we see the corresponding ASCII codes. What is this ASCII? ASCII stands for the American Standard Code for Information Interchange. It is a standardization established by the Americans for the users of the Computing System. Since the computers are used worldwide, there must be some standard system of representing the notations that we have on and off our keyboard. Languages like C, C++ use this ASCII coding system which could code only 256 (2^8). Since Java is a language user over the Internet, more characters are to be coded and hence the UNICODE (UNIversal CODE) system is preferred. It is a good thing if you remember the ASCII values of at least the Uppercase and Lowercase alphabets and Digits, since we will be later using them in string manipulation programs. Even if you cannot remember them, do not bother too much, since we always have options to get to know them.

When you make a keystroke 'A' from the keyboard, C language will interpret that as a number, more appropriately saying, as a binary number. The ASCII equivalent of 'A' is 65 and this 65 is converted to its binary equivalent, which is 1000001 (I would advice you to go through the chapter on Number Systems once again, in case you have any clarifications on conversion process). Thus, each one of the 256 characters will be coded in the binary format. Remember that each code is unique to itself.

Now that we are familiar with the character set of the language, we are ready to move on to the next block of the learning process, namely keywords and identifiers.

2.2.3 C Tokens—Keywords and Identifiers

Keywords Keywords are those, which the language designers have defined for a special purpose and shall be used only for that. As a user one cannot change the definition of a keyword. C keeps a small set of keywords for its own use. There are 32 keywords that, when combined with the formal C syntax, form the C language as defined by ANSI C. Some keywords were used along with the existing list and were used rarely in programming. To name a few asm, entry, fortran, etc. Look at the Figure 2.3 below to know the keywords. You can notice that all the keywords are in lowercase and none of them should be used for naming an identifier while writing/editing programs. The C compiler will flash with error message when we try to use any of these 32 keywords for naming an identifier. In simpler words, keywords are predefined for the compiler and cannot be redefined by the programmers.

auto	double	int	struct
break	else	long	switch
case	enum	register	typedef
char	extern	return	union
const	float	short	unsigned
continue	for	signed	void
default	goto	sizeof	volatile
do	if	static	while

Figure 2.3 Keywords in C

Identifiers Identifier is a technical term used to mean "name". An identifier in C could be the name of a variable, a constant, a function, a structure, etc. In simpler terms, they are nothing but labels. Identifier as the word itself suggests, an identifier would identify something for the compiler. It identifies what type of value is stored in a variable or a constant or what name a function or a structure has in a program. It is very important that we shall name all the variables and frequently used constants (like pi = 3.14) to make a prior intimation to the C compiler. Otherwise the compiler might find it difficult to recognize a variable or a constant or a function defined in a program. This is very similar to giving name to a new born baby in our homes.

2.2.4 Rules for Naming an Identifier

Before we give a name to the identifier that we use in a C program, we must know a set of rules that have to be followed, rather very strictly. These are given below:

1. The first character of the identifier must be a letter of alphabets (remember C is case-sensitive and hence there are 52 alphabets in C—26 lowercase and an equal number of uppercase) No other character should be used as the first character while naming a variable. For example, the name single for an identifier is acceptable, whereas the name 1single is not acceptable, since the first character is a digit. Also, the name single and the name SINGLE are treated completely differently by the compiler since they are of two different cases (the former being a lowercase name and the latter being an uppercase name).
2. There should not be any embedded space within an identifier. For example, the name reg no for an identifier is not acceptable, since it has a space in between the words reg and no. The C compiler does not accept a space. In fact, it will not accept any other character (other than digit) in between two words of an identifier. If you wish to distinguish two words of an identifier, C has a provision. It allows the programmers to use an Underscore (_) between two words of an identifier. In other words, an underscore is regarded as the 53rd alphabet in C language. So, the above name reg no could be written as reg_no.
3. When you name an identifier in C, you are allowed to use digits within it. For example, the name mem_2_scr is acceptable. One can use digits even towards the end of an identifier's name. But no other character shall be used either within or towards the end of an identifier's name.
4. When you name an identifier, be sure that you are not using any of the keywords defined in the language. This point has been already discussed in keywords. The name of the variable shall be unique to itself.
5. This rule tells you about the length of an identifier's name. The identifier could contain a maximum of 32 characters including digits and underscores for its name. If this limit is exceeded, the same will not be notified by the compiler, since it is not a syntax error. If you exceed this limit, the compiler will simply consider only the first 32 characters of the identifier and will skip the remaining characters.
6. The last rule is not exactly a rule, but a convention to be followed while naming an identifier in C. The convention is Be Suggestive when you name an identifier. This means, you must have a valid reason for naming an identifier. For example, if I am

writing a program to find the area of a circle, I can very well use names for the variables in my program as 'x' and 'y'. But when we have a look at this program at a later time, we cannot be sure what 'x' implies and what 'y' implies. In order to make a program easily understandable, we can be suggestive when we give names to the identifiers. In the above program, which I had mentioned, the variables could be given names like rad, area instead of x and y. So, we understand that the variable rad is going to store the value of radius and the variable area is going to store the area after it is calculated.

2.2.5 Data Types in C

Data types are an important feature of any programming language. Why do we have data types at all? Don't we know what type of input we are going to give to the computer or to the program that we have? It is not for our sake that we must learn the data types. It is necessary to know it so as to let the C compiler know what type of data we are giving as input and what type of data we want back in the output. This means whenever we use variables in our programs, we need to specify the type of value they will be holding throughout the execution of the program.

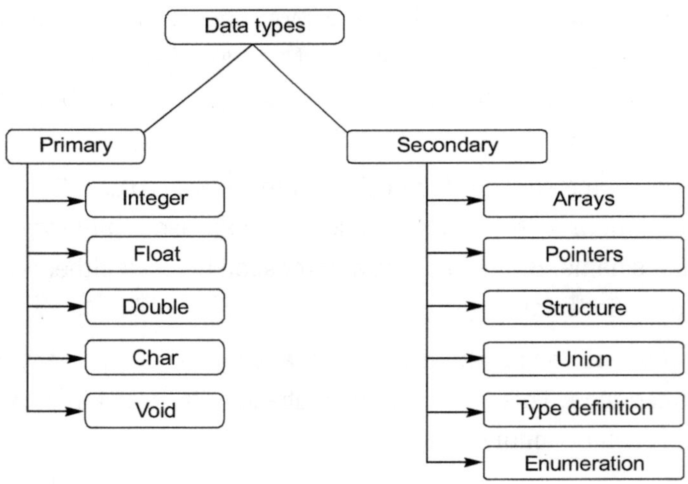

Figure 2.4 Data types in C

Data types are classified into two major categories namely, primary and secondary. The primary data types can again be classified into numerical data types and alphanumeric data types (character type). Numerical data types include integer, float (single-precision) and double (double-precision). Alphanumeric data type includes characters. Figure 2.4 shows the different data types available in the language.

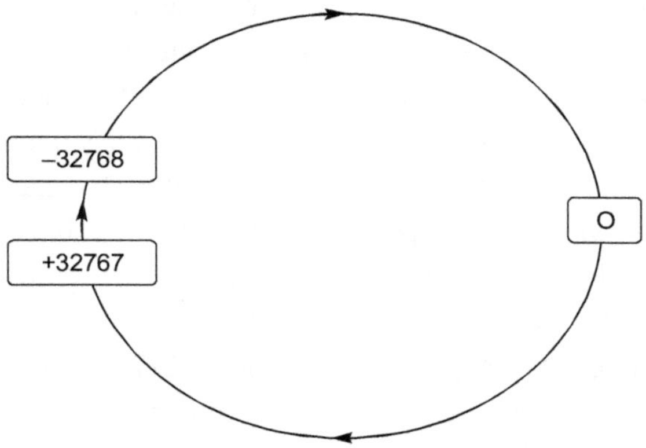

Figure 2.5 Integer data type representation

Data types are further classified into different subtypes based on their size and sign. Except the float data type, all the data types have their own qualifications. Integer has four subtypes, double has one and character has two. Put together there are seven numerical types and two character types. Each data type has its own size and range of values. Table 2.1 illustrates the different data types in the language with all necessary details. Do keep the table in memory, since the same will be required during programming. For the time being, do not bother too much about the last column of the table (format specifier). We shall come to that when discussing the input and output instructions.

Table 2.1 Detailed list of data types in C

Data type	Keyword used	Memory (bytes)	Range	Format specifier
Signed character	char	1	–128 – 0 – +127	%c
Unsigned char	char	1	0 to 255	%c
Signed short integer	int	2	–32768 – 0 – +32767	%d
Unsigned short integer	unsigned	2	0 to 65535	%u
Signed long integer	long	4	–2147483648 to +2147483647	%ld
Unsigned long integer	unsigned long	4	0 to 4294967295	%lu
Float	float	4	–3.4E38 to 3.4E+38	%f
Double	double	8	–1.7E308 to 1.7E+308	%lf
Long double	long double	10	–1.7E4932 to 1.1E+4932	%Lf
Void	void	-	-	-

Different data types have different ranges and hence we must be careful while we use variables pertaining to a particular data type. The C compiler will not check whether we have used the values within the range for a particular variable belonging to a particular data type. That is why we call C very often as a **Weakly Typed Language**. One can use any number of variables in a program. But all of them must be properly type declared using the **Type Declaration Instruction**. The compiler otherwise will report errors. What happens when we cross the limit for a particular variable of a specific type? For example, if a variable *x* is declared in the program as int, the compiler will treat that as 'signed short integer' (please see the table that the keyword used for a signed short integer is just 'int') and hence one can assign values only from –32768 to +32767. What will happen if we assign a value 32770 for *x*? We are crossing the limit and we expect the C compiler to flash an error message. But it will not do so. Errors of this type are termed as **logical errors**, since they arise out of wrong thinking. The compiler will notify only the **syntax errors** in a program. So, for the above situation, what will be the result? Imagine the integer data type's range to be a circle, as shown in Figure 2.5. We see that after the upper limit value 32767, we have the negative extreme –32768. Since, 32770 is three values more than the actual limit, three will be added with the negative limit and the result will be checked again for availability within the range. If it lies within the range, then that value will be the one for the value which we assigned. In the above case the value which will be recognized by the compiler will be –32766. In the table below, the columns size and range are purely architecture-dependent. The values indicated are the most common ones found in 32-bit systems.

Thus, the compiler's response will be quite different and we have to be very careful when we program in C. You can just imagine the same kind of Circle represented in Figure 2.5 for representing the ranges of all the data types and try to get a clear idea of them. Do not ever memorize all the columns of the above table, as you will be practicing them in all your programs and they will be registered in your mind.

2.2.6 Essential Operators in C

Why do we have operators in programming? Obviously the answer is to perform operations. What are the various operations you can perform? Basically we can do addition, subtraction, multiplication and division. Some of you even would be familiar with modulus (remainder out of division). But basically only two operations are there in Mathematics, namely addition and subtraction. Multiplication is nothing but repeated addition and division is nothing but repeated subtraction. But C language has plenty of operators. In fact, it is very rich language with regards to the operators. No other language has such a large number of operators. All the operators are designated by symbols (combination of one or two characters). Let us look out what are the different operators available in this language.

Before you could learn the different types of operators in C, I suggest, you be familiar with the jargon involved. Consider an expression like c = a + b. In this expression, a and b are called independent variables and c is called the **dependent** variable. Mathematically we may write the above expression as c = f (a, b) and we read it as c is a function of a and b. We see the above expression has two sides a **Left Hand Side (LHS)** and a **Right Hand Side (RHS)**. Any expression basically will have two sides, as indicated above. In these kinds of expressions, the RHS will be evaluated first and the result will be assigned to the LHS. In the above expression, the variables a, b and c are called as **operands** and = and + are called as **operators**. In other words, any expression will involve operators and operands. Basically, an operator could be one of two kinds, namely **unary** and **binary**. Any operator that needs only one operand for performing the operation is called a unary operator (una means one) and that which needs at least two operands is called a binary operator (bi means two). Let us now explore the different categories of operators in C.

Category of Operators

1. Arithmetic
2. Relational
3. Logical
4. Bitwise
5. Assignment
6. Shorthand
7. Modification
8. Pointer
9. Structure and Union
10. Function
11. Array
12. Sizeof
13. Ternary
14. Typecast

Let us start with arithmetic operators. The arithmetic operators in C are listed below:

1.	Addition	+ (plus)
2.	Subtraction	– (minus)
3.	Multiplication	* (asterisk)

4. Division / (forward slash)
5. Remainder % (percentage)
6. Unary Minus – (minus)
7. Parenthesis ( )

It is quite simple to understand the way the addition, subtraction and multiplication operators work, since they occur in the basic mathematics, which we have learnt. It is important to understand how the division and the remainder operators work.

For example, consider the following expression: X = 1/2. What do you think the value in X will be? I think everyone will say 0.5. It is not. The value stored in X will be 0. How does this happen? When you work out the same in your calculators or any other computing devices, you will get 0.5 and when you work out this in C, it gives you a zero. Why? The reasoning is quite interesting and deviates from the basic arithmetic you have learnt in schools. We have a numerator (1) and a denominator (2) in the above expression and we see that both are integers. Integers are nothing but whole numbers, which do not have a fractional part. 1 is an integer, whereas 1.0 is not. Even if the fractional part is a zero, we don't consider it as an integer. When we do arithmetic operations with integers as operands, we tend to get only integers and it is applicable to all the arithmetic operators. So, we can form a tabular column like the one shown in the Figure 2.6 for the basic arithmetic operations (+, –, * and /).

As said earlier, the division operation has to be looked into very carefully, as C does a bit differently from the basic division process that we do with pencil and paper.

The next operator to be discussed in this category is Unary Minus (–). It carries the same sign as the Binary Minus, except for the fact that it works completely differently. We can say in more simple words that Minus Operator plays a dual role in C. As far as the jargon is concerned, you know what Unary and Binary are. Lot of questions does arise in our minds. Why do we have a Unary Minus, when there is a Binary Minus? Why don't we have Unary Plus? or Is there anything called Unary Plus? The answers for these questions are very simple and easy to understand. What do you do when you want to express a Positive number (an integer or a real number)? Yes, you are right if your answer is we simply express the number without having a need to prefix the 'plus' (+) operator. For example, '8' is a positive number and we do not say '+8' under normal circumstances. But, what about negative numbers, say, if we want to express a negative '8'? Obviously, we have to prefix a negative sign to the number, that is, we have to express it as '–8'. This 'prefixed minus' sign is called the Unary Minus. It is a must that all the negative numbers shall be prefixed with this minus.

Operations	Operand-I type	Operand-II type	Result type
Addition (+), Subtraction (–), Multiplication (*), Division (/)	Integer	Integer	Integer
	Integer	Float	Float
	Float	Integer	Float
	Float	Float	Float

Figure 2.6 Results of the basic arithmetic operations with data types

The next operator in the list is the Modulus Operator (designated by a % symbol). Most probably, this operator would have been used by you in your higher classes, but in a different way. We term that mathematically as a "remainder operator". What this operator does is that, it finds out the remainder of the two integer values. Of course, it divides one number by the other (like the division operator), but leaves the remainder as the answer (unlike the division operator, which leaves the quotient). Then, what could be the answer for 8%5? Yes, the answer is 3, since upon dividing 8 by 5, the remainder that we get is 3. What could be the answer for 5%8? If your answer is 0, then it is wrong. Then, what else is the answer? It is 5. Wondering how!!! Work out this division with a pencil and a paper or just ask yourselves, what could be the quotient on dividing 5 by 8? Or how many times do 8 go in 5 exactly? Obviously 8 will not go in 5 or simply saying, 8 goes 0 times in 5. So the quotient you obtain is 0 and then to get the remainder, you multiply the divisor with the quotient and the product obtained is subtracted from the dividend. (In division process, the numerator is termed the dividend and the denominator is termed the divisor. Normally to get the remainder of the division, we first multiply the quotient with the divisor and subtract the product obtained, from the dividend). In the above example, when we multiply the quotient (which is 0) with the divisor, we tend to get 0 as the product and when we subtract the product, that is, 0 from the dividend, which is 5, we get 5. This is the justification for 5% 8 = 5. I can give you a blind way of understanding this. For any modulus division of the format m%n, the answer will be 'm' provided (i) m < n and (ii) m, n > 0. But, don't just keep this blind idea in your mind, since you may be required to justify in the way I have done above. Now, try the following: –5%8, 8% – 5 and –5%–8. What do you think the answers will be for the above expressions? If your answers are –5, 3 and –5, they are right. But, how? Again apply the same idea of division. When you evaluate the number of times 8 goes in –5, it will obviously be a 0. Hence, the remainder is –5. When you evaluate the number of times –5 goes in 8, it will be –1 and when this quotient is multiplied with the divisor (–5) we get 5 and when this product is subtracted from the dividend, we get 3. And finally when you evaluate the number of times –8 goes in –5 it will again be a 0 and hence the remainder is –5. Can you conclude anything from the above discussion, with regard to the % operator? Yes, the sign of the result of %

division is always decided from that of the numerator of the expression. So we can have a table for deciding the sign of the result of the % division as shown in the Figure 2.7.

%	+	–
+	+	+
–	–	–

Figure 2.7 Operation and results with % operator

What could be the result of the expression 5.85 % 9.4? Don't take too much pain in solving this expression, since % operator is applicable only to integer and its associated data types and not to any other data type. Thus, the above expression will give rise to an error and hence be ruled out.

The last one in the list of arithmetic operators is parenthesis ( ). This set of parenthesis also plays a different role with regard to the context in which it is used. When it is used in an arithmetic expression (an arithmetic expression is the one which has arithmetic operators alone operating on the operands), it shall be considered as an arithmetic operator. In any complex expression, the expression within the innermost parenthesis shall be evaluated first and then the one in the next innermost and so on. For example, consider the following arithmetic expression: a+b/(c*(d+e)). In this, we have an expression 'd+e' within parenthesis and another expression 'c*(d+e) also within paranthesis. The former expression is within the innermost parenthesis and the latter within the next innermost (or probably the outermost) parenthesis. So, when the entire expression is evaluated, 'd+e' is evaluated first and the result is multiplied with 'c'. This is the importance of using a set of parenthesis in an expression. We can give priority to a particular part of the whole expression thereby overriding the normal precedence rules. It is right time for me to explain the precedence rules associated with the arithmetic operators. The following list provides the rankings or priority to the arithmetic operators.

1. ( ), – (Unary Minus)
2. *, / and %
3. + and – (Binary Minus)

This is otherwise called the hierarchy of operators and all the expressions we use in C language are governed by such hierarchy rules. As we go on with this discussion, I will let you know the hierarchy of all the operators pertaining to their category.

There comes a question in our minds. If any expression contains both '*' and '/', which will be preferred first by the C compiler? Preference will be given to the operator that comes first from left.

For example, consider the following expression: c = 4*6/8%3–4+9. What will be the result of the above expression after you apply the above precedence rules? It is 5. How? When the C compiler starts to evaluate the expression on the Right Hand Side, it traces it from the left and when it does so, the first operator encountered is '*' and hence 4*6 will be evaluated first, whose result is 24 and this intermediate product is now divided by 8, since '/' appears next to '*'. So, we get a quotient of 3. This in turn is divided by 3 and the remainder of this division happens to be 0. The next operator encountered is '–' and when 4 is subtracted from 0, we get –4 and finally we have '+' using which we add –4 and 9 and get the result as +5.

What is the importance of knowing hierarchy? Why do we want to implement in languages? Let us understand the reason with an example.

Try to evaluate the following expression: c = 5/9*(f–32). This is the formula to convert the temperature in Fahrenheit to Centigrade. Using the hierarchy of arithmetic operators we get zero as the result, for any input for the temperature in Fahrenheit. This happens because according to the hierarchy, f–32 will be first evaluated, which will give a value and then 5/9 will be evaluated, which will give a zero. When Zero is multiplied with any value, obviously the result is Zero. Thus, even though the expression given above is mathematically correct, it will not give the right result. Some modification should be done in the above expression to get the correct answer—change the expression to c = 5.0/9*(f–32) or c = 5/9.0*(f–32) or c = 5.0/9.0*(f–32)—it will work.

Such is the importance of hierarchy of operators in C and it has to be followed when you are writing a program or developing an application in C.

The next category of operators to be discussed is the Logical Operators. The logical operators in C are NOT, AND and OR, designated by the symbols, ! (exclamation mark), && (double ampersand) and | | (double pipe) respectively. These symbols accept the inputs and process them in the same way as the arithmetic operators. But the way they treat their operands is different. For a better understanding of how the operands will be treated, you should know what **Boolean Values** are. In digital electronics, there are two numbers normally used, 0 and 1. We had discussed about these two numbers earlier as binary digits and now we will call them as Boolean Values. In C language, any non-zero number is treated by the compiler as a '1' (stands for true) and any zero as a '0' (false). All the three logical operators accept the inputs and internally convert them to either '1' or '0' based on their values. For example, consider the following logical expression (any expression that has logical operators in it): c = a&&b, where the values of a and b are 20 and 5 respectively. Since 20 and 5 are non-zero values, they are internally converted to '1' and hence the above expression reduces from c = 20&&5 to c = 1 && 1. Thus, we see that regardless of the values we give, they are either treated as a '1' or a '0'. Out of the three logical operators, NOT is a Unary operator and the other two are binary operators. Thus, it is not possible to have an expression like c = a!b.

The logical operators give the result also in Boolean format. This means the result of the operation will either be a '0' or a '1'. Let us consider first the operator AND (&&). Using 'double ampersand' for AND is a must. We may commit a typing mistake by just typing a single ampersand. It will lead to a different meaning and a different answer and this type of problems will not be notified to the users or the programmers. Hence, the programmers and the learners must be very careful when they use this operator. In digital electronics, this operation is achieved with the logical devices called GATES. There are two categories of gates called primary and secondary. The primary gates are AND, OR and NOT. They work based on the combination of the inputs given to them. For AND and OR gates we can give any number of inputs and for a NOT gate, we can give only one input. Let us consider for simplicity that we give two inputs for AND gate. There are obviously four different combinations. They are (0,0), (0,1), (1,0) and (1,1). They are formulated inside a table called the **truth table**. This table for a two-input AND gate is shown in the Figure 2.8.

Input a	Input b	Output c
0	0	0
0	1	0
1	0	0
1	1	1

Figure 2.8 Truth table of a two-input AND gate

Thus we can now generalize an AND operation and define it more formally in terms of C. An AND operation is defined as that which gives a true output if and only if all its inputs are true. See that we have used only two inputs for justifying this general statement. This holds good for any number of inputs. Also, AND logic is often called as Boolean Multiplication and is very useful for Boolean Algebra.

An OR gate works in a way just opposite to that of an AND gate in digital electronics. The truth table of a two-input OR gate is shown in the Figure 2.9.

Input a	Input b	Output c
0	0	0
0	1	1
1	0	1
1	1	1

Figure 2.9 Truth table of a two-input OR gate

Now we can define the OR operation from the outputs generated in the table as follows: An OR operation is defined as that which gives a true output if at least one of its inputs is true.

A NOT gate is a Unary gate and hence there can be only one input for it and it produces an opposite output for an input. For example if the input is true, it produces a false output and vice versa.

Thus a NOT operation is defined as that which produces a true output when the input is false and vice versa.

Thus, when you have an AND operation throughout an expression, the expression will be checked till a first '0' is encountered. When you have an OR operation throughout an expression, it will be checked till a first '1' is encountered. Suppose that there are 100 inputs in an expression and throughout we have an AND operator, then the expression will finally yield a '0' if the expression contains a zero either in the beginning, in the middle or at the end. The evaluation of the expression will stop once the first '0' is encountered. If in the same expression, we have an OR operator, the result will be a '1' if the expression contains a '1' in the beginning, in the middle or end. The evaluation will stop once the first '1' is encountered. This is the way the logical AND and OR operators work. One has to be very careful while performing operations with AND and OR operators, especially when they both occur in a logical expression. The hierarchy of logical operators is as given below:

1. NOT (since Unary)
2. AND
3. OR

The next category of operators is relational operators. They are otherwise called comparison operators. They are six in number and they are >= (greater than or equal to), > (greater than), <= (less than or equal to), < (less than), == (equal to) and != (not equal to). As the name suggests, they are used for comparing two inputs and all of them are binary operators. Probably, these operators would be familiar to you since they are available in the basic mathematics itself (only a few symbols are new to you). These operators accept the input and return either a true or a false (i.e., a '1' or a '0'). For example, if we have an expression X = 10>=3, it evaluates to true, since '10' is greater than '3'. Thus, there will be a '1' stored in 'X' after the evaluation. If we have, say, X = 10<3, the expression will evaluate to false and hence a '0' will be stored in 'X'. The operators '==' and '!=' are called otherwise as 'equality' and 'inequality' operators respectively. The main difference between the logical and relational operators is that, the logical operators accept the input in Boolean form and return the result also in Boolean form, whereas, the relational operators accept the input in Numerical form (Arabic Numerals) and return the result in Boolean form.

Relational operators are simple and easy to work with and they carry the basic meaning as in mathematics. We have to be very careful when we use '==' (double equal to or equality operator) in relational expressions, since we have another operator '=' (single equal to), which stands for 'assignment' operator. We will get unanticipated results when we use one in place of another. Before going to the discussion of the above difference, let us know about the assignment operator. An assignment operator is the one which is used in almost all the expressions, whether relational or logical or arithmetic. The whole expression will be called an assignment expression, which has two sides, a Left Hand Side (LHS) and a Right Hand Side (RHS). Usually, the RHS will be the first to be evaluated and the result will be assigned to the LHS. Also, in any assignment expression, the LHS will always have only one variable (not even a constant is allowed) and the RHS may have any number of variables or constants clubbed together by the valid operators of the language. For example, C=A+B+6 is a valid expression. But the expression 3=R+T is not valid, since we have a constant on the LHS. Also, the expression A+B=C is not valid since there are two variables on the LHS. Now let me come to the point of using an assignment operator in the place of an equality operator. Consider the expression C=A+B. Here we add the values of 'A' and 'B' and assign the result to 'C'. Consider the expression C==A+B. Here we are adding 'A' with 'B' and comparing the result with 'C'; that is, we are checking whether the sum of A and B and the value of C are the same. If so, the above result will yield a '1' and if not, we will get a '0'. Assuming the value of A as 8, the value of B as 3 and the value of 'C' as 11, the second expression will yield a '1', whereas the first will yield a '11' (will overwrite any previous value available in 'C'). Thus, one has to be very careful with the assignment operator and the equality operator.

The next category of operators is bitwise operators. The bitwise operators are a special category in C language, since no other programming language prior to C supports this kind of operators. In fact, it is the bitwise operators in C which have made the language so popular. There are six bitwise operators in C, namely one's complement operator (~), bitwise logical AND (&), bitwise logical OR (|), exclusive OR (EX-OR / X-OR - ^), shift left (<<) and shift right (>>). Among these operators, one's compliment operator is a unary operator and all the others are binary. When we use a bitwise operator in an expression, the operands in the expression are converted to their binary format and then the expression is evaluated. Let us first have a look at the binary bitwise operators and then go to '~'.

Consider the bitwise logical AND. It is designated by the symbol '&'. For a logical AND operator we use a '&&'. We have to be very careful again. For example, consider the expression C=4&&5 and another expression C=4&5. I would like to project the common mistake that is being made while typing a program. One may type '&' instead of '&&' and vice versa. This sort of error will not be notified to the users since it is a logical error. The result of both these expressions will vary and hence one has to take utmost care while using a '&' and '&&'. Let us see what happens to the above two expressions. The first one is a logical expression and the second one is a bitwise logical expression. In the first expression, both 4 and 5 will be converted to intermediate value '1' since they are non-zeroes and hence the expression will be reduced to C=1&1, whose result is a '1', according to the definition of AND operator. But when we look at the way the second expression is evaluated, it will show the difference between the two operators. In the second expression, both 4 and 5 are converted to their binary equivalents. Normally the conversion will be such that the binary format has 16 bits or 2 bytes of memory. The reason for taking 16 bits will be discussed when we do data types in C language. For discussion sake, let us confine ourselves to 3 bits. The binary of 4 is 100 and the binary of 5 is 101. When we use these two binary formats and operate them with '&', we have to consider each pair of the three pairs of bits available and perform the logical operation on them (see Figure 2.8). The result of the expression is 100, whose decimal equivalent is '4'. Thus, we get '4' in the expression C=4&5 and we get a '1' in the expression C=4&&5. The moral is: Use the operators appropriately and get the proper results.

Let us consider the bitwise logical OR. It is designated with the symbol '|'. For a logical OR operator we use a '||'. We have to be very careful again in using a logical OR and bitwise logical OR. Using one in place of the other will yield unwanted results. For example, consider the expression C=4||5 and another expression C=4|5. Again this may be a typing error as in the case of using a '&' in the place of '&&'. The error will not be notified by the compiler, since we have been using the right syntax only. The result of both these expressions will vary and hence one has to take utmost care while using a '|' and '||'. Let us see what happens to the results of the above two expressions. The first one is a logical expression and the second one is a bitwise logical expression. In the first expression, both 4 and 5 will be converted to intermediate value '1' since they are non-zeroes and hence the expression will be reduced as C=1||1, whose result is a '1', according to the definition of OR operator. But when we look at the way the second expression is evaluated, it will justify the difference between the two operators '|' and '||'. In the second expression, both 4 and 5 are converted to their binary equivalents.

The binary of 4 is 100 and the binary of 5 is 101. When we use these two binary formats and operate them with '|', we have to consider each pair of the three pairs of bits available and perform the logical operation on them (see Figure 2.9). We have the result of the expression as 101, whose decimal equivalent is '5'. Thus, we get '5' in the expression C=4|5 and we get a '1' in the expression C=4||5. Again, to get the correct results use the operators appropriately.

The next bitwise operator is Exclusive OR (otherwise termed as Ex-OR or X-OR) designated with the symbol '^' (caret). In the same way as we defined the logical AND, OR and NOT operators, we have to formally define an X-OR operator too. An X-OR operator could be defined as the one in which the output will be true if and only if we have ODD number of TRUE inputs. This could be justified with a truth table shown in the Figure 2.10. In this truth table, we have used just two inputs X and Y and so we have the outputs for four different combinations of the inputs. Had there been three inputs, there must be eight possible combinations. We could see from the table that only for odd number of TRUE inputs we get a TRUE output and for other combinations we get a FALSE output.

X	Y	Z = X^Y
0	0	0
0	1	1
1	0	1
1	1	0

Figure 2.10 Truth table of a two-input X-OR gate

I will just illustrate you how this operator functions when we write programs with an example. Consider, for example, the problem of swapping (interchanging) the values of two variables. For swapping, we may have umpteen solutions, like, we may go for using a temporary variable, we may just use the arithmetic operators, etc. Let us consider two variables X and Y and let their values be 2 and 5. When a temporary variable, say temp, is used, the following set of codes will result in swapping the values of two variables:

1. temp = X (2 transferred to temp)
2. X = Y (5 transferred to X)
3. Y = temp (2 transferred to Y)

The diagrammatic representation of the stepwise swapping process for easy understanding is given in Figure 2.11.

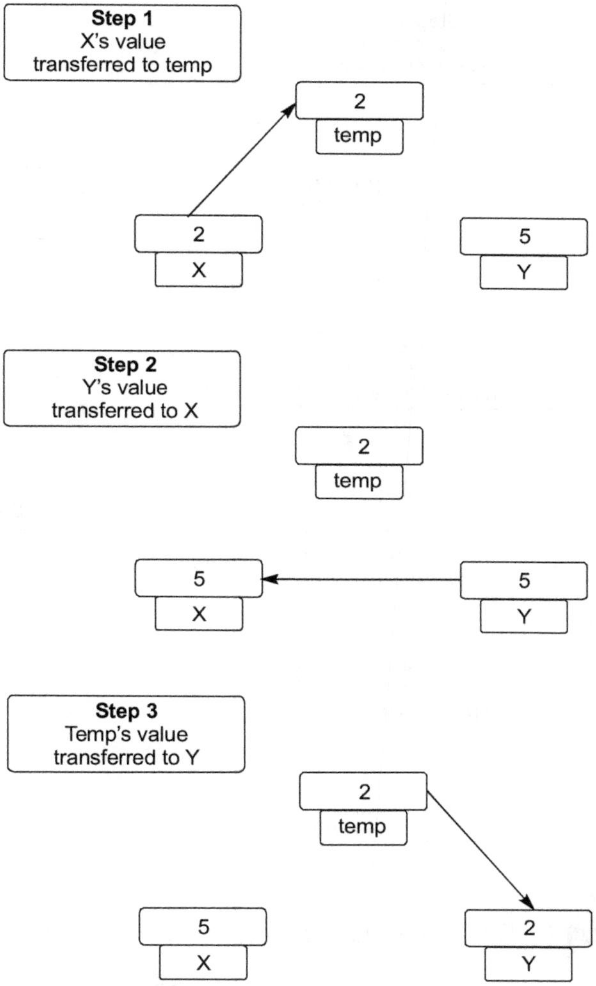

Figure 2.11 Swapping values using a temporary variable

The second method for swapping the variables values will not be using a temporary variable. Rather, we will be using the variables X and Y and use the basic arithmetic operations addition and subtraction on them and swap the values. The following are the steps:

1. X=X+Y (2+5=7, 7 replaces 2 in X)
2. Y=X–Y (7–5=2, 2 replaces 5 in Y)
3. X=X–Y (7–2=5, 5 replaces 7 in X)

At the end of these three steps, we see that X has 5 and Y has 2 and hence the values are swapped. This process is shown in the Figure 2.12.

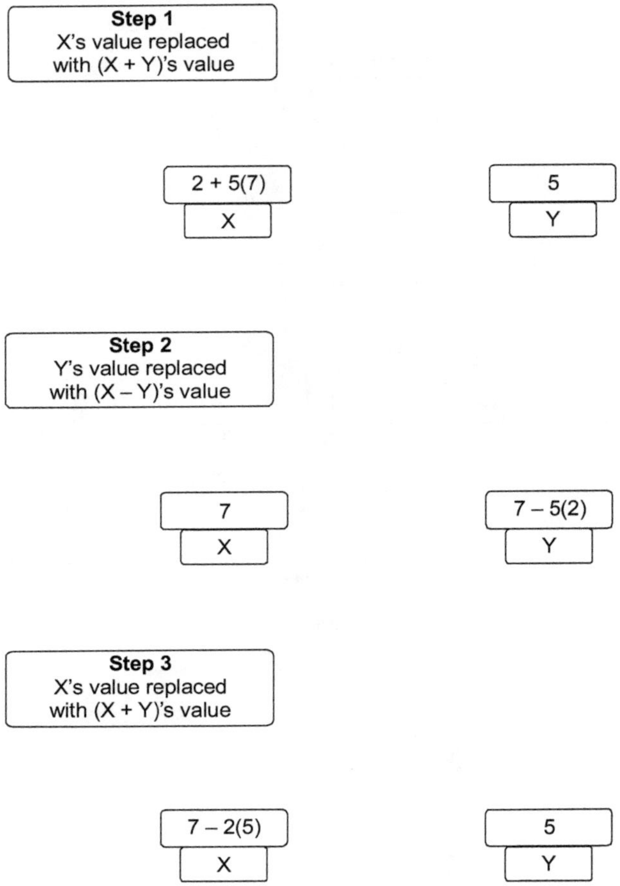

Figure 2.12 Swapping values without using a temporary variable

The same process can be accomplished with the X-OR operator and the steps are given below:

1. X = X^Y
2. Y = Y^X
3. X = X^Y

At the end of these three steps, the values would have swapped. Consider for example, the values of X and Y to be 2 and 5 respectively. So, to start with we will have 2 in 'X' and 5 in 'Y'. After the above three steps are performed, the value in 'X' will be 5 and that in 'Y' will be 2. Hence the swapping will take place. Consider the first step, where we have X = X ^ Y, that is X=2 ^ 5. Here the values 2 and 5 will be converted to their binary as usual and then operated at bit level. Hence, we have the intermediate expression as (considering three-bit pattern for the integers 2 and 5),

X = 010 ^ 101, which will yield an output 111 (refer Figure 2.10). The decimal equivalent of this binary is '7' and it is stored in 'X' and the previous value of 'X' is lost (that is 2 is lost). Consider now the second step, where we have Y = Y ^ X, that is Y = 5 ^ 7. The equivalent binary expression for the above is Y = 101 ^ 111, which will yield 010. The decimal equivalent of this result is '2' and it is stored in 'Y'. Hence the previous value of 'Y' is lost (that is, 5 is lost). Now after the execution of steps 1 and 2, we have, '7' in 'X' and '2' in 'Y'. Consider now the third step, where we have X = X ^ Y, that is X = 7 ^ 2. The equivalent binary expression for the same is X = 111 ^ 010, which will give 101. The decimal equivalent of this is '5' and it is stored in 'X'. Thus, we see that, at the end of the three steps, the value of 'X' is 5 and that of 'Y' is '2'. This means the initial values in 'X' and 'Y' have been swapped. Thus, we have another logic for swapping the values of two variables apart from those that we had already discussed. Where do we exactly use this operator? Since this is a bitwise operator, this finds application at low level where one can swap the information between two registers or two memory locations very easily.

The next operator for discussion is shift right operator (>>), which is also a binary operator like '&', "|' and '^'. The basic function of this operator is to shift the binary equivalent of the given number to its right by the number of bits specified. Consider, for example, the expression, a = 9>>2. Here, we mean that the binary equivalent of the number '9' has to be shifted to the right by two bits. Let us assume the 16-bit pattern for the above operation. In a 16-bit pattern, the leftmost bit is called as the Most Significant Bit (MSB) and the rightmost bit is called as the Least Significant Bit (LSB). So, when you count the bits, always count from the right and proceed to left and start the counting from '0' and not '1'. (As a convention, when you are learning C, do the counting from '0' and not from '1'.)

0	0	0	0	0	0	0	0	0	0	0	0	1	0	0	1

Figure 2.13 Binary pattern of '9' in the memory

Figure 2.13 shows the pattern (binary equivalent of '9') for 9 in the memory. The bits are shifted in the direction towards right (because of shift right operation). The first bit to be shifted will be the LSB. Since, we have shifted the LSB towards right side, we would be apparently losing that bit and hence that position will become vacant. To fill up the position, we shift the next LSB (the bit at the second position from right) to the vacant bit. Like this we will be shifting all the bits one by one till we reach the MSB.

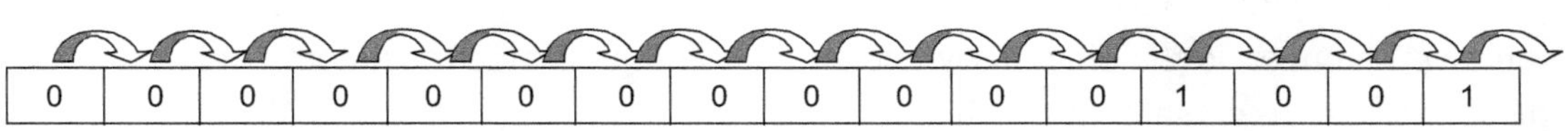

Figure 2.14 Methodology of right shift in the memory

	0	0	0	0	0	0	0	0	0	0	0	0	1	0	0

Figure 2.15 Status of binary '9' after shifting by one bit to right

This is indicated in the Figure 2.14. The resultant pattern after one complete shifting is shown in the Figure 2.15. We could see that the MSB becomes vacant at last. But, one cannot leave any of the bits vacant. Either it shall have a '0' or a '1'. So what could be the bit available in the vacant slot? Will it be occupied by the bit that is expelled from the LSB or will it be a different one? One has to keep in mind while shifting a bit towards right or left that, a shift is not a rotate. This means, the bit that is lost at the rightmost end (LSB) is lost forever. The vacant position at the leftmost end (MSB) is filled with a '0' (this is called a leading zero). Thus we have a new pattern as shown in the Figure 2.16. We see from the pattern that, after shifting the given pattern (decimal '9') by one bit, the value of the pattern has changed. The decimal value of the above pattern is '4'. In the expression that we have considered, we have to shift the given pattern by two bits. The above process is illustrated for only one bit. That is in the above discussion, we have shifted the given pattern by one bit towards right. So, we have to repeat the entire process with one more bit and for this, we have to take the above pattern as the source (that is the new value '4'). If we carry out the same procedure with the above pattern, we will finally get the pattern which is shown in the Figure 2.17.

0	0	0	0	0	0	0	0	0	0	0	0	0	1	0	0

Figure 2.16 Pattern after including a leading zero during the right shift

0	0	0	0	0	0	0	0	0	0	0	0	0	0	1	0

Figure 2.17 Pattern after shifting the binary of '9' to the right by two bits

The decimal value of this pattern is '2'. Thus we have shifted the binary value of '9' by two bits towards right using the operator '>>' and obtained the result as '2'. This result is stored in the variable 'a'. The intermediate result in the process was '4'. Hence, we can conclude that upon shifting the binary equivalent of any decimal number towards right side by '1' bit, we actually divide the number by '2'. Thus, we can generalize the result as, if we shift the binary equivalent of a decimal number 'd' towards right by 'n' bits, we actually divide the number by 2^n.

So far we have been discussing the bitwise operators, which are all binary operators. The operator '~' (pronounce this operator as tilda), which is used for getting the 1's complement of the given number, is the only unary operator among the bitwise operators. To have an in-depth discussion on this operator, we shall actually know what a "complement" is. We have many number systems in use like Decimal (10 numerals—0 to 9), Binary (2 numerals 0 and 1), Octal (8 numerals—0 to 7) and Hexadecimal (16 numerals—0 to 9 and A to F). Each number system has two types of complements called as "radix complement" and "radix-minus-one complement". For example, the decimal number system has 10's complement (radix complement) and 9's complement (radix-minus-one complement), the binary number system has 2's component (radix complement) and 1's complement (radix-minus-one complement) and so on. Consider a decimal number '43'. The 9's complement of this number could be found out by subtracting each of the digits from '9'. On doing this, we get '56', which is the 9's complement of '43'. To find out the 10's complement, we have to first find the 9's complement and add '1' to the result. For the number '43', 9's complement is 56 and the 10's complement will be '57'. Similarly, when we consider a binary number like 11011, the 2's complement could be obtained by subtracting each of the digits from '1'. The 2's complement could be obtained by adding one to the result of the above process. On subtracting each of the digits of 11011 from '1' we get, 00100 and adding '1' to it we get, 00101.

The operator '~' does the job of finding the 1's complement for the given number after converting it to bits. For example, consider the expressions a = 0 and b = ~a. The value in 'a' is '0' and it has to be converted to its binary equivalent in 16-bit pattern. (We are taking 16-bit pattern because we consider '0' as an integer and normally an integer will take 2 bytes—16 bits, in memory.) The 16-bit binary equivalent of '0' is shown in the Figure 2.18 (memory picture for 'a'). When we evaluate the expression b=~a, the 16-bit pattern of 'b' will be the one shown in Figure 2.19. Can you predict what will be the value of 'b'?

0	0	0	0	0	0	0	0	0	0	0	0	0	0	0	0

Figure 2.18 Binary equivalent of '0'

1	1	1	1	1	1	1	1	1	1	1	1	1	1	1	1

Figure 2.19 Pattern after the 1's complement of '0'

If you had followed a procedure that had been taught to you in converting a number from binary to decimal for getting the result of the above pattern, then you are wrong. If you had followed that method, then your answer would be 65535 and C does not yield this result. Instead, C will do the calculation in a different way. The decimal equivalent of the second 16-bit pattern noted above would be '–1'. Wondering how??? Let me justify the answer.

In subjects like Digital Electronics or in Computer System Architecture, you would have come across the concept of encoding and decoding and, most appropriately, we can call it as **sign-magnitude** representation of a number. In sign-magnitude representation, the leftmost bit, which is called the **Most Significant Bit (MSB)** is used for indicating the sign of the number. If it is set as '0', the sign of the number stored would be positive and if it is set as '1', the sign of the number stored would be negative. C uses exactly this idea in giving the result to you. It will store all the negative numbers in an encoded format (usually the format used is 2's complement approach). When we apply the '~' operator to the first 16-bit pattern, all the zeroes in it are converted to ones, due to the fact that the operator is a 1's compliment operator and it has to do like that.

Keeping the above logic in mind, let us try to analyse how the decimal equivalent of the second 16-bit could be '-1' and not '65535' as you would have calculated. From the second 16-bit pattern, we notice that the MSB is '1' and hence the number stored in the pattern is surely a negative number. But, we have to decode the pattern, since it is encoded in 2's complement format. In order to decode that, we have to find out its 2's complement. For finding 2's complement of a given binary pattern, we have to first find the 1's complement and then add a '1' to the Least Significant Bit (LSB) of the result. The 1's complement of the second 16-bit pattern is shown in Figure 2.20 (note that it is the same as the pattern in the Figure 2.18).

0	0	0	0	0	0	0	0	0	0	0	0	0	0	0	0

Figure 2.20 1's compliment of the pattern in the Figure 2.19

When we add a '1' to the LSB of the pattern shown in the Figure 2.20, we will be getting the pattern shown below in the Figure 2.21.

0	0	0	0	0	0	0	0	0	0	0	0	0	0	0	1

Figure 2.21 2's compliment (by adding a '1' to the LSB of Figure 2.20)

The decimal equivalent of the pattern in Figure 2.21 would be '1'. So, we decide the sign of the resultant pattern from Figure 2.19 and the magnitude from Figure 2.21, which together will give the result as '–1'. Many results in C language will not comply with that of the users of the language and one has to be thorough in the concepts of the language, to be able to work like a C compiler. With this operator '~', we have come to an end of the discussion related to the bitwise operators.

The next operator is Assignment Operator designated by the symbol "=". Note that we have already discussed "==" operator (called "equality" operator) in the Relational Operators category and one has to be extremely careful when they use '=' and '==' operators, since they both yield completely different results. For example consider the expressions, a=b and a==b, where the variables 'a' and 'b' have the values 3 and 4 respectively. In the first case, the value of 'b' is transferred to 'a' and hence we will lose the value already existing in 'a' and hence in the end, we will have the value '3' in both 'a' and 'b'. When we look at the second expression, it is obvious that we are comparing the values of 'a' and 'b' and checking whether they are equal. In our case, they are not equal and hence we will get a 'false' value returned from the evaluation (refer to the earlier discussion of relational operators). Thus, the result of the second expression would be a '0' (Zero). The beginners in C often commit the mistake of confusing the operators which are alike. In C, different operator symbols will be used for different contexts and one must be clear about using the symbol in the right context.

The next category of operators is Shorthand Operators. As the name suggests, they are really used in programming for cutting short an expression that may be lengthy. For example, consider the expression, a=a+b, where we have two variables 'a' and 'b'. In the expression, we are trying to update the value of 'a' by adding the value of 'b' with already existing value of 'a'. When we are doing so, instead of saying a=a+b, we can very well say 'a+=b' and we notice that the expression is cut short because of '+='. Thus, whenever we update a variable's value, we can use a shorthand operator. The following are some of the shorthand operators: −=, *=, /=, %=. A shorthand operator enables us to cut short the expression without losing the meaning. Some operators do not have their shorthand form, e.g. ">=", "<=", etc. since they already have two symbols in their notation.

We have so far discussed all the essential operators, which are important for basic programming in C. Let us temporarily stop the process of learning the operators here and we will learn more operators in the forthcoming sections, whenever one is needed. Let us move to the next building block of C language, namely, I/O instructions.

2.2.7 C Instructions/Statements

Type declaration instructions We are aware of the data types available in C (please refer 2.2.5). Once we start to write programs in C, we must follow certain regulations and structure. The very first part of any C program would be (not necessarily in all the programs) the declaration of variables and their corresponding types. This section is called as type declaration part. For example, consider the following C statements (As of now do not worry about the syntax of the statement since we will deal with the same later):

```
int a,b=10;
float c;
```

In the first statement, we have used two variables a and b and both of them are of type integer and in the second statement we have used another variable c, which is of type float. This section of the program informs the compiler in advance that there are going to be three variables in the program with the names a, b and c. All the variables that would be used in the program shall be necessarily declared in the type declaration itself and not anywhere else. This means, if we are using a variable in the 100th line of a program containing 500 lines, we cannot be in a mindset to declare that variable in 99th line of the program, since the compiler would flash that as an error.

This issue, however, has been fixed in C++, where the user is free to declare the variable at the place it is used for the first time. The style of type declaration that is used in C enhances the structure of the program and the program does not look clumsy.

When the variables are declared, the C compiler does three things. The following steps illustrate these three things.

1. Identifies memory location for each variable declared
2. Associates name for each memory location (as used by the user)
3. Assigns value for each variable (if it is given by the user, that would be used or a garbage value would be used)

For example, in the first statement above, both a and b are declared as integers and b is given a value of 10 by the user while it is declared. This way of giving an initial value while declaring the variable is called as initialization during declaration. Let us execute the above steps for the variables a and b. In the case of a, the compiler will first reserve a location in memory for the variable, say, 1000th byte of memory. Then it associates the name, 'a', with that memory location. Finally, it assigns a value to that variable. In this case, the value would be garbage, since the user has not given any value. This garbage value would be within the range of the corresponding data type and it would be surely decided by the compiler and not by the user. Also, the garbage value for a variable would be different for different users for the same kind of program that may be executed simultaneously in two systems. In this case, 'a' is an integer and more precisely, a short signed integer and hence the value assigned for it would be within the range of –32767 – 0 – +32767 (please refer the Table 2.1). In the second case, the variable 'b' is declared as integer and also given a value of 10. The first two steps of the above procedure is executed for 'b' as it was for 'a'. The third step alone is quite different for 'b' and it would be given a value of 10 instead of being given a garbage value, since it has been initialized as it is declared.

I/O instructions Input and output instructions or the functions are the heart-throb of any language, since the programmer would be able to communicate with the computer only through these instructions. This is exactly the counterpart of framing sentences in learning English (as depicted in the Figure 2.2 earlier). There are many input and output instructions in C, of which the user has necessarily got himself/herself familiar with two important ones, namely scanf() and printf(). In C language, the instructions (what we term as commands in some other programming languages) are mostly in the form of functions. We will see more about the functions in the chapter 8. But, for the beginners of the language, let me tell you that functions are nothing but that do some specific task assigned to them and they must follow the syntax rules defined by the language. C language is full of functions (any function name in C would be followed by a set of parentheses, that may or may not be empty and it is a very easy identity for them) and hence it is very often called as a compiled language or a procedure-oriented language.

The two important and essential input and output functions, as I had quoted earlier are scanf() and printf(). Let us discuss how and why we should use these functions.

The function scanf() is used to get the input from the input device (normally a keyboard in case of C, since C is a Command Based Language). The expansion for the word "scanf" is "scan format". It has the following method of usage, formally called as the prototype (or model).

```
int scanf(const char *format, ...);
```

Its prototype is available in the header file called stdio.h. It returns the number of input fields processed successfully. It processes input according to the format and places the results in the memory locations pointed to by the arguments. (stdio.h is nothing but a standard input and output header, containing prototypes of all the input and output functions. We will see more about a header file later in the chapter 2). The more general model of scanf() is as given below:

```
scanf("format specifier(s)", (list of) address(es) of the argument(s));
```

As an example, consider the following statement which uses a scanf().

```
scanf("%d%d",&a,&b);
```

In the above statement, we see that we had used "%d". Now, you can just go back to Table 2.1, where we had used one column called "format specifier". All the data types in C have their own format specifiers and one has to be very careful in using them while programming, especially in the functions scanf() and printf(). In the above example, I have used two %d format specifiers inside the function scanf() and this means I am trying to input two short signed integers. This information is easily recognized by the C compiler, since we had used the format specifier %d. Had we used two "%f" inside the function scanf() in the above example, then the compiler would expect two float values from the user. So, the only way to let the compiler know that we would be giving a float input is by providing proper format specifier.

Now, let us analyse the job of the function scanf(). This function is just doing the job of a postman. A postman will go through the postal address in the envelope and drop the letter in that address. Similarly the function scanf() gets a specified value from the input device (keyboard) and drops that value in the address or memory location of the corresponding variable.

We have seen in the previous section that when a variable is declared, three things happen at the compiler level. These steps invariably would have happened for the variables a and b used here. They both would be having their own memory locations and have a garbage value.

Inside the function scanf(), we see that we are using the arguments a and b along with a "&" (Ampersand) symbol. We had already used this symbol when we discussed the bitwise operators. The symbol "&" is a bitwise logical AND operator and it has its own meaning as discussed. But the same symbol is used here and it has a different meaning here. The function of "&" here is to fetch the "address of" the argument that follows it. So, it is called as "address of" operator. In the above case, the addresses of a and b are being fetched by the "&" operator and in the addresses, the values provided for them by the user are dropped. Thus the garbage values are replaced by the new values given by the user. Let me remind you that as long as the values for variables are not provided by the user, they would continue to have garbage values in the program.

Let me conclude this discussion by saying that the function scanf() instructs the compiler to fetch the value pertaining to a data type from the input device (keyboard) and drop the same in the address located by "&" for that corresponding variable.

Remarks

1. Another important thing to be noted here is that the values that are scanned are dropped according to the order specified in the list of arguments. In the above case, the first value scanned would be dropped in the address of 'a' and the second would be dropped in that of 'b' and so on. In more simpler words, when we use more than one format specifier in the function scanf(), the first one will associate itself with the first argument and its address in the list, the second one with the second argument and its address and so on.
2. In the above example, we could see that we have used the format specifiers inside the function scanf() without providing any space in between them. No special character, not even a comma, is supplied in between them. What would be the result had we used the same function scanf() in the following way?

```
scanf("%d,%d",&a,&b);
```

If you closely see the above version of the function scanf(), we have provided a comma (,) in between the two format specifiers "%d". This question would surely test the reasoning power of a C programmer. Normally the answer for this question, even by a professional programmer in C, would be, "an error". But it is not so. The above version of the function scanf() is absolutely permitted and error-free. I would justify this for sure.

When we execute this code of the program, the compiler would expect two integer values from the user. Normally when the user is keying in more than one value from the input device (keyboard), he/she would separate the two values or the two inputs with a space or a tab or an enter key. This argument goes by the common sense that when we write words, we separate them using a blank space. These three separators are called as the delimiters and they are invariably the default delimiters used by anyone who writes a program in C or in any other programming language. In the above newer version of the function scanf(), we see that the format specifiers are separated by a comma and hence when we give the two integer values from the input device (keyboard), we have to use a comma in between them. This means we have to overlook the three default delimiters discussed above. The delimiter in this case would be a comma. For that matter, any character could be used other than a comma and the same character shall be used as the delimiter when we execute the program by giving the inputs. We can even have a scanf() like the one below:

```
scanf("%d ^ %d*%d",&a,&b,&c);
```

In the above version of the function scanf(), we see that the first two format specifiers are separated with a '^' and the last two are separated with a '*'. This means, when we give inputs we have to use a '^' between the first two and use a '*' between the last two. I hope you would be satisfied with my argument. Just try this when you write programs later.

Another important I/O instruction is the function printf(). This function is used to print the values or the processed results or any message the user wants to print in the output device (normally a visual display unit, the screen in our case or a printer). Thus the function printf() would try to retrieve values from memory when the user wants to print some variable's value or just print a message as given by the user. The expansion for the word "printf" is "print format". It has the following method of usage.

```
int printf(const char *format, ...);
```

Its prototype is also available in the header file stdio.h. The function printf() formats a variable number of arguments according to the format, sending the output to stdout (output device). It returns the number of bytes output. In the event of error, it returns EOF (end of file). The more general model of scanf() is as given below:

```
printf("control string with or without format specifier (s)", <list of arguments>);
```

For example, one can have a function printf() like the one shown below:

```
printf("The values of a and b are %d,%d",a,b);
```

In the above case, we see that the format used satisfies the general format given above. Thus the result of the above printf() would be "The value of a and b are 5 and 20" (had we given 5 and 10 respectively for a and b). Thus we see that first %d in the expression associates with the first argument in the list and the second %d with second argument in the list. Thus we can generalize this result by saying that if there are 'n' format specifiers in the function printf(), there must be 'n' corresponding arguments and always the format specifiers will be associated with the arguments in a particular order (as stated above) and never in a random manner.

Now, a big question would arise in our minds. It is "what would happen if there is a mismatch, that is, either there are more format specifiers than the actual number of arguments or otherwise?" Let me tell you what will happen.

Let us consider the first case, where we have more number of format specifiers than the actual number of arguments as in the following example:

```
printf("%d,%d,%d",a,b);
```

In this case, we have three format specifiers and only two arguments. For the first two format specifiers, the compiler would associate a and b. But what will happen to the third format specifier inside the function printf() above. Since the third format specifier could not find any variable for getting itself associated, it prints a garbage value for it (the value would be within the range of the short-signed integer and is randomly decided by the compiler and not by the user). So the output could be like the one given below:

```
10, 20, 2515
```

(You as a user may get a different value for the third value here in this output.)

Let us consider the second case, where we have the number of format specifiers less than that of the arguments. In this case, the output generated would be 10, 20 (since there is no format specifier in the control string, the compiler does not bother to print a value).

In the general format of the function printf(), we have stated that the presence of the format specifier is optional. This means we can have another version of the function printf() like the following:

```
printf("Hello, how are you?");
```

You may see that the above version of the function printf() does not have any format specifiers and hence any arguments. This means the presence of format specifier necessitates the presence of an argument. If there is a format specifier and there is no argument, then we would get a garbage value in the output and if there is no format specifier and there is an argument, we would not get anything in the output. And as a last case, we may have the function printf() without any proper control string or even a format specifier. It is absolutely valid and would work without any problems. Thus, the following are the different versions of the function printf() a programmer might use.

1. printf(); This would not print anything and is of little use, but syntactically valid.

2.	printf("%d",a);	This would print the value of 'a' by getting the same from the memory, but you can see that there is no control string accompanying.
3.	printf("The value of a is %d",a);	This would also print the value of 'a' but along with the control string given by the user.
4.	printf("");	This would not print anything and of little use in the program, but syntactically valid.
5.	printf("%d,%d");	This would print two garbage values for the format specifiers belonging to the type short-signed integer, since there is no argument accompanying them.
6.	printf("%d,%d",a):	This would print the value of 'a' and a garbage value, since there is only one argument and two format specifiers and the first format specifier will associate itself with the first argument.
7.	printf("%d",a,b);	This would print the value of 'a' alone, since there is only one format specifier, even though there are two arguments.
8.	printf("%d");	This would print a garbage value as there is no argument accompanying the format specifier.
9.	printf("",a);	This would not print anything, since there is no format specifier and only an argument.

When we discuss the function printf(), it will be apt to discuss the **escape sequences**. Escape sequences are those which are highly useful when formatting our programs. The escape sequences which are very frequently used are:

\n	Newline character
\t	Horizontal tab
\a	Audible alert
\v	Vertical tab
\b	Backspace
\r	Carriage return
\f	Form feed
\\	Backslash

\" Double quote

\' Single quote

Any character that is accompanying the backslash will be printed along with the control string in the format specifier. The escape sequences are specially designed for the function printf(). Normally they are not used in the function scanf() and even if we use them, it does not serve any purpose.

The newline character (\n) when used in the function printf(), would generate a new line with the cursor blinking at the beginning of the line. This character works just like the carriage handle of a typewriter, which when pulled and moved towards the left, generates a new line and positions itself so that the user could start typing from the beginning of the new line. Assume that the function printf() is written as printf("hello\nworld\n"). The output would look like the screenshot shown in the Figure 2.22.

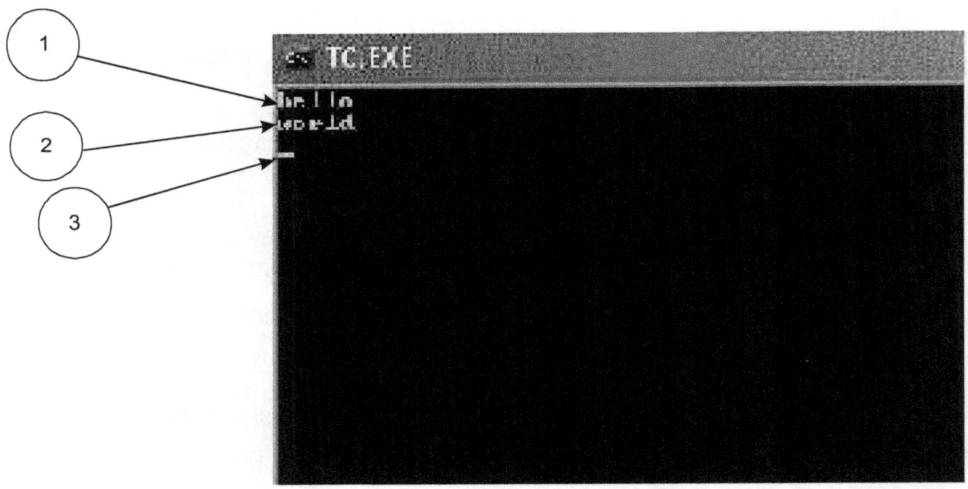

Figure 2.22 The newline character (\n) in action

In the Figure 2.22, we have three indicators 1, 2 and 3. The indicator 1 denotes the first word in the control string of the function printf(), 2 denotes the second word and see that these two words are available one after another, since there is a "\n" in between them in the control string. Indicator 3 denotes the position of the cursor after printing the words inside the control string of the function printf(). The cursor is positioned at the third line of the output since there is a "\n" after the word "world" in the control string of the function printf(). Now you would probably know what must be the coding to generate an output as that is expected as in the Figure 2.23. Yes, you are right. The coding shall be printf("hello\n\nworld\n"). There are two "\n" characters between the words "hello" and "world", so that a blank line is left between them in the output. There is only one

"\n" character after the word "world" and hence the cursor is positioned in the line immediately after the word "world" in the output.

Figure 2.23 The newline character (\n) in action in a different manner

The horizontal tab character (\t) when used in the function printf(), would generate a tab space (normally a tab space corresponds to 8 blank spaces, which can be modified by the programmers). This is shown in Figure 2.24.

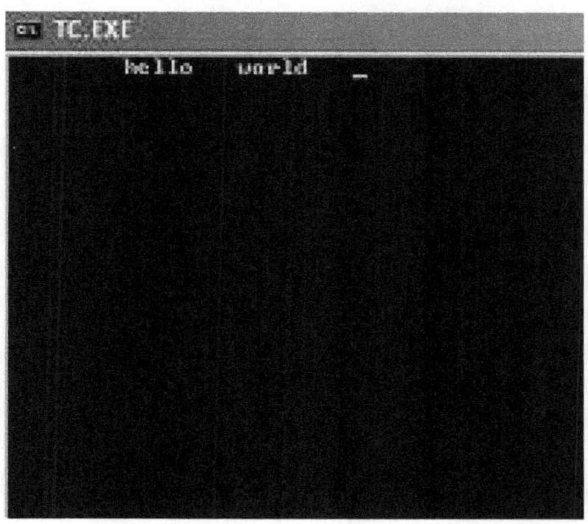

Figure 2.24 The tab space character (\t) in action

We can see from the Figure 2.24 that there are three tab spaces used in the function printf(), one before the word "hello", one after that and one after the word "world". That is the function printf() would look like printf("\thello\tworld\t")

The audible alert character (\a) would generate a beep sound (provided there is an in-built speaker) along with the other contents of the function printf(). This cannot be exemplified using a figure.

The vertical tab character (\v) generates a tab space vertically, unlike the horizontal tab. In most of the recent compilers, this would not work properly.

Figure 2.25 The backspace space character (\b) in action

The backspace character (\b) would make the cursor to position itself at the last character of the output generated by the function printf(). If there is any character left out from being printed in the control string of the function printf(), this character will erase the last character of the previous word and would start printing the first character of the next word. The Figure 2.25 shows a typical '\b' character in action. The following is the way the function printf() would have been used for getting the above output.

```
printf("hello\bworld\b");
```

We had used a '\b' character after the word "hello" and there is one more word "world" following "hello" in the control string of the function printf(). So, after printing the first word, the cursor would move one space backwards and start printing the second word and finally would position itself at the last character of the second word.

The carriage return character (\r) would place the cursor at the beginning point of the same line, unlike the '\n' character. Consider the version of the function printf() as printf("hello world\r"). The output of this is shown in the Figure 2.26.

Figure 2.26 The carriage return character (\r) in action

Since we had used a carriage return character (\r) after the word "world" in the control string, we see that the cursor is positioned at the beginning of the same line. What would happen if the function printf() looks like printf("hello\rworld");

The output would be like the one shown in the Figure 2.27. The justification for the output could be given by you.

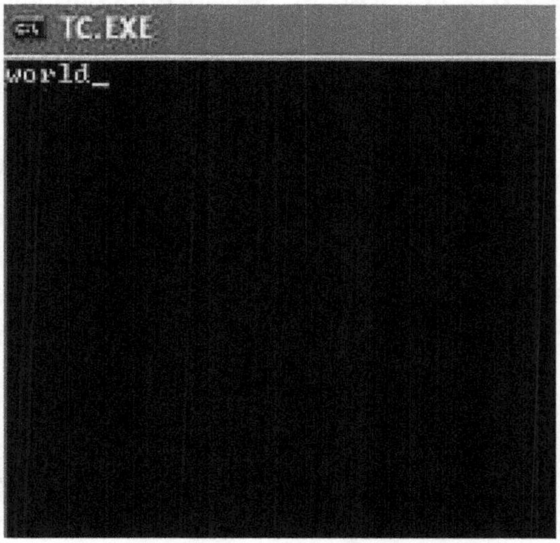

Figure 2.27 The carriage return character (\r) in action in a different manner

The form feed character (\f) would be used while working with a printer, and if we want to print a page or an output, we can send the same to the printer and programmatically, we can tell the compiler when there shall be a page break and when there shall not be while a program or its output is getting printed.

Any character that is used along with '\' will be printed in the output. For example, let us assume that we had used "\\", then we can anticipate a '\' at the output. Also, if we want to print a character like '$' in the output, then we may include a '$' along with a '\'. This is shown in Figure 2.28.

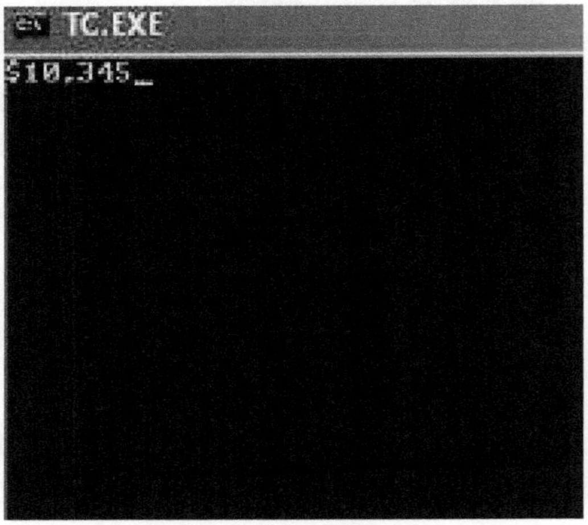

Figure 2.28 $ character is printed along with the output

The escape sequences, I hope, are now clear to you. You will probably have to work more with these sequences to get to know the real implications of these.

Arithmetic instructions Arithmetic instructions will involve the arithmetic operators. For example, consider the following expression: c = a + b. This is a very simple example for the arithmetic instruction, where we have the operator '+' between the operands a and b. In basic mathematical principles, we would express the same instruction as c = f(a,b) and this would be read as "c is a function of a and b". Thus, a and b are independent variables and 'c' is a dependent variable, whose value depends on the values of a and b.

The following is also an example of an arithmetic instruction: c = a+b/d%e. Here, we notice that apart from the operator '+', we have the operators '\' and '%'. Since we had already learnt about the hierarchy of arithmetic operators, you must know the order in which the above arithmetic instruction would be executed. All the arithmetic instructions have two sides, a Left Hand Side (LHS) and a Right Hand Side (RHS). Always the LHS will have a variable and that too only one variable.

A constant is not allowed on the LHS. The RHS of the instruction will have any number of variables and constants. If we write the expression c=a+b as a+b=c, it will not work in programming. Even though a+b=c serves the purpose of the operation, it is not accepted since it is not an assignment statement but an equation.

Control instructions Control instructions are used in a program to control the flow of the program. The control instructions are of two types in general. They are conditional control and loop control. Conditional controls work on the basis of conditions and loop controls work on the basis of loop cycles. We will discuss more about the conditional and loop control instructions later in this chapter.

2.2.8 The Function main()

We know that C language is that which has many built-in functions that are categorized into different headers and included as and when required in a program. We had seen two such functions earlier when we discussed the I/O instructions, namely printf() and scanf(). The function main() is one of the functions in C and the most important of all the available functions.

When a program begins running, the system (operating system) calls the function main, which marks the entry point of the program. Every program must have one function named main(). No other function in the program can be called main(). The function main() has one of two forms:

int main (**void**) *block_statement*

int main ( ) *block_statement*

You will probably come to know about the blocks later when we introduce you to programming ideas. Now, take for granted that the function main() has two forms of usage as represented above.

As said earlier, without the function main(), no C program would compile or execute. The name of the function itself implies that it is the core function in any program.

2.2.9 Conditional Branching

There are three conditional control instructions available in C language and they are listed below:

- if
- if-else
- if-else-if

Let us have a detailed discussion on all these controls.

if This is the simplest condition of the three. The following is the syntax of the if condition:

```
if (<expression[s]>
{
Statement-1;
Statement-2;
......................
......................
Statement-n;
}
```

→ **Block of Statement (s)**

The expression in the above syntax may be the one that has a single condition or multiple conditions clubbed together with the logical operators. Following the expression, we see that we have a block of statements. In C programming, when any set of statements is included within the curly braces ({ and }), we could say that a block is created. The advantage of having a block is that we can execute a set of statements in a single go, upon meeting the condition(s) in the expression. The statements inside a block will all be executed or will all be skipped together. It is not possible (it is not sensible too) that we can select a particular statement from the block and execute or skip it (unless and otherwise there is a nesting of *if*). Also, one cannot execute statements inside a block at random. The block statements are always executed in the order of their appearance inside the block. The block following the expression inside *if* is also called as the **body** of the *if* statement. The body will be executed if the return value of the expression is non-zero (true) (we know that the logical and relational operators and the expressions involving them will return a Boolean zero or a Boolean one as the result). Thus when the return value is true, the block following it will be executed and when it is false, the block will be simply ignored. The following fragment exemplifies the usage of *if* statement in programming code.

```
if (x>0)
{
printf("Good Morning");
printf("Have a Nice Day");
}
```

Thus, if the input for the variable 'x' is greater than 'zero', then the messages inside the function printf() will be executed in that order and the block will be terminated after that. If the input value for x is less than 'zero' or 'zero' itself, then the entire block will be skipped.

The simplest form of the above *if* is as given below:

```
if (<expression[s]>) Statement;
```

Here, we see that there is only one statement and not a block as in the syntax given already. When the expression returns a true, the only statement available will be executed, otherwise, we will not get any output. When there is only one statement after the expression, then the set of curly braces is optional. This means that even though the curly braces are provided, the expression will just look lengthy and the curly braces are redundant. The following piece of code exemplifies this idea.

```
if (x>0)
printf("Good Morning");
```

We see that only one statement is executed if the expression 'x>0' returns a non-zero value. But see that in both the cases, the set of parentheses is essential and one cannot skip them. For example, the following piece of code is invalid.

```
if (x>0)
printf("Good Morning");
```

The 'if' statement that is explained above has very least applications in real life programming, the reason behind this being no alternative option available once the expression inside the set of parentheses returns a "zero" (false). Thus, the designers of the language have introduced another conditional control, which is an advanced version of the *if* statement. This is discussed in the following section. Figure 2.29 is the representation of *if* structure in the form of a flowchart.

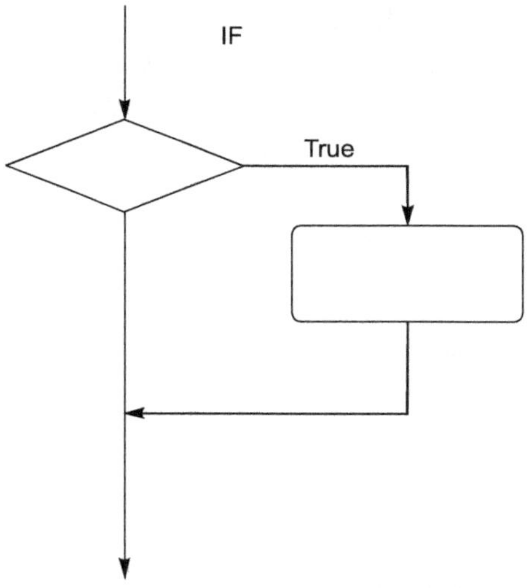

Figure 2.29 Flowchart representation of *if* structure

if-else The following is the general syntax of *if-else* structure:

```
if (<expression[s]>)
{
Statement-1;
Statement-2;
.......................            ──► True or if block
.......................
Statement-n;
}
else
{
Statement-1;
Statement-2;
.......................            ──► False or else block
.......................
Statement-n;
}
```

We see from the general syntax that there are two alternatives for the expression given within the set of parentheses. This means, one of them will always be executed. The expression inside the parentheses may return a true (non-zero) or a false (zero). If it returns a true, then the block following *if* will be executed and hence it is called a true or if block. If the expression returns a false, then the block following *else* will be executed and hence it is called a false or else block. In no case, will both the blocks be executed or be skipped. The following piece of code is an example for an *if-else* fragment used in programming.

```
if (x>0)
{
printf("Good Morning");
printf("You have entered a positive value");
}
else
{
printf("Good Evening");
printf("You have entered a negative value");
}
```

Thus, if we enter a positive value for 'x', the block following *if* would be executed, otherwise that following *else* would be executed. As in the previous case, the set of parentheses is essential and one cannot skip it.

The following are the alternative forms of the *if-else* structure used in programming.

Form – 1

```
if (<expression[s]>)
{
Statement-1;
Statement-2;
.......................
.......................        ──► True or if block
Statement-n;
}
else
Statement;
```

Form – 2

```
if (<expression[s]>)
Statement;
else
{
Statement-1;
Statement-2;
.......................
.......................        ──► False or else block
Statement-n;
}
```

Form – 3

```
if (<expression[s]>)
Statement;
else
Statement;
```

In the first form, we see that there are multiple statements for execution in the true block and there is only one statement in the false block. Since there is only one statement following the *else* statement, we had skipped using the set of curly braces (as it would be redundant, as discussed earlier in the previous section). The following is the example code for this.

```
if (x>0)
{
printf("Good Morning");
printf("You have entered a positive value");
}
else
printf("You have entered a negative value");
```

In the second form, we have multiple statements for execution in the false block and only one statement in the true block. Again, the usage of a set of curly braces after the *if* is redundant and hence is skipped. The following fragment exemplifies this.

```
if (x>0)
printf("You have entered a positive value");
else
{
printf("Good Evening");
printf("You have entered a negative value");
}
```

In the last form, we have only statement after *if* and one after *else*. This is the simplest way in which we can use the *if-else* structure in our coding. The following is a piece of coding that tells us this.

```
if (x>0)
printf("You have entered a positive value");
else
printf("You have entered a negative value");
```

While programming, all the three forms of *if-else* would be used frequently. Figure 2.30 shows the flowchart representation of *if-else* structure.

At this juncture, I would like to introduce you a new operator. Recall that you had one category of operator called **ternary operator**. The ternary operator is called so because it has three sections (or arguments). The following is the general syntax of a ternary operator.

Argument-1? Argument-2 : Argument-3;

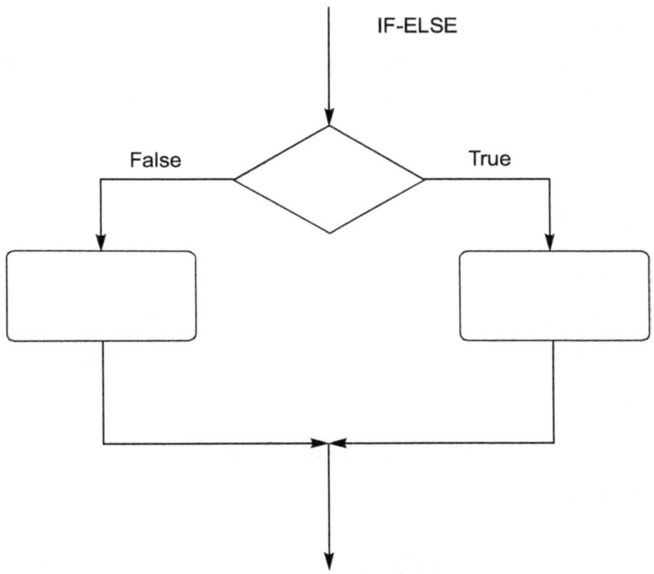

Figure 2.30 Flowchart representation of *if-else* structure

Normally, argument-1 would be an expression or a combination of expressions. Argument-2 is the statement that would be executed upon the expression returning a true (non-zero) value and the argument-3 is the statement that would be executed upon the expression returning a false (zero) value. Now, I believe you would have understood the reason why I introduced this operator at this point. Yes, you are right. This operator is just a replacement for the third form of the *if-else* structure. Just go back and see the general syntax of the third form of the *if-else* structure and it will just be the same idea that we had discussed.

Assume that we have the following piece of fragment in some part of our program.

```
if (x>0)
printf("You have entered a positive value");
else
printf("You have entered a negative value");
```

The above piece of coding could be replaced with the ternary expression, whose general syntax is given above. The following will be the replaced code for the piece of coding given above.

```
x>0? printf("You have entered a positive value") : printf("You have entered a negative value");
```

It is obvious from the expression above that the simplest form of the *if-else* could be replaced with a ternary operator and we call it a ternary expression. Another advantage of using a ternary expression instead of the simplest *if-else* is that we need not use the set of parentheses for the expression, as we do in the *if-else* structure. Even if we use them, it is just redundant and its presence or absence will not be a big issue with respect to the coding.

Consider the following piece of coding:

```
x>9 && x>0? printf("Good") : printf("Bad");
```

Suppose that we enter a value of '10' for 'x', then the output for the above piece of coding would be 'Good'. We notice that even if we have two expressions clubbed with an AND operation, we could skip the usage of the parentheses. This is the advantage of ternary operator over the simple *if-else* structure.

Another wonder that we can do with a ternary operator is that we can nest them to any number of levels and execute. For example, consider the piece of coding given in Figure 2.31.

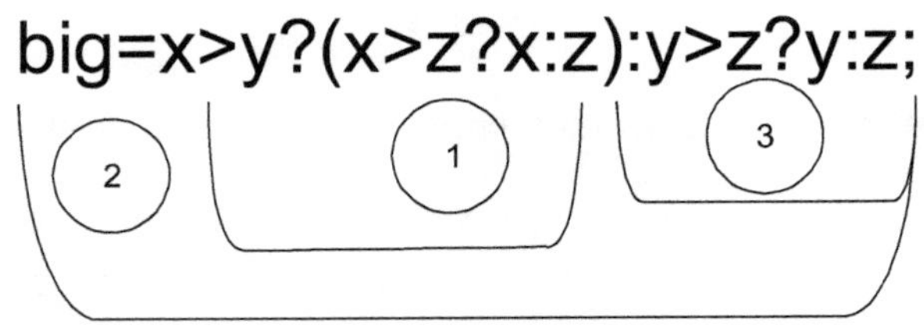

Figure 2.31 Nesting of ternary operator

In the above piece of coding, we see that the ternary operator is used thrice and this is what we call as nesting. We have nested the ternary operator up to three levels (the indicators 1, 2, and 3 show the three ternary expressions). When we give values for x, y and z, they will be tested accordingly and the result (the biggest of three values) will be stored in the variable 'big'.

To begin with, the ternary expression indicated by 2 will be in action. The first argument for the expression indicated by 2 will be the condition x>y, the second argument will be that indicated by 1 (another ternary expression by itself) and the third argument will be that indicated by 3 (yet another ternary expression by itself). Let us consider the following cases for the above nested ternary expression.

When the condition x>y returns a false, obviously it means that the reverse condition y>x is true. Now the control comes to the third argument (indicated by 3) of 2. Here the condition y>z is checked and if it is true, then the biggest value will be the one in 'y'. Otherwise the biggest value will be the one in 'z'.

When the condition x>y returns a true, then the control would go to the second argument (indicated by 1) of 2. There the condition x>z is checked and if it returns a true, then the control will store the value of 'x' in the variable 'big' and that will be the biggest of the three. If the condition returns a false, then the value of 'z' will be stored in the variable 'big' and hence that will be the biggest of the three.

As an example, consider the values for x, y and z as –5, 4 and 98 respectively. To start with, the value '–5' will be compared with '4' and obviously –5 is less than 4 and hence the condition will return a false (Boolean zero). So, the control will skip the second argument of the ternary expression indicated by 2 and will go to the third argument indicated by 3, which is another ternary expression by itself. Here, 4 and 98 are compared (remember that already –5 and 4 are compared in the first argument and it has been confirmed that 4 is bigger among them) and again it will return a Boolean Zero, i.e., a false. This means, the control would now go to the third argument of the ternary expression indicated by 3 and there we have the value of z and it is nothing but '98' and it will now be assigned to the variable 'big'. This way, you may shuffle the values and test the validity and the way the above expression works.

There is a shortcoming with the ternary expressions, even though it has many advantages. There can be only one argument after the question mark (?) and after the colon (:). This means, we cannot create blocks (*if* block or *else* block) after the question mark or the colon. In other words, only the simplest form of the *if-else* could be replaced with a ternary expression and not the other forms that are discussed earlier.

if-else-if The first conditional structure (*if*) has no optional statement or block to be executed, except the block that follows the *if* statement, when the conditional expression returns a false. The second conditional structure (*if-else*) has two blocks (one following the *if* block and one following the *else* block) and hence there is always an alternative block that is executed even when the condition returns a false. If we need more than one condition to be tested and hence execute different blocks according to the one that returns a true, then we are provided with another structure in C and it is *if-else-if*. Here the conditional expressions are tested one after another and the corresponding block of code is executed and the rest are skipped. The following is the syntax of the *if-else-if* structure.

```
if (<expression-1>)
{
Statement-1;
Statement-2;
........................          ──► Block-1
Statement-n;
}
else
if (<expression-2>)
{
Statement-1;
Statement-2;
........................          ──► Block-2
Statement-n;
}
else
if (<expression-3>)
{
Statement-1;
Statement-2;
........................          ──► Block-3
Statement-n;
}
else
if (<expression-4>)
{
Statement-1;
Statement-2;
........................          ──► Block-4
........................
Statement-n;
}
```

```
........................
........................
else
if (<expression-n>)
{
Statement-1;
Statement-2;
........................        Block-n
Statement-n;
}
else
{
Statement-1;
Statement-2;
........................        Default Block
Statement-n;
}
```

We could see from the syntax that there are 'n' different expressions and corresponding blocks for each of them to get executed. The control starts evaluating the expression from the very first *if* and proceeds according to the return value of that. That is, if it gets a return value of false for the first conditional expression, then it will skip the first block and proceed to evaluate the second conditional expression and if that returns a true, then it will execute the block following that. When all the conditions in a particular *if-else-if* structure fail, then the default block available as the last one will be executed and it is purely optional to have such a block. But it will be sensible to provide a default block in such cases, as the user will not know the outcome if all the conditional expressions fail. But what happens then? Do you think the control will skip the entire structure and come to the subsequent part of the program? The answer is no. The control will check all the remaining conditions, regardless of whether a particular condition returns true and the corresponding block gets executed. In this aspect, the compiler is said to waste the execution time and it is built in such a way.

For example, consider the following fragment of a program that contains *if-else-if* structure, which checks whether the given character is a vowel or a consonant.

```
if(a=='a' || a=='A')
printf("Apple");                              → Block-1
else
if(a=='e' || a=='E')
printf("England");                            → Block-2
else
if(a=='i' || a=='I')
printf("Ice Cream");                          → Block-3
else
if(a=='o' || a=='O')
printf("Owl");                                → Block-4
else
if(a=='u' || a=='U')
printf("Umbrella");                           → Block-5
else
printf("The character is not vowel");         → Default Block
```

When the program gets executed, if the conditional expression 2 returns a true, then the control will execute the block corresponding to that, but still will check the remaining conditions for their return values, before it leaves the entire structure to the subsequent part of the program. We see that there is a default statement following all the formal conditional expressions. If the user inputs a consonant, none of the available conditional expressions will return true and hence the default statement (The character is not a vowel) would be the output and the user could understand that the input he/she has given is processed and an outcome is given. If this default statement has not been provided, then the user must be unaware of what is happening around if he/she has supplied a consonant as the input.

The following Figure 2.32 represents the *if-else-if* structure in the form of a flowchart. The structure can be extended beyond the specified number of levels in the Figure 2.32, as required for the program. The *if-else-if* structure is also termed as *if-else-if* chain.

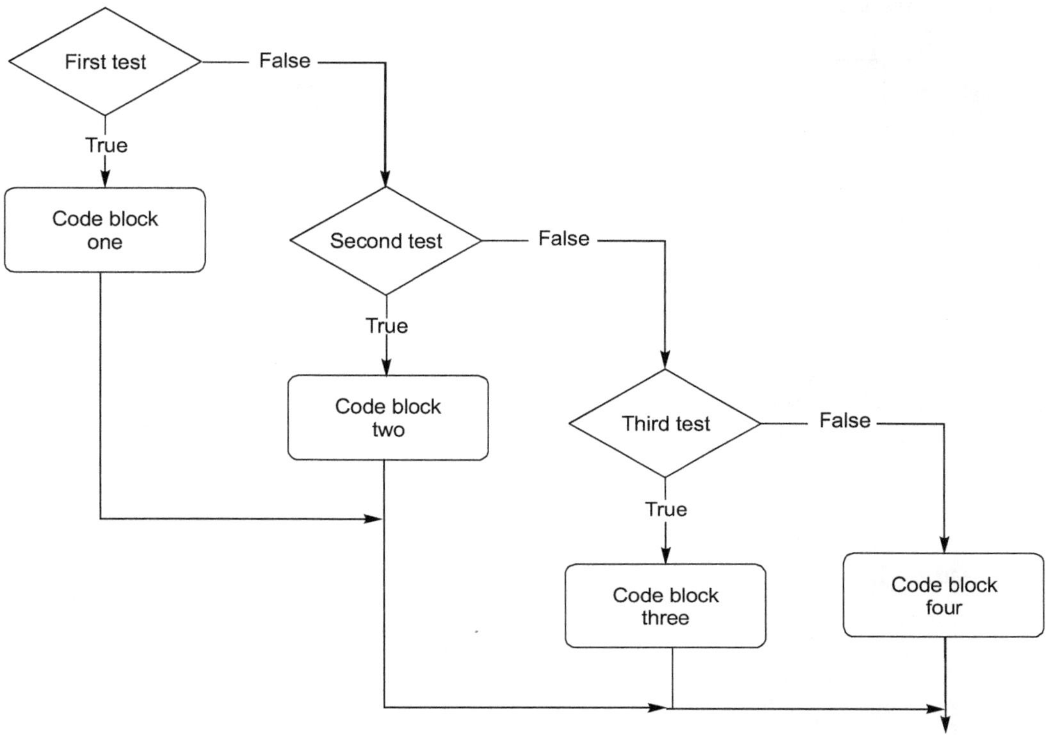

Figure 2.32 Flowchart representation of *if-else-if* structure

When we discussed the syntax of *if-else* structure, we noted that there are three different forms of the structure that could be used in programming. Similar to that, we have umpteen forms for the *if-else-if* chain, since one block or the other would be provided with only one statement and hence we can skip the curly braces for that block. As an example, two of the alternate forms of *if-else-if* structure are provided below. Based on this, you may derive many similar forms by applying a similar logic.

In the example provided above, we had used the simplest form of the structure, where each block contains only one statement and hence we could skip the curly braces that are provided in the general syntax.

```
if (<expression-1>)
Statement-1;                 ──────► Block-1
else
```

```
if (<expression-2>)
{
Statement-1;
Statement-2;
......................                  Block-2
Statement-n;
}
......................
else
if (<expression-n>)
Statement-1;                            Block-3
else
{
Statement-1;
Statement-2;
......................                  Default Block
Statement-n;
}
```

We see from the above alternate form of *if-else-if* that the first and the nth conditional expressions are provided with only one statement following them and the second conditional expression and the default block are provided with more than one statement for execution. Hence, we could skip the curly braces, when we have only one statement following the conditional expression. Even if we provide curly braces, we will be just lengthening our code and they will be redundant.

We had enough information about the conditional structures and their different forms of execution and usage. We will now discuss another control structure in C language, which is Case Control Structure.

2.2.10 Multiway Branching—Case Control Structure

The case control structure (otherwise called as switch-case control structure) is a multiway branching structure, which may be termed as an alternate to the *if-else-if* chain. Similar to *if-else-if* chain, one can have many options in case control structure. It is otherwise a far better and tidy way of representing the *if-else-if* chain. The control will execute the option based on the match it does before getting in the switch-case block. The following is the syntax for the case control structure.

```
switch (expression)
        {
case item-1:
        Statement-1;
        Statement-2;
        ......................
        Statement-n;
        break;
case item-2:
        Statement-1;
        Statement-2;
        ......................
        Statement-n;
        break;
        ......................
        ......................
case item-n:
        Statement-1;
        Statement-2;
        ......................
        Statement-n;
        break;
default:
        Statement-1;
        Statement-2;
        ......................
        Statement-n;
    }
```

Case item | **Case item** | **Case item** | **Default case** | **Switch-case block**

In each of the cases above, the case item can be only a constant and shall not be a variable. Even the constant that is used must be a numerical or a character constant and not anything else (like a float or its associated data types). This is in contrast with the *if-else-if* structure, where one can use any type of constant or a variable. We will later do a comparative study of the case control structure and *if-else-if* chain, so as to better analyse which to use when.

In order to terminate the *switch* control once a particular *case item* is executed, we supply a *break*, so that the control comes out of the *switch-case* block. Consider for example, the following fragment of a program, wherein we print the given single digit numbers in words, using case control structure.

```
switch(m) {
case 0: printf("Zero",m);  break;
case 1: printf("One",m); break;
case 2: printf("Two",m); break;
case 3: printf("Three",m); break;
case 4: printf("Four",m); break;
case 5: printf("Five",m); break;
case 6: printf("Six",m); break;
case 7: printf("Seven",m); break;
case 8: printf("Eight",m); break;
case 9: printf("Nine",m); break;
default: printf("\nThe number %d is not a single digit number",m); }
```

For a given value of 'm' (which would be scanned using the proper scanf() in the program), the individual cases inside the *switch-case* block is matched and the output is generated accordingly. If the given value does not match with any of the cases inside, then the default case would be executed and here we could see an analogy between the default block in the *if-else-if* structure and the default case inside the *switch-case* block. Both are provided for the same purpose: in the former structure, to provide the user with some output when none of the conditional expressions return a true and in the latter structure, when none of the cases match with the input given by the user.

We had used the *break* statement in all our cases (except the default) for obvious reasons that are mentioned earlier. What will happen if we do not use *break* for a particular case item? What will be the outcome if we use *break* for the default case?

If we do not use *break* statement for a particular case item, then the statements related to that case along with the subsequent case will be executed. For example, consider the following program fragment that prints all the single digit numbers in words.

```
switch(m) {      case 1: printf("One"); break;
         case 2: printf("Two");
         case 3: printf("Three");break;
default: printf("The number is not a single digit number");
}
```

We see that there is no *break* used in the second case item. Suppose that we enter a value of '2' for 'm' from the keyboard, we expect only the output "Two" to be printed. But along with that, we get the output "Three" also, since according to the control, the block of the case item starts at 'case 2' and ends at the *break* statement of the case item 'case 3' and hence whatever statements available after the case item '3' till the next *break* statement will be executed. We understand that *break* statement is the one that acts as a 'terminator' for a given case item. What do you think the output will be if there is no *break* statement in the cases 2 and 3? The output will have both "Two" and "Three". So, it is concluded that *break* statement is essential for terminating the execution of the *switch-case* block and bringing the control outside it.

In the above program fragment, the variable 'm' can be of any type related to integers or characters. Float and its associated data types are not allowed within the switch as well as in the case items.

Can you guess how we can rewrite the program fragment that we had written already for illustrating the *if-else-if* structure? (The program using which we have printed a word for the given vowel). The following would be the corresponding program fragment.

```
switch(m) {
case 'A': case 'a': printf("Apple"); break;
case 'E': case 'e': printf("Elephant"); break;
case 'I': case 'i': printf("Ice"); break;
case 'O': case 'o': printf("Owl"); break;
case 'U': case 'u': printf("Umbrella"); break;
default: printf("The character is not a vowel");}
```

We see that, unlike the logic that is used in *if-else-if*, we have not used the logical OR operator for checking the nature of the alphabet that we enter. The reason is that the *switch-case* structure does not allow the usage of operations of any kind in the case. The logic, however, is achieved in a different way.

Let us list out the advantages of case control structure over *if-else-if* structure. See that the advantages also give a fair idea on how the case control differs from *if-else-if* structure.

1. This structure is useful in case the user wants to terminate the control upon matching a particular case item and bring the control outside the *switch-case* block, unlike the *if-else-if* structure. Remember that in *if-else-if* structure, there is no way to bring the control outside the structure, even when a particular conditional expression returns a true value and the block following it is executed. In case control, it is achieved through the *break* statement.

2. When the statements of a particular case item are written, we do not enclose them within curly braces, unlike the *if-else-if* structure. It will be an error if we do so in *if-else-if* structure. But, in case control, the presence of the colon (:) denotes the beginning of case block and the *break* statement denotes the ending of a block. So, there is no need for enclosing the set of statements of a case item within curly braces. This is true for any number of statements for a particular case item.

3. The *if-else-if* structure, when professionally written, needs indentation of the program lines. Indentation will make the program to have a clear look and the users can easily identify the sections of the program. For example, consider the following fragment of a program written using case control structure.

```
switch(m) {        case 1: printf("One"); break;
     case 2: printf("Two"); break;
     case 3: printf("Three");break;
     default: printf("The number is not a single digit number"); }
```

The same program, when written using *if-else-if* looks like the one below.

```
if (m==1)
            printf("One");
                        else
                            if(m==2)
                                    printf("Two");.....
```

The structure needs indentation to be done in each line of the code as shown above. If we proceed with this sort of indentation, then the coding will creep towards the right side of the screen and will not be fitting within the working area of the editor. But the case control does not require indentation to be done for the blocks following the case items, since the provision of colon (:) and the *break* statement does the job of identifying the correct block and it is matched with the correct case item.

4. We need not use the set of curly braces to create a block of statements, as in *if-else-if* structure, since the presence of colon (:) and *break* satisfies this one too.

5. We can use all sort of operations inside the case control as in the *if-else-if* structure, but with a caution. We can perform the operations only with the values pertaining to integer and its associated data types.

6. Another advantage of the case control structure is that, one can use any order for the case items. That is the programmer need not follow any particular logical order or a sequence. For example, consider the following fragment:

```
switch(a)
{
case 3: printf("Three");break;
case 2: printf("Two"); break;
case 1: printf("One"); break;
default: printf("Bye");
}
```

The output for the above program would be fine, even if the order of the cases is jumbled. Let us now slightly modify the above fragment and present the same as shown below:

```
switch(a)
{
default: printf("Bye");
case 3: printf("Three");break;
case 2: printf("Two"); break;
case 1: printf("One"); break;
}
```

What do you think will happen? Since we have told that the cases could be in any order, in the above fragment, the default's position is also jumbled unlike the earlier version of this fragment. The output for this will be quite different from that of the earlier one. If we give a value of 1 or 2 or 3, then the output will be fine. But if we give any values other than the above, we are supposed to get the result "Bye", and of course we will get it, but along with that we will also get "Three". This is because, the default is not terminated with a 'break' statement and hence the printf() statements associated with both the default case and the succeeding case are executed. Thus the position of the default case could also be changed, provided we give a *break* statement following the default case.

There are many disadvantages of case control structure, which makes the usage of *if-else-if* structure inevitable. The following are some of the disadvantages and shortcomings with case control structure.

1. We cannot use data types other than integer and its associated types and character type constants (only unsigned character) inside the *switch-case* block for the case items. For example, the following case would be invalid and hence will get you an error message.

   ```
   case 1.4: printf("Good"); break;
   ```

   What will happen if the variable is declared as a float and the case item is an integer? Consider the following example to know about this situation more clearly.

```
float m;
scanf("%f',&m);
switch(m)
{
case 1: printf("One"); break;
case 2: printf("Two"); break;
default: printf("Good"); break; }
```

In this case, if the given input is '1.5', what will happen to the output of the program? Will any one of the cases be executed or will we get an error? The answer is quite simple. The rule is that we shall not use constants other than integer (and its types) and character in the case items and we are never restricted from declaring a variable of any type and passing it to the switch statement. This means that when the above value (1.5) is given for 'm', it will be accepting and matching the value (1.5) with all the cases and it is matching with case 1 (the value that is given by us will be truncated to an integer value) and hence that case will be executed. When we give values like 3.6, the default case will be executed.

2. What will happen if we have cases like the one given below?

```
case 1+2: printf("Good"); break;
case 2>1: printf("Big"); break;
case 'a' && 'b': printf("AND"); break;
case 'a' & 'b': printf("Hello"); break;
```

Again, the answer is very simple to understand. The operation will be performed and the result will be assigned for each case item. In the above three cases, the case items have the results as 3, 1 (Boolean), 1 and 96 respectively. These values can be realized by applying the suitable operations to all of them, since we knew all the operators used in them. This proves the fact that any kind of operation is possible with the case items. What about the following one then?

```
case a&b: printf("True");break;
```

where 'a' and 'b' are variables of type integer. This type of case is not allowed, due to the reason that a case item cannot have anything other than constants.

3. We cannot use the statement *continue* (we will discuss about this statement and its usage at a later stage when we discuss loops) inside the case control structure, unless there is a loop used within a particular case item. Since the loops allow the usage of the statements *break* and *continue*, the programmer will be tempted to use both of these in

case control also. But they must know that continue is not allowed in case as a stand-alone statement.

4. Even though we are allowed to use character constants for a case item, we cannot use strings (which are a collection of two or more characters). For example, the following case item is invalid.

```
case '>=': printf("This is greater than or equal to operator"); break;
```

This means, any operator that contains more than one character for designating it cannot be used as the case item. A few other examples are the operators like ?:, <=, ==, ++, —, >>, << and so on. But do remember they can be used along with the numerical and character constants in the case item.

We will be using a lot of ideas that are learnt in this section of the chapter later while learning advanced concepts in the language.

2.2.11 Unconditional Branching—*goto* Statement

goto is a statement found in almost many computer programming languages. It is a combination of the English words go and to. When used and executed, it causes an unconditional transfer of control (a "jump") to another statement in the same program. The statement is specified using some kind of label, which may be an identifier or a line number depending on the language in use. At the machine code level a *goto* is a form of branch or jump statement.

In some languages, the functionality of *goto* may be present without explicit use of the keyword *goto*, such as where a *break* or *continue* keyword may be present (we have seen the usage in *break* in the previous section).

Different programming languages impose different restrictions with respect to the jump location of a *goto* statement. For example, in the C programming language it is not allowed to jump to a label contained within another function. For the control to unconditionally jump, the label must be provided in the same program and that too in the body of the same function. The following illustration explains the function of *goto* statement.

```
scanf("%d",&opt);
goto aa;
switch(opt) {
case 2<=2: printf("Three");break;
default: printf("good"); }
aa: printf("You are here");
```

If we give a value for the variable 'opt', say, 2, the function scanf() will accept the value and it will not pass the control to the switch statement, since there is a *goto* statement available immediately after the function scanf(). So, the control will abruptly transfer to some part of the same function (in this case within the function main()) and execute the coding in that.

Why one shall not use goto statement? The *goto* statement has been the target of much continued criticism and debate, with the primary negative claim being that use of *goto* results in unreadable and generally difficult-to-maintain type of coding called as a **spaghetti code**. Spaghetti code is a pejorative term for source code which has a complex and tangled control structure, especially one using many *goto*'s, exceptions, threads, or other "unstructured" branching constructs. It is named such because program flow tends to look like a bowl of spaghetti, i.e., twisted and tangled. Structured programming greatly decreased the incidence of spaghetti code, and is widely regarded as one of the most important advances in the history of programming languages. C, is a structured programming language and the usage of statements like *goto* will surely spoil its structure. Java, which is one of the programming languages, that uses many of the syntax from C, does not use *goto* statement. Professional programmers in C, very rarely or do not at all use this programming structure. Probably, in our book, this would be the last place, where you will be seeing the usage of this statement.

2.2.12 Loop Control Structures

Loops are used to repeat a block of code. Being able to have your program repeatedly execute a block of code is one of the most basic but useful tasks in programming. You could have seen many programs or websites that produce extremely complex output (such as a message board) are really only executing a single task many times. (They may be executing a small number of tasks, but in principle, to produce a list of messages only requires repeating the operation of reading in some data and displaying it.) Now, think about what this means: a loop lets you write a very simple statement to produce a significantly greater result simply by repetition. In C programming, loops are important and unavoidable.

In C we have three kinds of loops, namely, *while*, *do-while* and *for*. Before proceeding, make sure that you have understood the concepts of C's true and false (Boolean Values—1 and 0), because it will be necessary when working with loops. All the loops in C have **Loop Counter**, which keeps track of the number of cycles of a loop. A loop counter is used only for this purpose and it does not server any other purpose in program coding. In some programs, one of the variables used in the program for calculation itself will be used as a loop counter, where it has to play a dual role. The loop counters are normally used in the conditional expressions that appear in the loop fragment.

While Let us start from the while loop and understand the way it functions. The syntax of the *while* loop is given below:

Initialization of Loop Counter;

```
while (<conditional expression [s]>)   ----->  Loop Header
{ ----------------------------+
Statement-1;                  |
Statement-2;                  |
........................      +----------->  Loop Body
Statement-n;                  |
Loop counter modification;    |
} ----------------------------+
```

Basically, there are four sections in any loop. They are initialization, conditional expression, loop body or set of loop statements, modification. If you have a closer look at the above syntax, you may notice that all the said sections are available in it. The general procedure for executing the above structure in programs is stated below.

1. Initialization of Loop Counter
2. Checking the Conditional Expression
3. Executing the body of the loop
4. Modifying the Loop Counter

The execution of the loop begins with the first step, proceeds to the second step and based on the return value (true or false) of the expression, the control proceeds to the third step. Once the third step is executed, it goes to the next step and modifies the loop counter and then goes to the second step where the conditional expression will be checked again. Thus, we see that the first step of the loop is executed only once and the remaining loop cycles make the control to move with the other three steps. Once the conditional expression returns a false, the loop will terminate at that step. So, to conclude, the control starts at the first step and terminates at the second step (unless there is a *goto* statement somewhere inside the loop). For example, let us suppose that we are writing a program to print numbers from 1 to 100 on the output screen. The following would be the program fragment for achieving that using the *while* loop.

```
x=1;
while(x<=100)
{
printf("%d\t",x);
x=x+1;
}
```

We see that the above fragment complies with the general syntax in all aspects and will be executed based on the four steps that we had discussed above. So, it begins with the initialization, where the variable 'x' is initialized as '1'. Then comes the conditional expression, 'x< = 100'. This expression will return a **true** value as long as x's value does not exceed 100. Since, now, the value of 'x' is '1', the control will get into the body of the loop and execute the same (step 3). When it is executing it we could understand that '1' is getting printed on the output screen. Then, we see that the value of 'x' is incremented by '1' (since we want the output to contain numbers from 1 to 100 in steps of 1). After incrementing the value of 'x' by '1', the control will proceed to check the value of 'x' (conditional expression) and proceed to execute the loop based on the return value of the expression.

Thus the loop is executed 100 times and when the value of 'x' becomes 101, we can see that it goes beyond the limit of 100 and hence the loop will be terminated. Figure 2.33 gives the screenshot of the output that is generated for the above piece of coding.

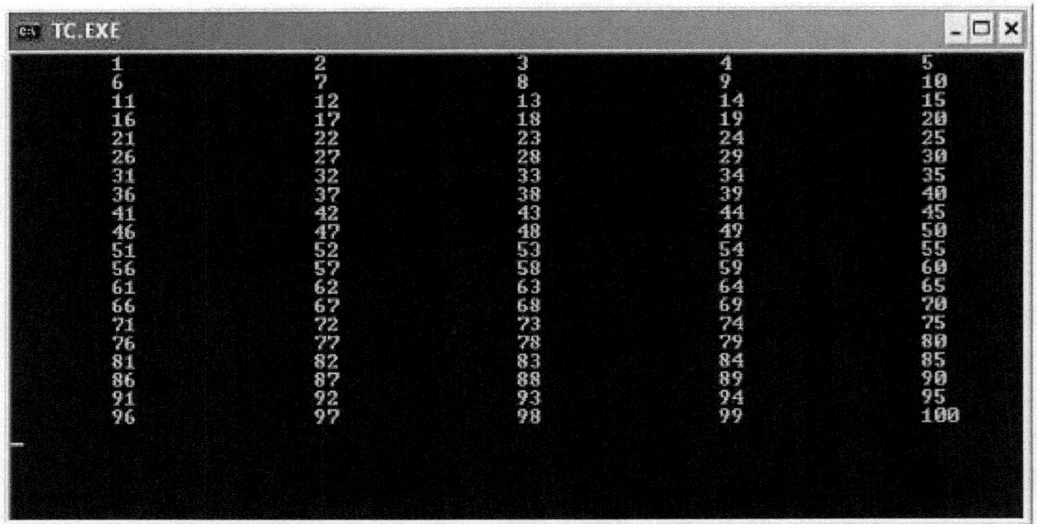

Figure 2.33 *while* loop in action—printing numbers from 1 to 100

See that we had used one of the escape sequences '\t' in order to get an output, in which we have enough space between the numbers 1 to 100. What would you do in case the program requires you to print numbers from 100 to 1? Just initialize the value of 'x' as 100 and decrement the value of x in steps of 1 as you proceed and while doing so, check whether the value of 'x' is not going below '1', i.e., when the value of x becomes 0, the loop will terminate. Thus, working with the while loop is quite simple.

do-while This is another loop control, like the *while*. This serves the same purpose as the *while* serves in programming. But the fundamentals of *do-while* are much the same as *while* loop the following is the syntax of *do-while* loop.

```
do  ------------------------------------------> Loop Header
{ ------------------------------+
Statement-1;                    |
Statement-2;                    |
.......................         |-------------> Loop Body
Statement-n;                    |
Loop counter modification;      |
} ------------------------------+
while (<conditional expression [s]>);  -------> Loop Footer
```

The general procedure for executing the above structure in programs is stated below.

1. Initialization of Loop Counter
2. Executing the body of the loop
3. Modifying the Loop Counter
4. Checking the Conditional Expression

Can you identify any difference between *while* and *do-while* loops, on comparing the syntaxes of *while* and *do-while*. Not too many know? Of course, there are two differences, if you closely observe. The primary difference is that the conditional expression in while loop is checked at the entry of the loop and the same is tested here at the exit of the loop. The second difference is due to the first one. Since the condition is checked at the entry of the loop, the loop body will not be executed at all, once the expression returns a false. But, in the case of *do-while* loop, the body of the loop is executed first and only then the condition is checked and hence, even when the expression is going to return a false, the loop body would have been executed at least once. As an example, consider the following program fragment, whose outputs are shown as screenshots in the Figures 2.34 and 2.35.

Program Fragment-1

```
int x=0;
while(x!=0)
{
printf("%d",x);
}
printf("Nothing inside the loop is printed");
```

Figure 2.34 Output of the Program Fragment-1—*do-while* loop in action

Program Fragment-2

```
int x=0;
{
printf("%d",x);
}while(x!=0);
printf("\n\nThe value of 'x' is printed");
```

Figure 2.35 Output of the Program Fragment-2—*do-while* loop in action

We could see that the initial value of the variable x in both the cases is one and the same. The conditional expression also checks for the inequality of the variable x with zero in both the cases. But in the case of *while*, since the conditional expression is checked at the entry itself and returns a false, the control will never get inside the loop and execute the loop body. But in the case of *do-while*, since the condition is checked at the exit, the body of the loop is executed once. We find that the conditional expression however, returns a false in this case too. But still we are able to get the output value of x, unlike the earlier case. Thus we can conclude that the *do-while* loop will execute at least once in its lifetime, regardless of whether the condition is true or false.

This loop is normally used in creating menu-driven programs, where we would like to have the menu of items to do on the screen, regardless of whether we are executing a particular menu or not. But remember the fact that, whatever can be done with the *while* structure, can be done with the *do-while* structure too.

for This is the most versatile and handy-to-use loop in C language. The syntax of the loop is also very simple that, the user would not forget to provide any particular component when he/she is writing a program. In the other two loops, we see that the initialization is done outside the loop structure and the modification is done inside the loop body. The *for* loop has such a syntax that all come under one line. The following is the syntax of the loop.

```
for (initialization of LC; Conditional Expression[s]; Modification of LC)
{
Statement-1;
Statement-2;
........................;
Statement-n;
}
```

Loop Body

Loop Header

We notice from the header of the loop that the three essential components are available under one roof and the user may not forget to supply them when the code is written. Also, we see that the three components of the loop header, namely, the initialization of loop counter, conditional expression and modification of loop counter are separated with a semicolon symbol, which is also essential. Also, if the loop body has only one statement, then the set of curly braces that are provided in the program are not needed and the loop will still work fine and also be error-free.

The general procedure for executing the above structure in programs is stated below.

1. Initialization of Loop Counter
2. Checking the Conditional Expression
3. Executing the body of the loop
4. Modifying the Loop Counter

As an illustration for the working of *for* loop, consider the following program fragment. We are printing the numbers from 1 to 100, as we did using the *while* loop earlier.

```
int x;
for(x=1;x<=100;x=x+1)
printf("\t%d\t",x);
```

The output of the program is shown in the Figure 2.36.

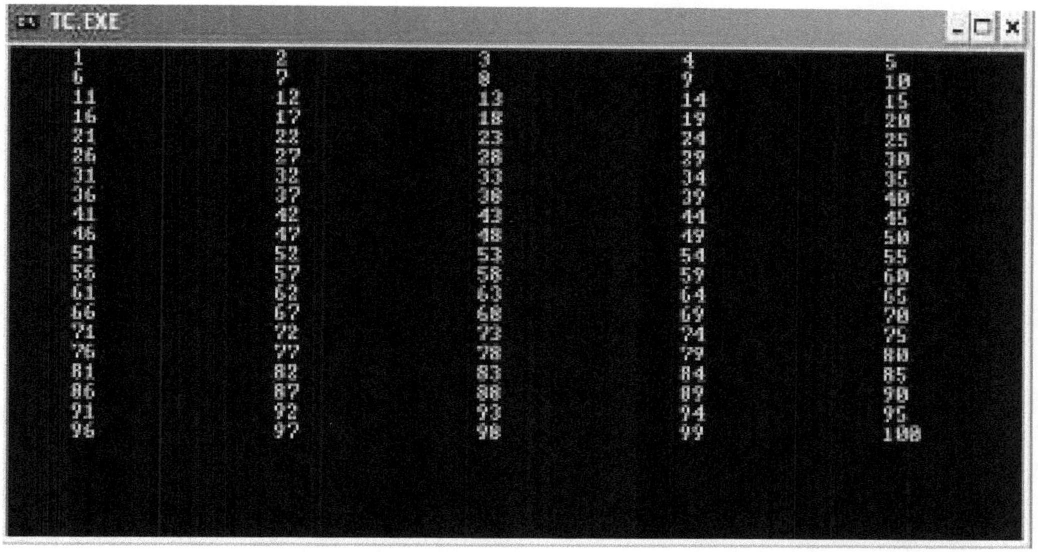

Figure 2.36 *for* loop in action—printing numbers from 1 to 100

There are many forms of *for* loop, which are called as odd loops. These are just like the different structures for the conditional controls we had discussed earlier (*if*, *if-else* and *if-else-if*). The syntaxes of the various odd loops are given below.

Odd loop-1

```
initialization of Loop Counter;
for (; Conditional Expression[s]; Modification of LC)
{
Statement-1;
Statement-2;
........................;
Statement-n;
}
```

Loop Body (the statements between { and }) — Loop Header (the for line)

Here, we see that the initialization of the loop counter is done outside the loop, like the way we do it in the *while* and the *do-while* loops. But still the two semicolons are necessary within the set of parentheses and the first column is left blank.

Odd loop-2

```
for (initialization of Loop Counter;; Modification of LC)
if (Conditional Expression[s])
{
Statement-1;
Statement-2;
........................;
Statement-n;
}
```

Loop Body

Loop Header

In this structure, we see that the conditional expression is used within the *if* structure and the column corresponding to it is left blank and the two semicolons are still there.

Odd loop-3

In the third form of the odd loop, we can see that the modification of the loop counter is done within the body of the loop and the corresponding column is left blank in the structure. Still we could see that the two semicolons are present.

```
for (initialization of Loop Counter; Conditional Expression[s];)
{
Statement-1;
Statement-2
........................;
Statement-n;
Modification of LC;
}
```

Loop Body

Loop Header

Odd loop-4

In the structure shown below, we could see that all the three components are provided outside the structure. But, this kind of odd loop is of rare usage in programming and it does not sound sensible if we use this kind of *for* loop, since this structure defeats the basic purpose of the loop that all the components shall come under one roof.

To conclude, do know that there are four versions of *for* loop (called odd loops) and we use them very rarely in our programs, as they do not comply with the purpose of the *for* loop.

```
initialization of Loop Counter;
for (;;)                                   <------------ Loop Header
if (Conditional Expression[s])
{ ---------------
Statement-1;
Statement-2
.........................;                 <------------ Loop Body
Statement-n;
Modification of LC;
} ---------------
```

We will be using the three loops that we have discussed so far, in our programming later and you will see and appreciate how much these loops would be useful to you in performing repeated operations.

SUMMARY

After having grasped the fundamentals of C programming in this chapter, you will probably be confident of programming. In this chapter, we had started with the history and development of C language and the people who were instrumental for its emergence as a powerful language. Then followed the significance of learning C and implementing the same in Bioinformatics.

Following the introductory ideas, we discussed the building blocks of C language, in which we had learnt the essential ideas like character set of the language, keywords and identifiers, data types, a few category of operators, types of instructions, the importance of the function main(), conditional and unconditional branching, multiway branching and loop control structures. We had seen in detail the way in which the ternary expression is used and also learnt that they can be nested at any number of levels.

We also discussed how to use the functions printf() and scanf() effectively along with the escape sequences. The different data types and the kinds of operators available in the language would have very clearly justified the powerfulness and the efficiency of the language. The intricacies of the case control structure and how it is advantageous over the *if-else-if* chain is also discussed in this chapter and above all, the reason why the professional C programmers avoid using *goto* is also learnt in detail.

I am sure that you would have got more ideas from this chapter and I shall tell you that preserve all those you have learnt in this chapter and you will need them very often in the forthcoming chapters. There is no wonder why C is called as God's Programming Language.

CRITICAL THINKING QUESTIONS

1. Try the following exercise and find out which are valid names for an identifier and which are invalid. Also, try to justify why some of the names given are invalid
   1. Long
   2. printf
   3. 43F
   4. F_9
   5. value$sum
   6. exit flag
   7. 3lotsofmoney
   8. _valid_flag
   9. Jerry7
   10. _foo42
2. There is a variable 'a' declared as integer in a program. Then, what will be the result of the following expression

   a = 300*300/300
3. Try to evaluate the following expressions with the help of the precedence rules
   1. 5*5/5+4*6%4–6+8
   2. 7–4+6*7+8–4*2/12%6/3
4. Try and evaluate the following logical expressions
   1. a = 10&&4
   2. b = 5&&3
   3. c = 56||4
   4. c = 67||0

   5. c = 9&&0
   6. a = 0&&4
   7. x = 34&&23&&0
   8. x = 33||0||4
   9. x = 1&&0||4
   10. x = 3||5||6

5. Try and evaluate the following relational expressions
   1. X = 10>5
   2. Y = 4<=4
   3. X = 3>=3
   4. X = 4<3>–2
   5. Y = 10>6<7
6. What would happen if the binary equivalent of '9' is shifted towards left by two bits? i.e., consider the expression a = 9<<2 and find out the result.

SELF-LEARNING EXERCISES

1. Use the *if-else* conditional structure to check the odd or even nature of a number.
2. Write a program to find whether a given integer is positive or negative.
3. Use the *if-else* conditional structure to check whether a number 'm' is a multiple of another number *n*.
4. Use *if-else-if* conditional structure to classify the operators given as arithmetic, logical, etc.
5. Use the case control structure to implement the above logic.
6. Write a program fragment to find the reversed form of a four-digit integer using *while* loop.
7. Write a program fragment to find the factorial of an integer using *for* loop.
8. Write a program fragment to find the sum of the digits of a four-digit integer using a *do-while* loop.
9. Write program fragments to find out the area, volume, surface area and perimeter of different two- and three-dimensional figures (Find out the formulae for various geometric figures).

10. Write a program to print the Fibonacci series consisting of 'n' terms (A Fibonacci series has the first two terms as 0 and 1 and the remaining terms are obtained with the addition of the preceding two terms).
11. Write a program to find the Armstrong numbers between 1 and 1000 and print them (An Armstrong number is that which is the sum of the cube of digits of a given number, e.g. $153=1^3+5^3+3^3$).
12. Write a program to print all the prime numbers in a given range of values (get the lower and upper limits from the keyboard).
13. Write a program to find the sum of the first 'n' natural numbers (get the value of 'n' from the keyboard).
14. Write a program to perform the basic arithmetic operations using *switch-case* structure.
15. Write a program to print a Pascal's triangle.
16. Write a program to print a Floyd's triangle.
17. Write a program to find whether the given number is a palindrome or not.
18. Write a program to print all the multiples of a number 'n' in a given range of values.
19. Write a program to calculate the simple interest.
20. Write a program to calculate the compound interest.
21. Write a program to calculate the force for the given mass and acceleration due to gravity.
22. Write a program to display the size of all the basic data types using the size of operator.
23. Write a program to read a character from the keyboard and print its ASCII value.
24. Write a program to display all the characters used in C along with their ASCII values.
25. Write a program to calculate the pH value for a given H^+ ion concentration.
26. Write a program to calculate the average molecular weight of a given double-stranded DNA sequence with n base pairs.
27. Write a program to calculate the Body Mass Index value of a person.
28. Write a program to calculate the Relative Centrifugal Force.
29. Write a program to find the nature of a solution with a given pH value.
30. Write a program to find all the possible roots of a quadratic equation.
31. Write a program to check whether a person is obese or not from the given value of BMI.

32. Write a program to find and print whether the given nucleotide base is a purine or pyrimidine.
33. Write a program to swap two values without using a temporary variable.
34. Assume that we enter numbers from 1 to n, not in any particular logical order. We enter the numbers randomly. In doing so, we enter only n–1 numbers. After entering the last number from the keyboard, the program that you write shall flash with the number that you had missed to enter.
35. Write a program to implement a menu driven programming with the help of *switch-case* structure and *do-while* loop (choose any 10 options and write the program).
36. Write a program to print only the uppercase and lowercase alphabets on the screen (use *continue* statement for the program).

3 BASIC PROGRAMMING TECHNIQUES

3.1 EDITING, SAVING, COMPILING AND EXECUTING A C PROGRAM

Anyone who works with a computer and developing a script (whether it is a document or a presentation) need to have a good platform to work with. For example, we schedule our day-to-day activities, make it as a record in our diaries or a scribbling pad and then execute the same. Similarly when we learn and write programs in any programming language, we must script the programs in a an editor and then execute the program to check its validity. Before we start using any editor, we must know the background details of it and learn to be convenient with it. This chapter deals with how to make use of the editor for typing the C programs and all related aspects. I have given a simple program in C and a line by line explanation is also provided for the syntax used. I have also elaborated on how the program is saved, compiled and executed. At the end of the chapter, the most important concept in C language called Debugging is dealt with using a simple example and explanation.

3.1.1 What is an Editor and What is Editing?

An editor is a program where the user types and does all the formatting and editing processes. A notepad is a typical editor used in a Windows mode. When we work in a DOS mode, we use DOS editor and all that we can do in a Windows editor, could be done in DOS editor (options like save, copy, paste, exit, etc. are possible). Another famous editor still used by many of those who work in Windows mode is a WordPad. Those who develop web pages use the default editors like notepad or dedicated editors like FrontPage Express or Microsoft FrontPage. Different programming languages use dedicated editors for editing the programs typed using their syntax. For all the programming languages, one can very well use a notepad or even a DOS editor. But still the very purpose of having a dedicated editor would not be solved, if they are not used by the programmers.

As for example, programming languages like FORTRAN and COBOL use a very famous editor called a Norton Editor, shortly called as NE. Java uses a different editor called Kawa. Visual Basic has its own editor (built-in with the package), UNIX uses vi editor for typing either C programs or shell scripts. Likewise, C uses a dedicated editor called Turbo Editor, shortly called as TC. Figure 3.1 shows how a TC editor would look like.

Editing is a very cumbersome process, if it is done professionally, especially in languages like FORTRAN or COBOL. But in languages like C, C++ or Pascal, the process of editing is very easy and any individual could do the editing work, without having much expertise on the editor.

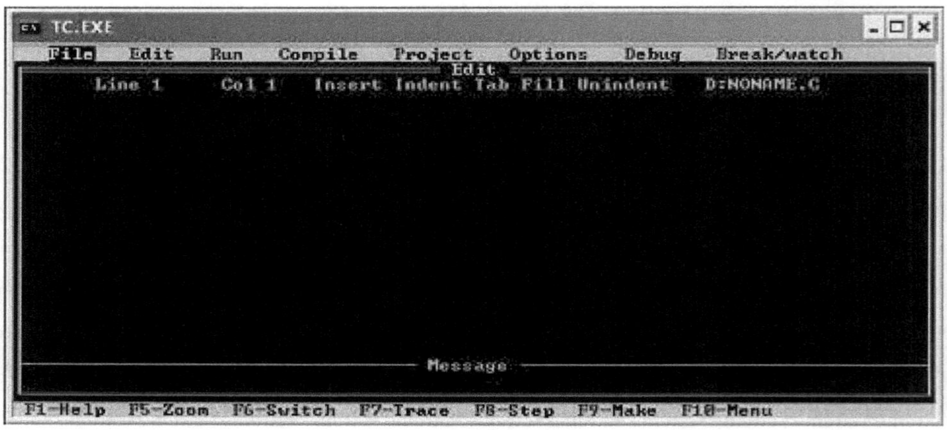

Figure 3.1 A Turbo Editor or TC

When a user is editing with a word processor, with FrontPage Express or with Microsoft Word, he/she must be an expert in using them, since they have such intricate designs and tools for the user to learn and to use. Even, when the user is editing with the help of DOS editor, he/she must know about the options available with the editor, so that no difficulty is faced during the process of editing. When a programmer is using a Norton Editor for editing a FORTRAN or a COBOL program, he/she must know the regulations that these languages impose on the user in using them.

When one uses a notepad editor in Windows, there is absolutely no expertise needed, since it is an easy-to-use editor. Similar to that, when we are using a TC, no difficulties arise during the process of editing a C program. In Chapter 2, we had been discussing the significant features of C and if you can recall, one such feature is that C is a free-form language. There we had mentioned that it is very easy to use a Turbo Editor. We had also mentioned a point that if the user once opens the editor, he/she can start typing the program from any point of the editor (whether fitting or not fitting within the visible part of the screen). Figure 3.2 gives a sample program fragment typed in the editor and Figure 3.3 shows the same program typed in a different style and these two figures imply the point that the Turbo Editor could be used without any difficulty. The output of both the styles will be one and the same, as shown in Figure 3.4.

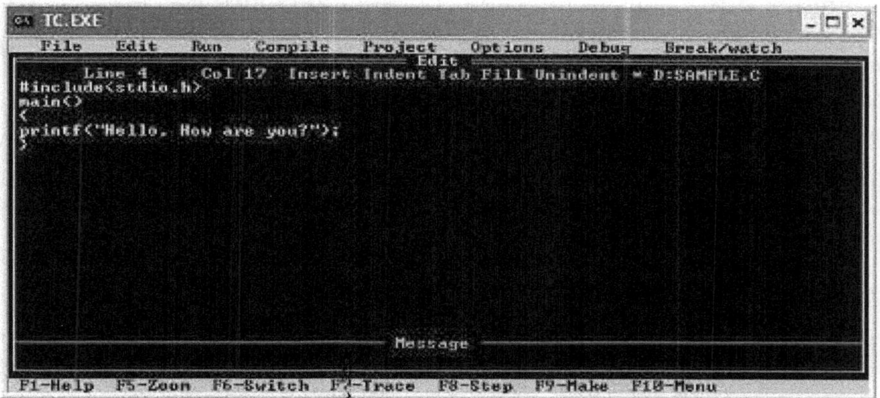

Figure 3.2 A simple program written in TC Editor

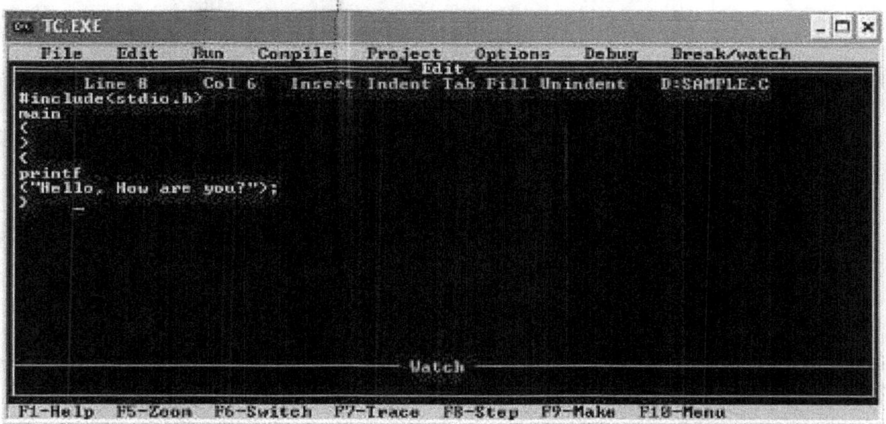

Figure 3.3 Program in Figure 3.2 written in a different style

Figure 3.4 Output for the programs in Figure 3.2 and Figure 3.3

Now that we know the way a TC editor looks and is used, we must know the different parts of the editor. As usual, the editor has the menu bar that contains options like file, edit, etc. When we would like to use a particular menu item, we must use the hotkey combination (as we do in DOS or in Windows). Remember that the roll of mouse is very minimal in TC editor. We can even say that mouse could not be used at the level of TC, since it is totally command based. So, to activate the option File from the menu, we have to use the key combination Alt+F from the keyboard. Once we use this combination, the corresponding menu pops down as shown in Figure 3.5.

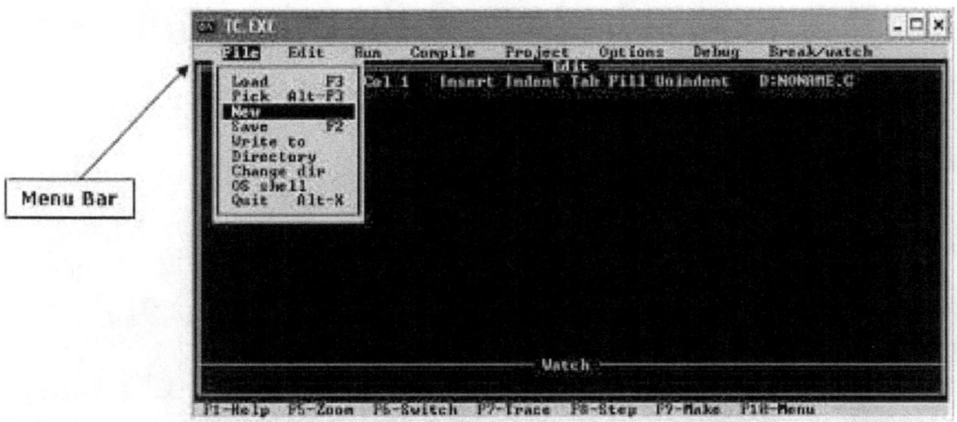

Figure 3.5 Menu bar and the file option in the menu bar

Normally, the Edit option will not work and will be disabled. But still we could perform the editing operations like cut-paste and cut-copy, as we do with other editors. Turbo Editor inherits these functions from its predecessor, the WordStar editor. When one wants to copy and paste a particular block of code to some other part, the following is the procedure.

Procedure for copy-paste and cut-paste of a block of code

1. Place the cursor at the beginning point of the code which you want to copy and holding the control key down, press the keys 'k' and 'b'.
2. Take and place the cursor after the last character of the end of the block of code that you want to copy and holding the control key down, press the key 'k' twice.
3. On completing the above two steps, you will get a highlighted block as shown in the Figure 3.6. Now the block is ready to be copied or moved (cut and paste) to another part of the editor. Figure 3.7 shows how it is pasted and Figure 3.8 shows how it is moved (cut and paste). For pasting it to another location of the editor, take the cursor to that particular position and by holding down the control key down, press the keys 'k' and 'c'. For moving (cut and paste) to another location from the present location, take the cursor to the specified location and by holding the control key down, press the keys 'k' and 'v'.

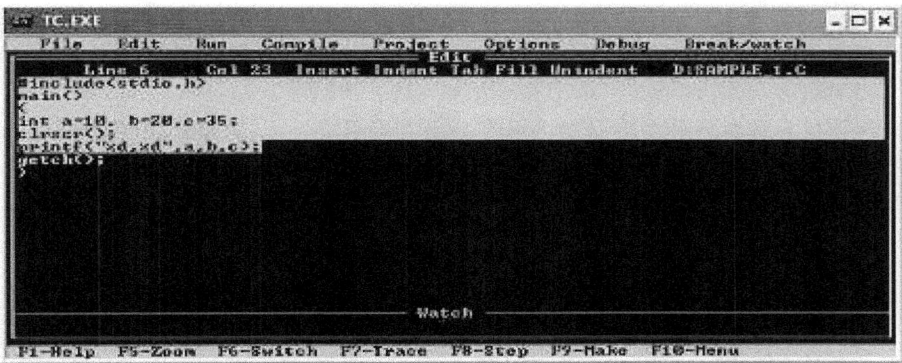

Figure 3.6 Block of code to be copied or moved to another location

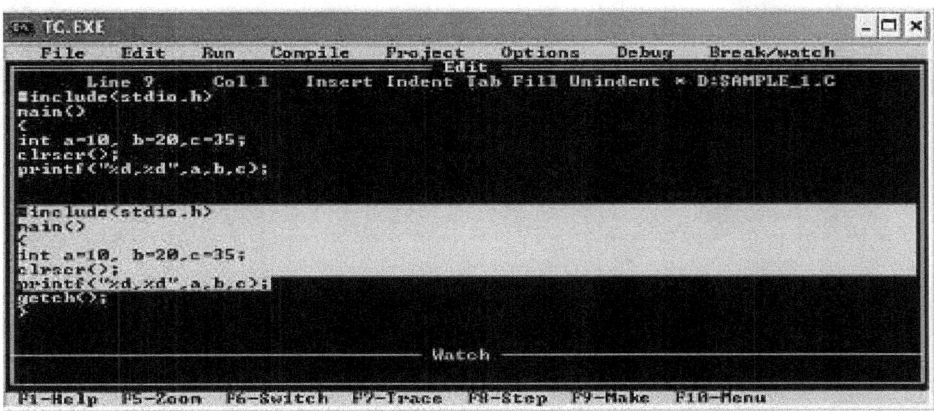

Figure 3.7 Block of code copied to another location in the Editor

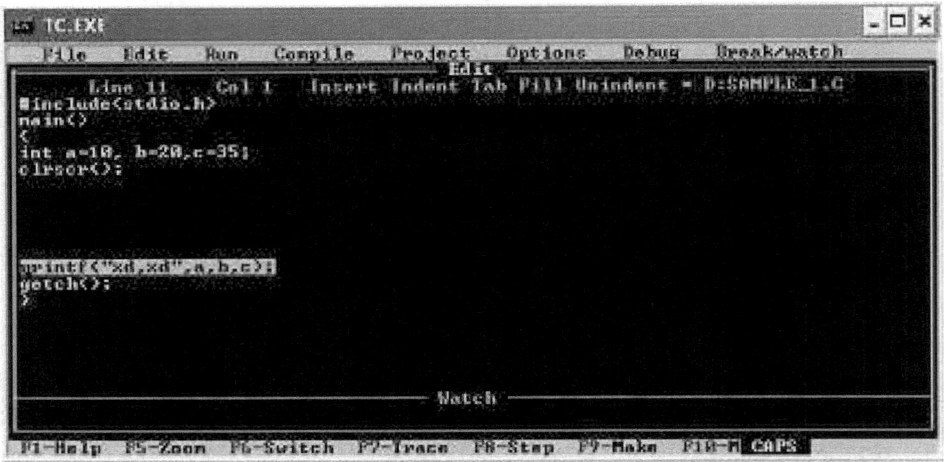

Figure 3.8 Block of code moved to another location in the Editor

If you need more help on the commands and operations that could be performed in the Turbo Editor, you may press the key **F1** by keeping the cursor on the editing part of the screen. All the cursor movement commands could be viewed by further pressing Page Down/Page Up keys. This is shown in the Figure 3.9.

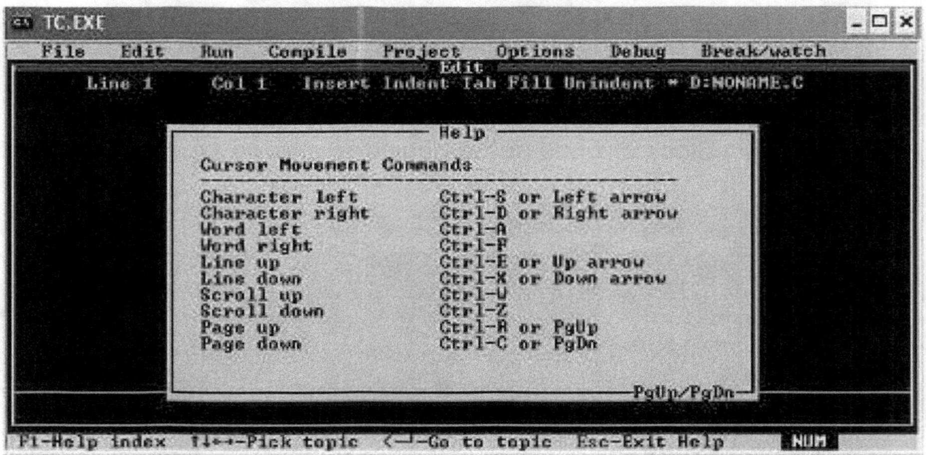

Figure 3.9 Cursor movement commands

If we want to have help regarding the **File** option in the menu bar, then use the combination Alt+F and then press F1. You will get all the information with respect to **File**. This is represented in the Figure 3.10.

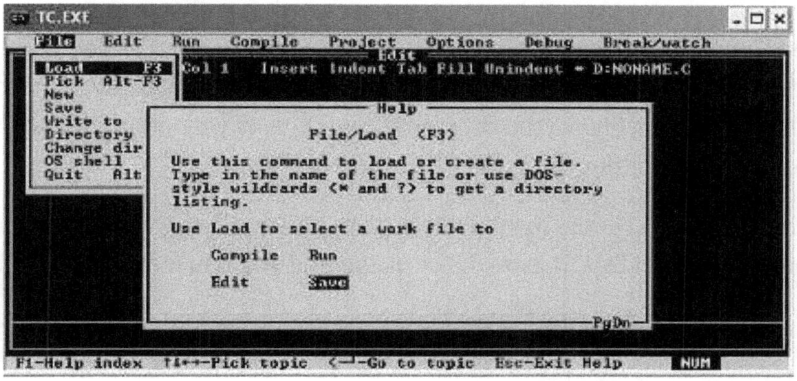

Figure 3.10 Help on File option

The next option available after the **File** is **Edit** and normally it will be disabled as said earlier and still we can get the help on it by highlighting the option using the key combination Alt+E or F10 and then pressing F1. Figure 3.11 represents the help corresponding to the **Edit** option.

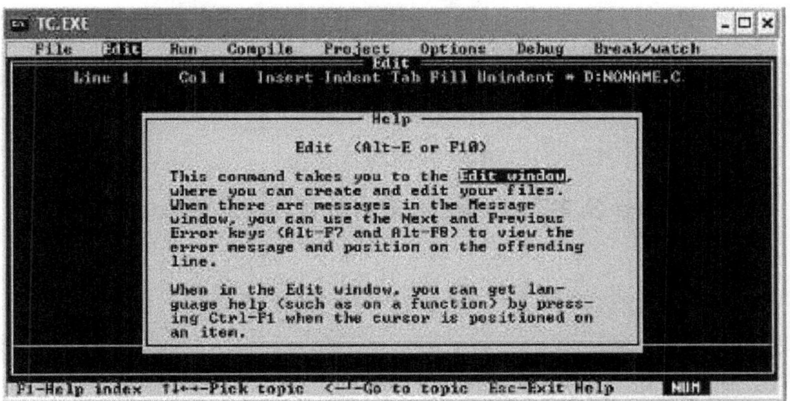

Figure 3.11 Help on Edit option

The same way, you can get the help for all the options on the Menu Bar and know the way they work. As you do programs, you will be better knowing them, since most of them will be used by us during the different contexts of a program.

3.1.2 Installing the C Compiler

The installation of the C compiler is quite simple. Just open Google in the Internet Explorer or Netscape and type Turbo C in the keyword area. You will get a screen that is shown in the Figure 3.12.

Click the first link provided and get into the screen that is shown in the Figure 3.13. In that, go to the section How to Download Turbo C 2.01? and click the link provided there. You will get a zip file from which you can extract the files to your hard drive. Carefully go through the read me document and extract the files in the corresponding drive of your computer and now you are ready for installation of the compiler.

Follow the steps given in the read me document and do the installation in your local hard drive. Once the compiler is installed, it is ready for usage and you can start writing programs.

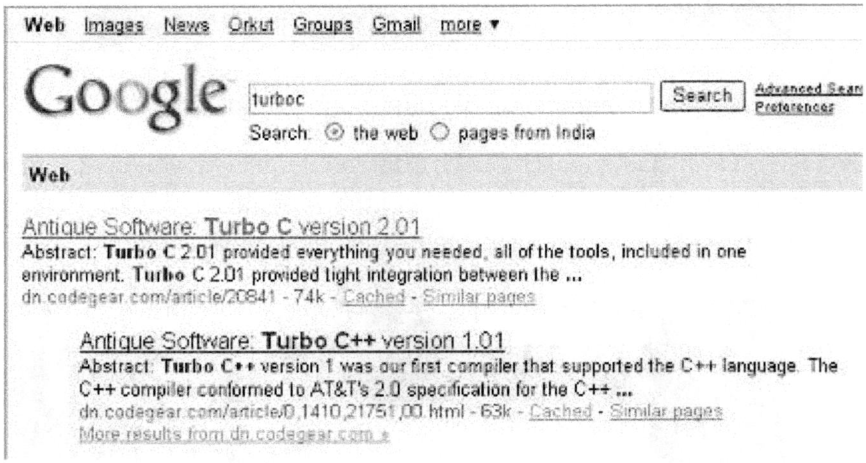

Figure 3.12 Link for Turbo C compiler on the Web

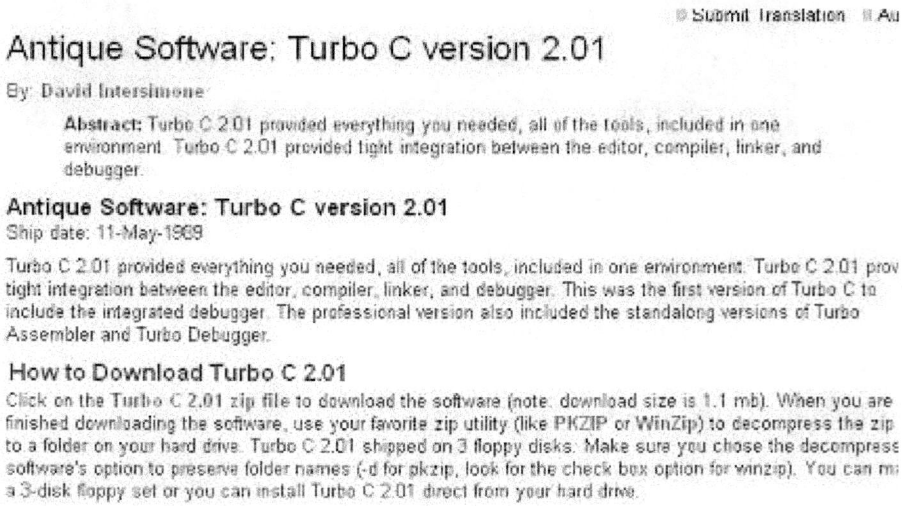

Figure 3.13 Download page for the Turbo C compiler

3.1.3 Editing and Saving Your Work

Create shortcut on your desktop for the C application file, so that you can always open the editor for typing your work from the desktop and you don't have to go to the drive for opening the editor each and every time.

Once the editor is opened, you can start typing the program coding in it and you do not have to follow any rules (as mentioned earlier, C is a free-form language). Let us type the following simple program in that.

```
#include<stdio.h>
main()
{
printf("Welcome to the world of C Programming");
}
```

Figure 3.14 shows the way it is typed in the editor and you can see that the file is not named (the default name noname.c could be seen at the right corner of the editor screen).

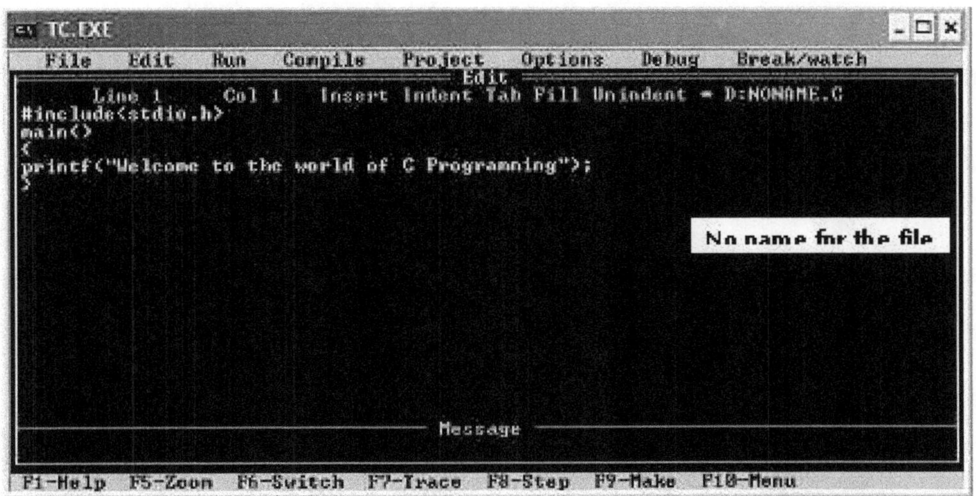

Figure 3.14 A program typed in the Editor

In order to name the file, we have to save our work. For saving the work, we have two options. We can use the key combination Alt+F and get the File menu to pop down and from it we can choose the option Save. You can see that there is a shortcut key assigned beside the option Save in the File menu, which is F2. So, we can directly access F2 instead of the other way around. When we choose either of the above methods, we will get a small window (a Save dialog box) that will expect a file name to be typed. Figure 3.15 represents the appearance of a dialog box for saving the program that is typed. Just start typing the name of the program in the dialog box and it is not necessary even to give the extension. Normally, C programs are saved with an extension .c and the case (uppercase or lowercase) is not a big deal in saving the file.

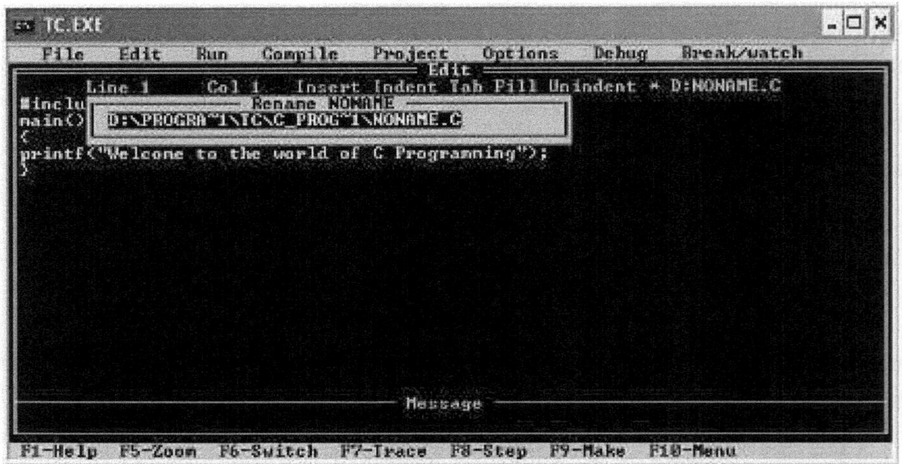

Figure 3.15 Save dialog box (obtained by using the key F2)

In Figure 3.16, we can see that the file is named as sample (skipping the extension) and in Figure 3.17, we see that the name noname.c is changed to sample.c.

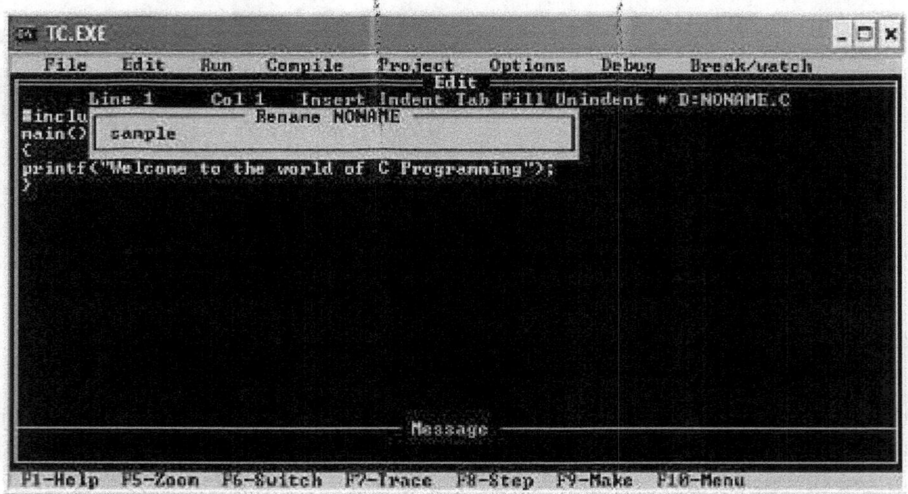

Figure 3.16 File name typed in save dialog box without the extension

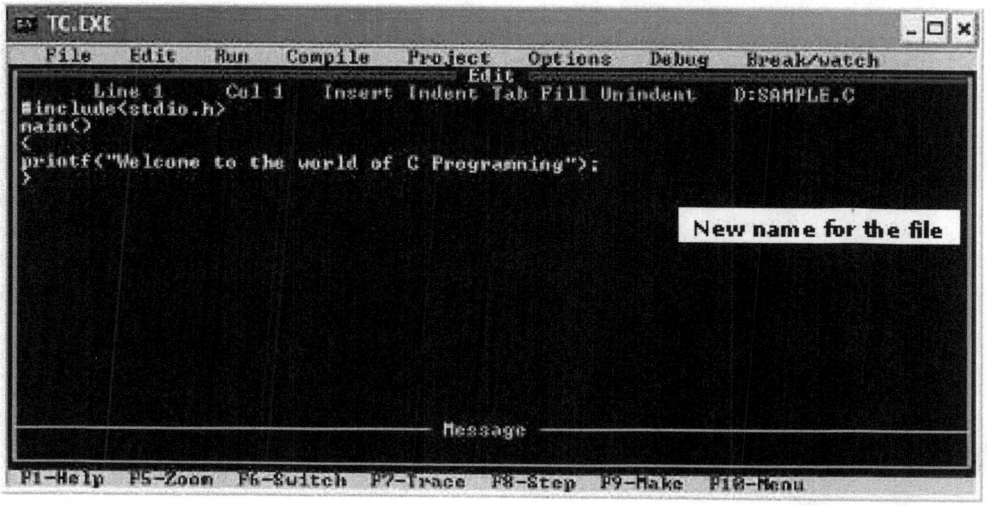

Figure 3.17 New name (sample) given for the file typed in the Editor

Each time we make any alterations in the program, we just have to update them using the key F2 and the change will be updated properly in the corresponding file.

Now that you have saved your work, it is time to compile and trace the errors in the program. We will learn that in the following section.

3.1.4 How to Compile a C Program?—The C Compilation Model

The process of compiling a C program is a very essential part of programming, which allows the user to proceed further with the validation. The compilation process will generate the **executable** or .exe format of the corresponding .c file. Without the proper compilation of C programs, it is not possible to execute and check the validity of the programs. On compilation, the user actually checks for any syntax or linker errors and fixes the same during the process. Compilation has three substages, namely, compiling, linking and making, all done one after another. The final stage, **making**, would create the .exe format of the C file.

For compiling the C program, use the key **Alt+F9** and a window will appear showing the status of the compilation process. Figure 3.18 represents the process of compilation. We see that the compilation is **success** and this implies there is no syntax error in the program and we are ready to link our program. For linking and making .exe file, we use the key **F9** alone and the corresponding screenshots are shown in Figures 3.19 and 3.20.

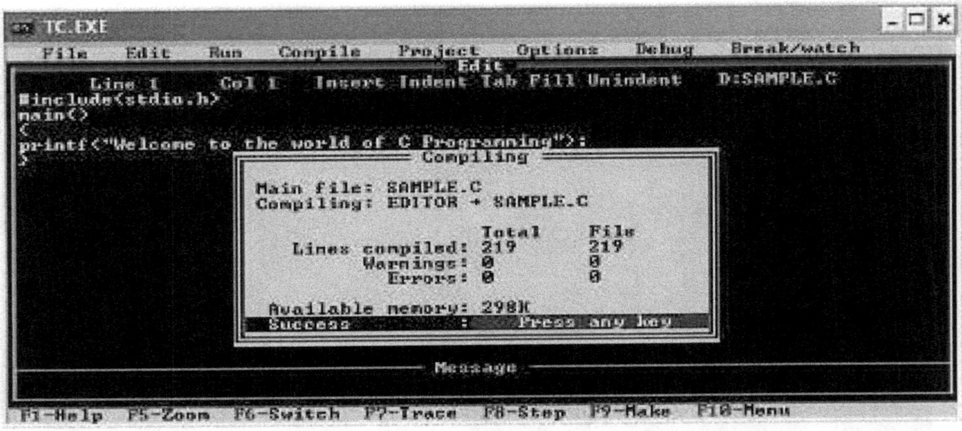

Figure 3.18 Compilation stage

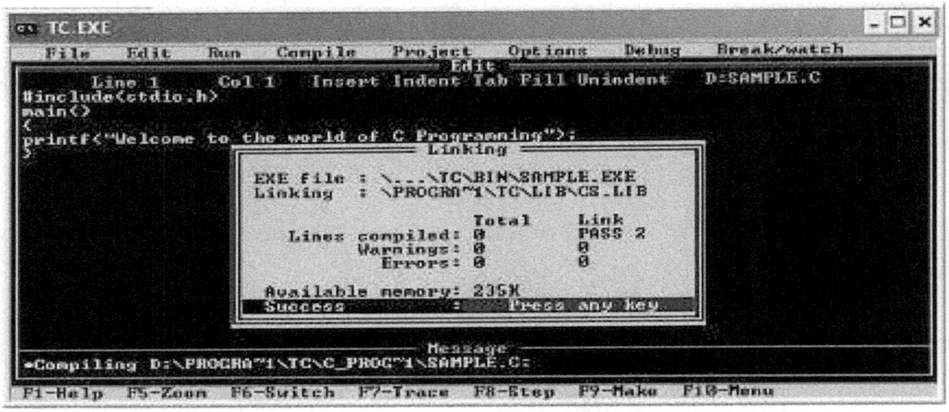

Figure 3.19 Linking stage

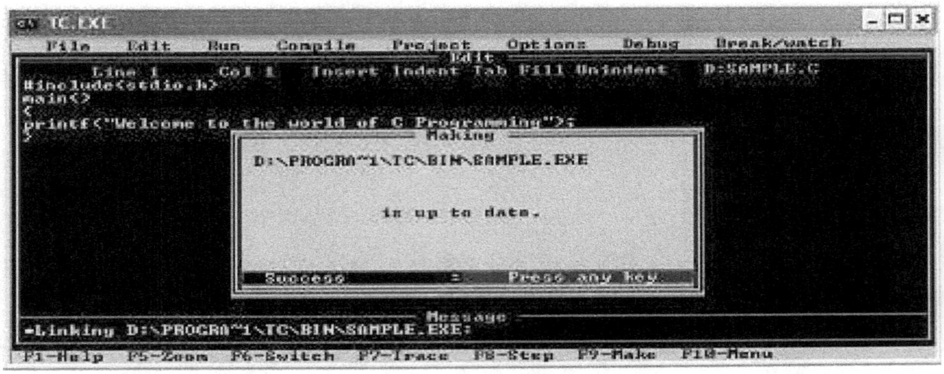

Figure 3.20 Making stage (making .exe file)

If the program is free from any syntax errors or linker errors, the above three stages of the compilation process would be success as shown in the figures above. Suppose the program has syntax errors or linker errors, it will be flashed in the space allocated for that down the editor (the space titled Message). Figure 3.21 flashes that a syntax error has occurred in the program and Figure 3.22 shows the details of the error. Figure 3.23 shows a linker error that might creep into your programs.

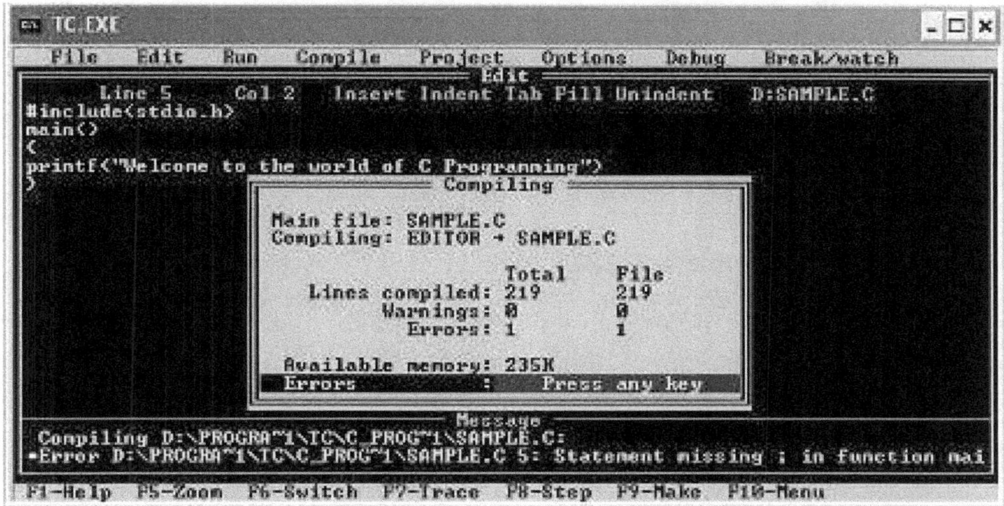

Figure 3.21 Error message flashed due to syntax

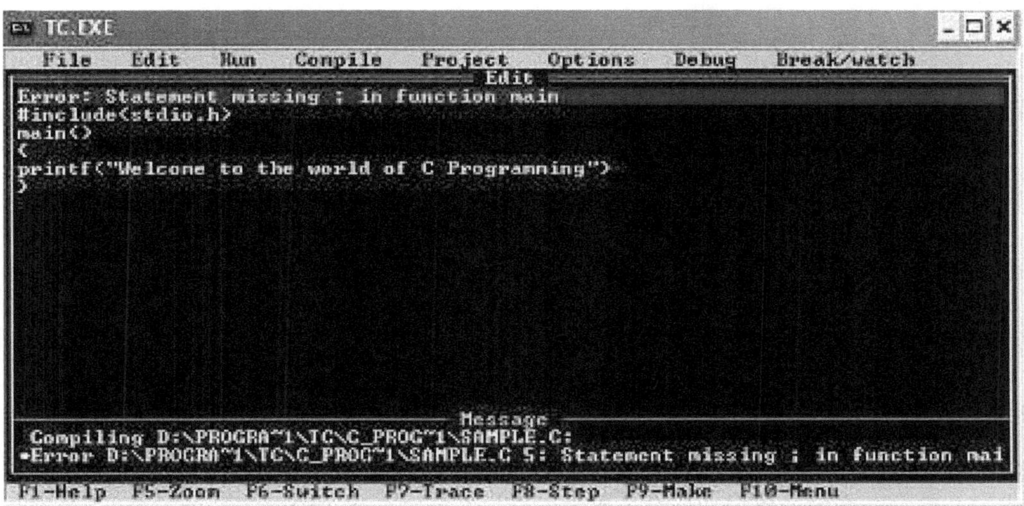

Figure 3.22 Details of the syntax error

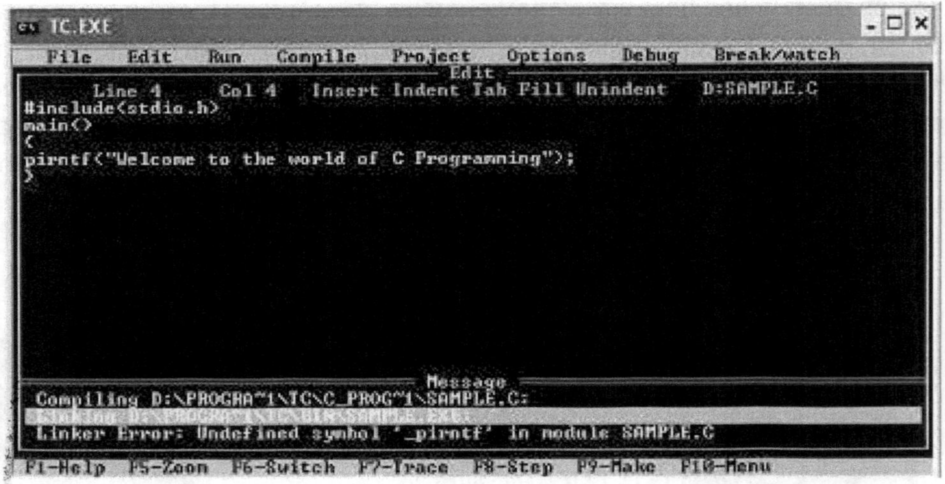

Figure 3.23 Linker error in the program

It is the programmer's responsibility to take care that no logical errors are appearing in the program, since only syntax errors and linker errors will be noticed and flashed by the compiler. For example, consider that in one of our programs, we have divided a given number by 'zero'. Since divide by zero is not allowed, we would anticipate an error while we run the program. But that will not happen, since it is a logical error and is due to the programmer. So it will be skipped from being traced by the compiler.

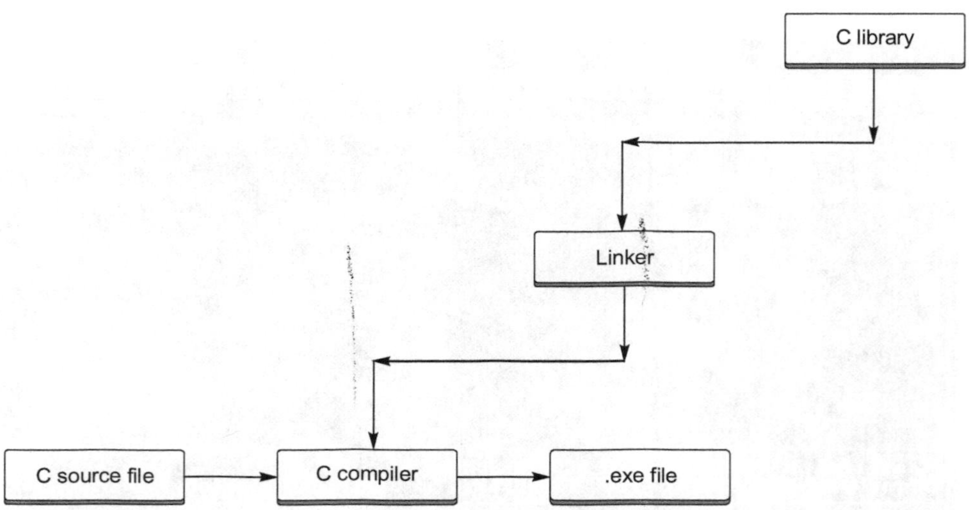

Figure 3.24 C compilation model

The C compilation model is depicted in the Figure 3.24, where the compilation process is indicated as a two-stage process. All the syntax errors are notified by the compiler and the other type of errors are reported to the compiler by the linker. The linker communicates with the library, if there is any issue in the program related to the library functions or built-in functions and reports to the compiler. Thus, even when the program is error-free with respect to the syntax, the linker might have reported some error due to any unrecognized functions or function names. The error shown in the Figure 3.23 is one such error, where the spelling for the function printf() is typed wrongly and hence the linker is unable to recognize the word, as it won't be available in **stdio.h** or in any other headers.

Also, if there is any error in a program, do fix all the errors and try to analyse why the errors come in the program. If you can do this, you will be surely a master in Programming. One good practice is to keep a database of errors with you and this way, you can easily trace the errors and fix them easily.

3.1.5 How to Execute a C Program?

Executing a C program depends on whether you get a successful compilation and hence **.exe** file. If we are able to get the executable form of the source code, we are ready to execute our program. Consider the example program that we have been discussing so far (the file named **sample.c**) and look at the Figure 3.20. We can notice that the **making** of **.exe** file is successful and the user could validate his/her program. For executing the file, use the key combination **Ctrl+F9**. You will be provided with another screen, which is called as the **user screen**. So far, we have been working with the editor and the corresponding screen is called as the **editing screen**. In the user screen, we can give the inputs for the program and check the validity of the program logic. In our example, we need not give any inputs, since we are just displaying a message. When the program is bug-free, we will get the message in the user screen. Figure 3.25 shows the way the message appears in the user screen. But in actual practice, when you press the key combination **Ctrl+F9**, you will not be able to view the output in the user screen properly, since the control will be available only for a moment in the user screen and immediately comes back to the editing screen. In order to view the output, the user has to use the key combination **Alt+F5**. The users might find it very difficult to use the said key combination every time they run a program and shall also want the control to remain in the user screen as long as they wish.

The question is whether we can achieve this idea? The answer is yes. C has console I/O functions for achieving this, which we will discuss in the following section of this chapter.

Figure 3.25 User screen displaying the output

3.2 USING CONSOLE I/O (INPUT/OUTPUT) FUNCTIONS

As said in the previous section, C provides console I/O functions if the user wants to view the output and make a delay in getting the control back to the editing screen from the user screen. One such function is **getch()**. The function is expanded as **get character**. This means the user, after viewing the output, has the option to make the control remain in the user screen itself and after entering a **character** from the keyboard, the control would be taken to the editing screen. The character entered may be any character available in the keyboard. Normally the function **getch()** would be used before we close the program while writing the coding, in the penultimate line of the program. The above program would appear as shown below.

```
#include<stdio.h>
main()
{
printf("Welcome to the world of C Programming");
getch();
}
```

When you run the above program, the status of the user screen would be like the one shown in the Figure 3.26 in which you could see that the control is waiting for the user's input character (the function **getch()** works here) to get back to the editing screen.

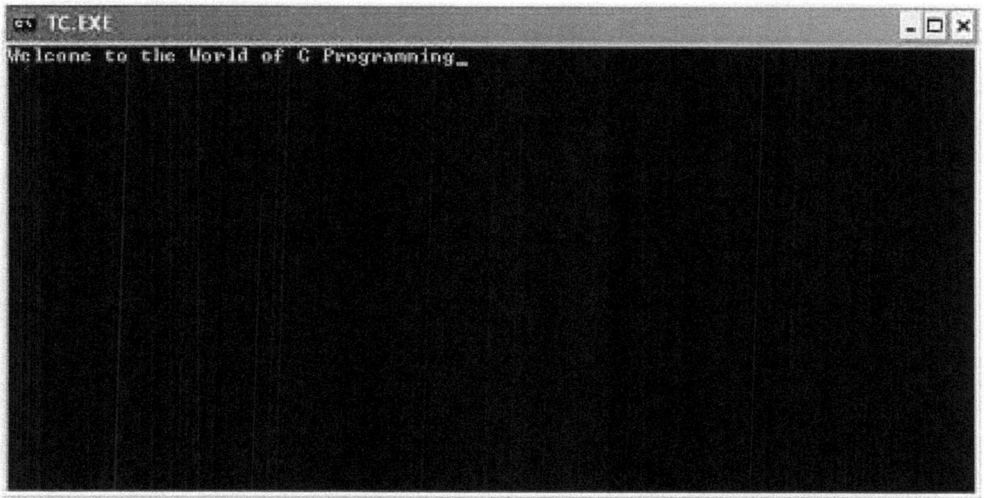

Figure 3.26 Function getch() in action

There are other versions of the function **getch()** in usage. They are the functions **getchar()** and **getche()**. We will be dealing with these two functions in detail later in our Chapters. These functions could also be used instead of **getch()** in our program to make a delay in getting the control from the user screen to the editing screen. But they will have a different effect when used, giving rise to the same kind of result.

For example, when we use the function **getchar()** instead of **getch()**, then the character that is typed by the user to get the control from the user screen to the editing screen will also be displayed along with the message. Figure 3.27 proves this fact. We see that an extra character (e) is displayed with the message.

This means after typing the character, the user has to press the enter key or any similar key to get the control back to the editing screen. This would be an additional work for the user of the program and so the programmer would naturally prefer using the function **getch()** instead of the function **getchar()**. When the function **getche()** is used in place of the functions **getch()** or **getchar()**, the result would be the same. But the way in which **getche()** behaves is quite different from that of the function **getchar()**. When we use the function **getche()**, that will also display the character that is typed along with the message displayed, but on typing the character, it will get the control back to the editing screen and one need not press any other keys like enter or escape from the keyboard, unlike that in using the function **getchar()**. But, try to understand the fact that both **getchar()** and **getche() echo** the character that is read from the keyboard, whereas the function **getch()** will just read the character and would not echo that on the screen.

Figure 3.27 Function getchar() in action

All these functions are grouped under the category conio.h (header file corresponding to the console input and output functions). There is one more function in the header conio.h that is used quite frequently in all our programs and that is the function clrscr(). This function clears the screen during each run of a program. For example, consider the program, which we have written above. In that we have not used the function clrscr() and hence, each time the program is executed, we will have a junk of the same kind of output in the user screen and this is true even if we run two different programs, one after another. The Figure 3.28 represents a user screen in which we have the outputs of two different programs and this has resulted since we have not used the function clrscr() in either of the program codes.

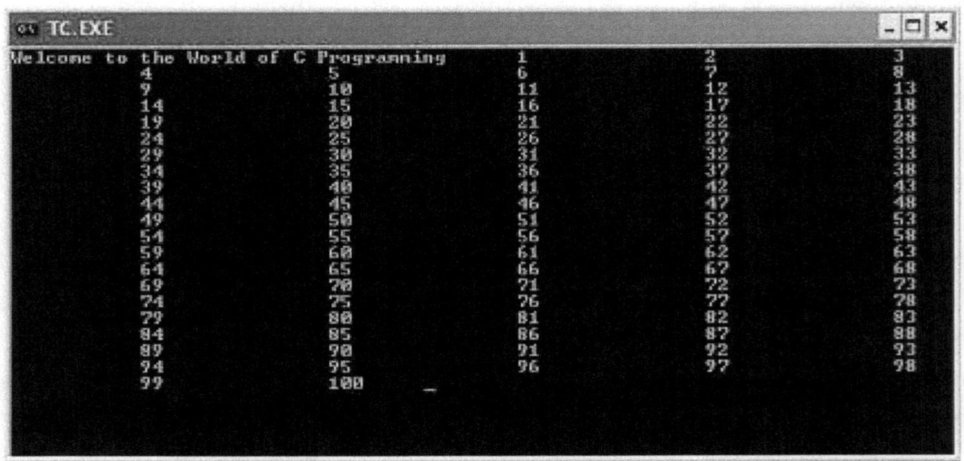

Figure 3.28 Outputs of two programs, where clrscr() is not used

Figure 3.28 has the outputs of the program showing the message and the numbers from 1 to 100 printed. When we open another program and execute it, we will get the third output along with the first two. This surely will make the user screen to have a clumsy look, which any professional C programmer would not like to have. The usage of the function **clrscr()** immediately after the **variable declaration section** (and not before that) of the program would solve this issue and each time a program is executed, the earlier output would have been cleared and the output of the program presently executed would be available in the user screen. Figure 3.29 shows the way the function **clrscr()** is used in the coding of the program. In this program, we have used a variable named 'x' and we had used the function **clrscr()** after that. If there is no variable declared, then the function could be used straightaway in the beginning without any reservations.

The functions **getch()** and **clrscr()** are two of the console I/O functions, which would be used in all our programs. The function **getche()** would be very rarely used and the function **getchar()** would be used when we discuss character arrays.

Even though these functions are grouped under **conio.h**, they can be used without even including the header in the program, as it is purely optional for the programmers to do so.

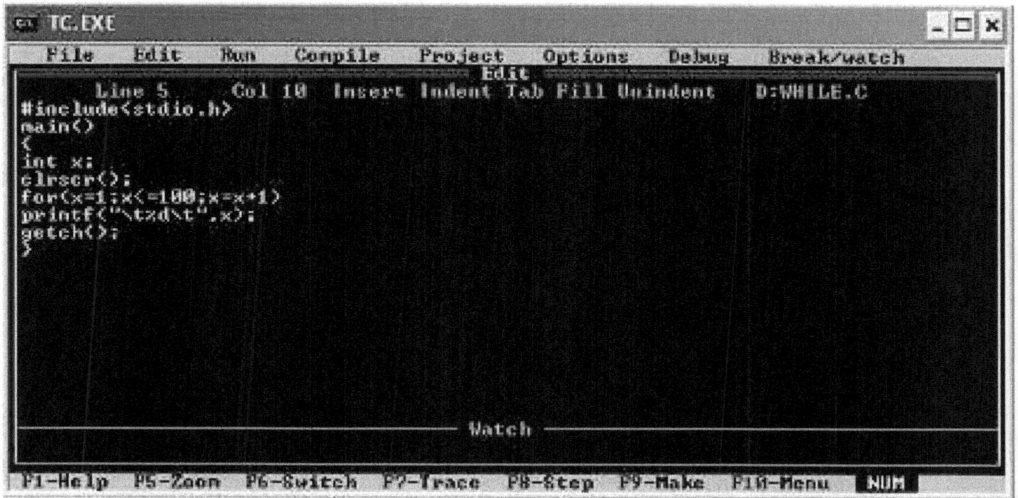

Figure 3.29 Usage of the function clrscr() in coding

3.3 FIRST PROGRAM IN C

After learning the building blocks and the essentials, we are now ready to write simple programs using the syntax of C. The following will be the very first and full-fledged program that we are going to discuss.

```
1.  /* program to add two integers and display the result */
2.  #include<stdio.h>
3.  main()
4.  {
5.  int x,y,z;
6.  clrscr();
7.  printf("Enter the values of x and y : ");
8.  scanf("%d%d",&x,&y);
9.  z=x+y;
10. printf("\n\nThe sum of %d and %d is %d",x,y,z);
11. getch();
12. }
```

We can see that the program has used some features, which we have not discussed at all. Let us consider and analyse the program line by line. The **first line** of the program has a brief description about the program and it starts with /* and ends with */ (a slash and an asterisk and an asterisk and a slash). The C compiler will simply ignore anything that it finds within these two pairs of symbols. This marks the **comment** line in the program and the programmers are free to use comments whenever they feel like using. In fact, the above style of comment is **single line comment**. If the programmer wants to use a comment running over more than one line, then he/she has to follow the **multiline comment**, which starts with the same pair of symbols and ends with the same pair of symbols. Consider the following code which illustrates the way a multiline comment appears in the program code.

```
/* Name     :   Program to add two integers and display the result
Author      :   XYZ
Date        :   03-07-2008 */
```

We see that the information regarding the author of the program and the date on which the program is written is also provided along with the name of the program. This way we can provide a multiline comment in our programs. One thing which the programmers must keep in mind is that the comments can spread over any number of lines, but they cannot be nested. Consider for example following comment.

```
/* Name   :  Program to add two integers /* and display the result*/*/
```

The above comment is invalid and the user would get syntax error, if such a comment is provided in the program. Figure 3.30 shows such an error. Another important point to be followed by the professional C programmers is that they must not use comments unnecessarily in programs and too much of their usage would make the program inelegant. The programmer has to take care that the number of comments in a program is not exceeding the actual number of lines of coding. This will surely spoil the look of the program and the program would definitely lose the professionalism.

If we analyse the **second line** of the program, we see that we have used #include<stdio.h>, which is called as a **preprocessor** in C language. Actually, the programmers include standard **header files** that are available in C library in order to use the built-in functions. For example, in our program we have used functions like printf() and scanf() in the lines 7, 8 and 10. These functions are already defined by the language designers and the user has to just use it and need not bother about the way they are designed.

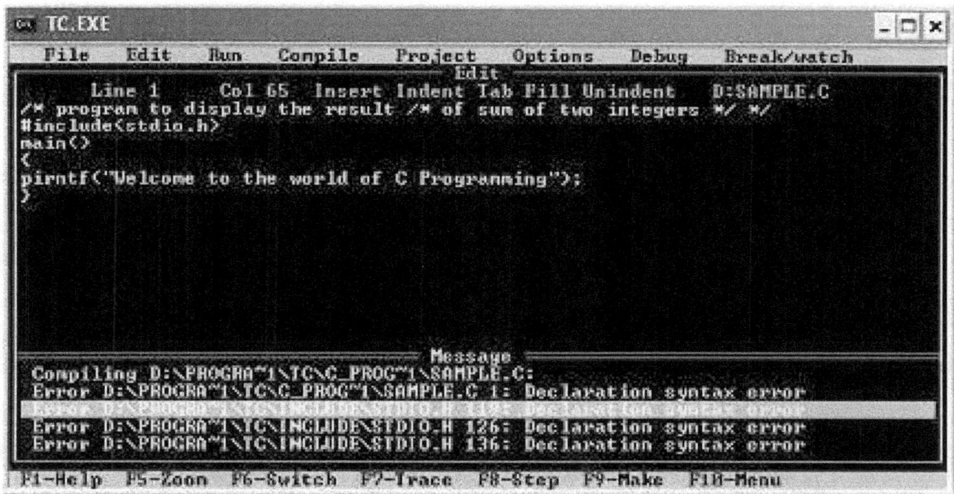

Figure 3.30 Error due to nesting of comments

The language designers have designed many such functions and they have grouped them into different categories. As many as 30 such categories are available in C language and each one is called as a header file and they have the extension .h (h stands for the word header). When the programmer wants to include a particular built-in function, he/she must be aware of the header corresponding to it and include the same in the program before the program's main() is coded. In the previous section, we had seen some of the console I/O functions, which are categorized under the **conio.h** header. But still we had stated that it is not necessary to include that header, since the same I/O functions could be defined in **stdio.h** as well. But some of the functions like **strcat(), sin(), cos(), etc.** could be used only when their header files are included in the program.

Otherwise they may not work properly. As for example, consider the following program in which we are trying to find out the square root of a given number using the built-in function sqrt(), which will take a value corresponding to double data type.

```
#include<stdio.h>
#include<math.h>
main()
{
double x=9.0;
clrscr();
printf("%lf",sqrt(x));
getch();
}
```

The output of the above program would be as shown in Figure 3.31.

Figure 3.31 Output for the program finding the square root (header included)

When the header file, namely math.h, is not included in the above program, the operation of finding the square root would not have been performed at all and hence we would get a different answer (in the case we are discussing we get 0.000000). This is shown in Figure 3.32. We must note another point here that there will not be any error message flashed during the compilation stage of the program and hence the programmer has to be very careful when using the built-in functions.

Thus, it is important to include the header files in programming and avoid getting answers like the one we had discussed and shown in Figure 3.32.

Figure 3.32 Output for the program finding out the square root (header not included)

The third line of the program indicates the beginning of the main section. One of the highlighting features of C language is that it is a compiled language containing many built-in functions. Users can also define their own functions and these functions are called as user-defined functions. One of the functions that would be used in all our programs is the function main() and we had discussed the importance of this function already in the Chapter Two (section 2.2.8). I would suggest the programmers to recall the corresponding section to know about the function main() before proceeding.

The fourth line of the program contains an opening brace ({) and it marks the beginning of the coding section of the program and also the beginning of a block of statements. When we were discussing the topics like conditional branching, loop control instructions, etc. we used the set of braces (opening and closing braces) and we denoted the lines of coding inside the set of braces as a block. Whenever we write a function too, we must include the lines of coding inside the block and any coding outside the block will not be considered as the part of the function and will be compiling to an error and hence execution of the program will get affected. Figure 3.33 shows this fact.

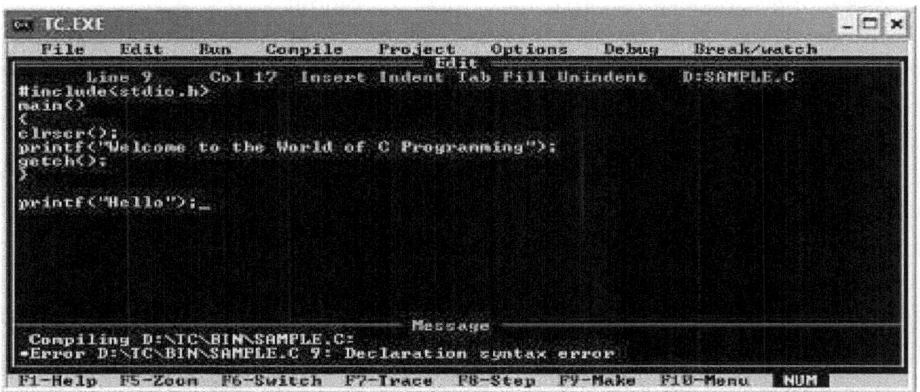

Figure 3.33 Coding outside the function main()—error generated

In the **fifth line** of the program, we are declaring the variables that are used in the program and the programmer must keep in mind the fact that the variables that are used in a program have to be declared in advance, regardless of the point of their usage in the program. The section where we declare the variables of the program is called as **Type Declaration Section**. We call the corresponding line(s) of coding that we use for declaring the variables as **Type Declaration Instruction(s)**.

The **sixth line** of the program has the function **clrscr()**, which is one of the console I/O functions and we had a detailed discussion on this in this Chapter in the section 3.2.

The **seventh line** of the program has the function printf(), which interacts with the user and prompts him/her to enter the values of the variables x and y. The detailed discussion on the function printf() could be recalled from the section 2.2.7(2).

The **eighth line** of the program scans the input given by the user and we already knew from the section 2.2.7(2), how the function scanf() works and here in this program we see that we are getting input for two integers and hence we are using two %d format specifiers and two corresponding addresses of arguments.

The **ninth line** of the program has the coding for the **addition logic** and we see that it is an assignment expression having two sides, the Right Hand Side and the Left Hand Side. The values that are given by the user for the variables 'x' and 'y' are added on the RHS and the result is stored in the variable 'z' on the LHS. The variable on the LHS is always the target and the variables on the RHS are always the source.

The **tenth line** of the program displays the result of addition with the function printf(). We notice that the input values are also displayed along with the actual result, to make sure that the inputs given by the programmer are accepted and stored properly. This way, the programmer can have a check on any logical errors that he/she could commit.

The **eleventh line** of the program uses another console I/O function **getch()** whose details we have seen in this Chapter, in the section 3.2.

The last and **twelfth line** of the program has the closing brace, which indicates the end of the block and no more coding could be added to main after this. This point has been already discussed while we were discussing the **fourth line** of the program.

We see that a simple program has these many details. We will discuss more programs of such kind and as we do so, we will learn a lot of new things and wonders in C.

3.4 SIMPLE PROGRAMS INVOLVING CALCULATIONS

As a beginner in programming, one would always like to learn simple concepts and implement them as programs. If one wants to be a professional programmer in C and have a better understanding of the functions such as printf() and scanf(), the data types, format specifiers and console functions, it is better to write more programs on basics and practise. As they use the above functions and data types, they would get more errors, both syntax and logical, and they would start applying their technical mind to solve the errors and slowly they would start mastering the language. In this section, we are going to discuss the basic programs in C, that would involve calculations and we would be using the basic formulae that we had learnt in our elementary and high schools. I have given the list of programs that I would like you to practice and I have written two of those programs here.

List of programs that will make you to become familiar with the syntax of C:

1. Finding the areas and perimeters (or circumferences) of the following two-dimensional figures:

   Square, circle, rectangle, triangle, parallelogram, rhombus

2. Finding the surface areas and volumes of the following three-dimensional figures:

   Cone, cylinder, sphere, cube, cuboid

I have written the programs corresponding to the area and perimeter of a square and volume and surface area of a sphere below.

```
/* Area and Perimeter of a Square */
#include<stdio.h>
main()
{
float side,area,perimeter;
clrscr();
printf("Enter the side of tbe square : ");
scanf("%f",&side);
area=side*side;
perimeter=4*side;
printf("Area = %f\tPerimeter = %f",area,perimeter);
getch();
}
```

The output for the above program is shown in Figure 3.34.

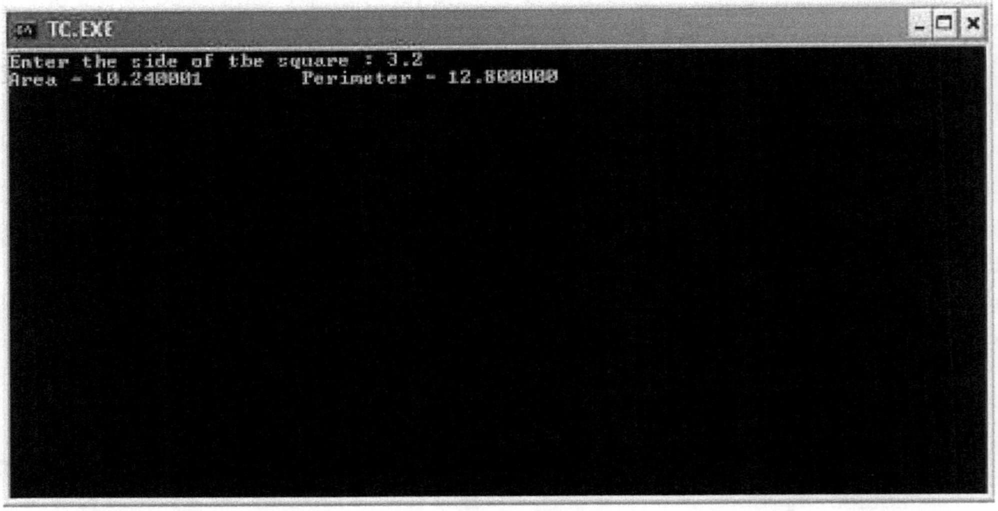

Figure 3.34 Program to find the area and perimeter of a square

The program for finding the surface area and the volume of a sphere is given below:

```
/* Surface Area and Volume of a Sphere */
#include<stdio.h>
main()
{
float surface_area,volume,radius;
clrscr();
printf("Enter the radius of tbe sphere : ");
scanf("%f",&radius);
surface_area=4*3.14*radius*radius;
volume=(4/3.0)*3.14*radius*radius*radius;
printf("Surface Area = %f\tVolume=%f",surface_area,volume);
getch();
}
```

The output for the above program is shown in the Figure 3.35.

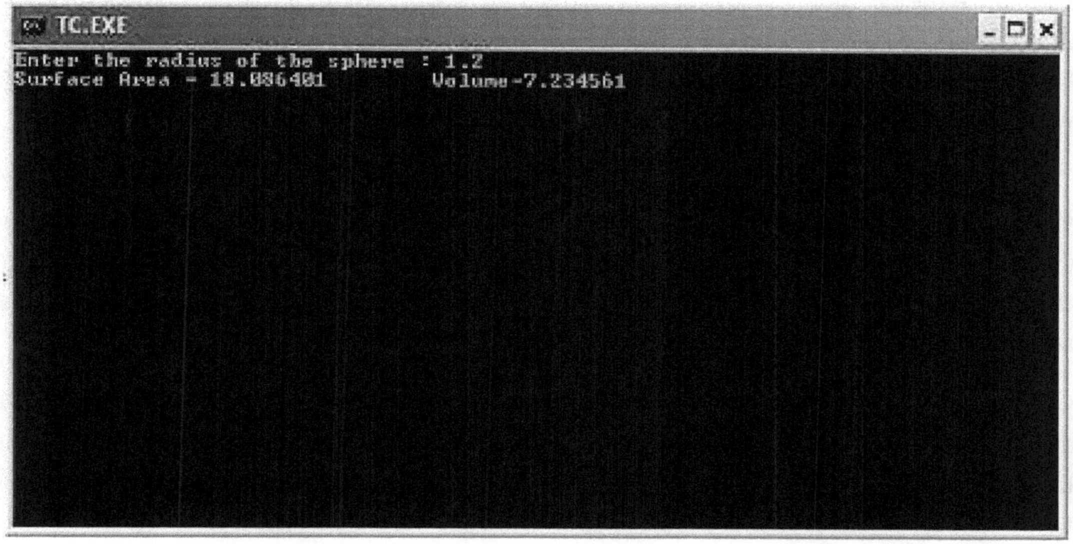

Figure 3.35 Program to find the surface area and volume of a sphere

3.5 UNDERSTANDING AND DEBUGGING A C PROGRAM

Understanding how a C program works, and analysing the technical details involved is quite difficult initially. But as we do more and more programs and fix the errors, we may find the task much easier than ever. I had mentioned in the section 3.1.4 of this chapter that one has to maintain a good database of errors, which will surely facilitate the process of debugging the programs. We have given a few exercises too, at the end of each section of each chapter, only to test the technical maturity of the programmers. For example, consider the following program fragment.

```
float a=0.7;
if(a<0.7)
printf("Good");
else
printf("Bad");
```

In the above program, we are just using a simple conditional instruction and the conditional expression is also very simple. What would you as a user and a programmer anticipate as the output? A typical fresher or a beginner in programming shall surely say that the output is Bad and the reason, he/she would say is that the value of 'a' that is checked in the condition is equal to 0.7

and hence the conditional expression returns a **false**. But, what if I say the answer is not **Bad** but **Good**. Yes, of course the answer is **Good** and this obviously means that the conditional expression has returned a **true** much against the anticipation of the programmers. Let me justify the answer with a detailed explanation.

When we declare a variable of type float and store some value (in this case it is 0.7) in it, the value will be stored in the form of binary number. Remember that the value we had given for 'a' (0.7) is not binary and is a decimal. So, while storing it, the compiler follows the logic of converting that to its equivalent binary. (For knowing the conversion of a decimal fraction to a binary fraction, refer to section 1.7.3).

When we convert the number 0.7 to its decimal, we will get 0.1011 approximately. We have approximated the result to four bits as it will be convenient for our discussion and even otherwise the actual will be a lengthy sequence of binary digits. In our case the data type that we have used is a float and it will occupy 4 bytes (32 bits). Hence the conversion process will go on till all the 32 bits are filled up with either a 0 or a 1.

This conversion process happens internally when the variable 'a' is declared and initialized. When the value of 'a' is used in the conditional expression, its value is retrieved from the memory and it is converted back to its decimal value (since the user has given a decimal value during the initialization of 'a'). We know how to convert the value in binary to decimal. When we convert the above binary number (the approximated number 0.1011) to its decimal, we would get a value of 0.6875 and if we had considered all the 32 bits for the given decimal number, then upon converting the entire sequence of binary number, we would get 0.69999 and obviously it is less than 0.7 when we do the comparison in the expression. Thus the result returned by the expression would be **true**. This leads to the output **Good**.

The conclusion that one can make out of this discussion is that floating point values always create significant issues like this and hence it is advisable not to use them at least in the conditional expressions.

From a very small program fragment, we could get so many ideas about the compiler and the way it behaves. Thus debugging a program fragment needs a lot of practical expertise and patience. You will get a few questions at the end of this chapter which will really make you think and understand the way the C compiler works.

SUMMARY

The idea behind this Chapter was to introduce the basic techniques in programming and using the editors for programming languages. By now, you would have gained a good amount of knowledge on the Turbo Editor and would appreciate its advantages when compared with the editors of other languages.

Installation of the C compiler in the hard drives of our computer systems was also discussed and the compiler is available at free of cost at the specified web link. Also, that the installation process is easy and consumes less time and does not require much expertise.

Editing the source code and the freeness in using the editor is also understood. Saving the file and updating the modifications after saving the file is also discussed.

The C compilation model was also discussed and you must now be aware of the compilation process and would have understood the fact that the compiler alone does not perform all the tasks during the process of compilation and that the linker has also an equivalent role to play.

Execution, which is the final stage of any program validation, was discussed elaborately and the details like editing and user screens were clearly explained.

Following the basic things related to the creation of programs and source codes, we had a brief discussion on console I/O functions like clrscr(), getch(), getchar() and getche() and their significance in the program.

Then we had the very first full-fledged program, where we discussed some additional programming techniques like using a comment and regulations in using it. We had also discussed the header files and their importance in any C program. The importance and necessity of using the source codes within a block was also understood.

Then we had a couple of simple programs involving calculations that might help you in knowing the syntax of the language. Some exercises too have been given in this section for you to practice.

We have concluded the chapter with a hint on debugging and I have given a sample program and debugged the same. You would be surprised to know that C language has umpteen tricks and wonders like the one we had discussed and for you to master them; do a lot of practical work and apply your technical mind.

In all the sections of this chapter, I had used a lot of screen shots to make you understand the concepts under discussion in the clearest possible way. It also enables you to check your work and cross check the details that are included throughout the Chapter. I wish the readers of this chapter shall surely appreciate the technical efficiency of C language.

CRITICAL THINKING QUESTIONS

Try to work out and get the outputs for the following programs:

1. main() { int a=10, b=20;printf("%d\t%d\t%d\t%d",a=b,a>b,a<b,a==b); }
2. main() { printf(); }
3. main() { printf("%d\d%d\t%d\t%d",72,072,0x72,0x72); }
4. #include<stdio.h> main() { int i=10 ; #include<stdio.h> ; printf(«%d",i); }
5. main() { int i; printf("%d",i); }
6. main() { int a=32000+1536+10*1; printf("%d",a); }
7. main() { printf("Hello\n"); main(); }
8. main() { long a; a=65536+300; printf("%d",i); }
9. main() { int x; x=4%5+6%5; printf("%d",x); }
10. main() { int x=100; if(!!x) printf("%d",!x); else printf("%d",x); }
11. i=2+3,4>3,1; printf("%d"i);
12. printf("%u",–1); what is the value printed out?
13. main() { int x,j,k; j=k=6;x=2; x=j*k; printf("%d", x); }
14. main() { unsigned int x = –1; int y; y = ~0; if(x == y) printf("same"); else printf("not same"); }
15. unsigned m=32; printf("%x",~m);
16. printf(5+"facsimile");
17. printf("%d,%d,%d", sizeof('3'), sizeof("3"), sizeof(3));
18. for(i=0; i<=10;i++,printf("%d",i));

SELF-LEARNING EXERCISES

1. Compare the characteristic features of C compiler with that of C++ and try to understand the key differences between them. Get various editors available for C and analyse their efficiency in editing the program and make a comparative study with the Turbo C Editor.
2. Since, C was instrumental in designing the entire UNIX kernel, get the information about C coding in UNIX environment and also explain the way it is compiled and executed in UNIX. Which one do you think is the best platform to work, with respect to C language—DOS or UNIX? Substantiate your points.

ARRAYS IN C

4.1 WHAT ARRAYS ARE AND WHY ARRAYS?

In principle, arrays in C are similar to those found in other programming languages. As we shall shortly see, arrays are declared and accessed slightly differently in C language when compared with others. Also, in C, arrays have a closer link with the pointers and we shall explore this later in the Chapter Six.

An array is defined as a structured data type that is capable of storing more than one value at a time, pertaining to a particular data type and these values are stored in contiguous memory locations. If you recall the Figure 2.4. from the Chapter Two, an array is listed in the secondary data types. The main advantage of using an array from the point of view of a programmer is that it can be referred with a single name.

In the programs that we had discussed in our earlier Chapters, we had used variables of different data types and all of them could store only one value at a given time. But in reality, many applications require the processing of multiple data items that have common characteristics. In that case, the programmer cannot use the concept of a single variable. Even if the concept of a single variable is used by the programmer by declaring more number of variables in the program, he/she ultimately makes the program coding look chaotic. The programmer has to remember more variable names in such cases and obviously this is not possible in complicated applications. For example, if one is writing a program for storing the marks of a particular subject for a class of 50 students, if the concept of single variable is used, then the programmer has to declare and use 50 different variables and above all, he/she has to remember which variables represents whom in the class of 50 and this would be very tedious for the programmer. What if the programmer wants to remember and use just one common name for the entire class for that particular subject and store all the marks of that subject one after another? There comes the provision of using arrays in programming.

As said earlier, the programmer has the advantage of using and remembering just one name, called as the array name, and that would be shared by all the members of that particular array. Each data item of the array could be of any basic data type like integer, character, float, etc.

4.2 DECLARATION, USAGE AND CLASSIFICATION OF ARRAYS

While programming, the arrays, like other variables, shall be properly declared and accessed. This section would let you know how to declare an array and access the data items.

The following is the general syntax for declaring an array:

<Storage Class> <Data Type> <Array Name[Size of 1st Dimension] [Size of 2nd Dimension]...[Size of nth Dimension]>;

The idea of using the storage class is beyond the scope of this chapter and hence we will defer the discussion of the same here. Apart from that, we see that an array is declared just like an ordinary variable and the size of the array must be known to the C compiler well in advance and it must be given within square braces as given in the syntax. The following is how one may declare a one-dimensional integer array.

```
int a[100];
```

The above declaration tells the compiler that there is an integer array by name 'a' and it is capable of storing 100 elements (of type integer). Once we declare the array and execute the program, the compiler reserves memory space for the 100 elements (200 bytes in this case, since each integer occupies 2 bytes of memory). The way the above array would look like in the memory is shown in the Figure 4.1.

a[0]	a[1]	a[2]	--------------	a[99]
23	54	–98	------------	19

Figure 4.1 Memory picture of an array

Another important thing that the programmers must keep in mind while declaring the array is that they must always use a constant value within the square braces for the size of the array and they shall not use a variable there. The following declaration of an array is invalid and would generate an error while compilation.

```
int a[x];
```

The above declaration uses a variable 'x' and its value is not known to the compiler and hence would generate an error while compiling the program. The following declaration of an array is valid, even one may argue that the value of 'x' is known to the compiler.

```
int x=100, a[x];
```

The error generated is shown in the Figure 4.2. How do we access the individual elements of the array? This is done by using the **subscript** or **index** of the array. The subscripts are normally unsigned integers. The C programmers shall know one fact before indexing the array or before learning the concept of indices that the indexing shall start from '0' (zero) and not from '1'. Initially, this would be difficult for the programmers who would have used FORTRAN or BASIC, but as they do more programs they would find it easy to use the idea.

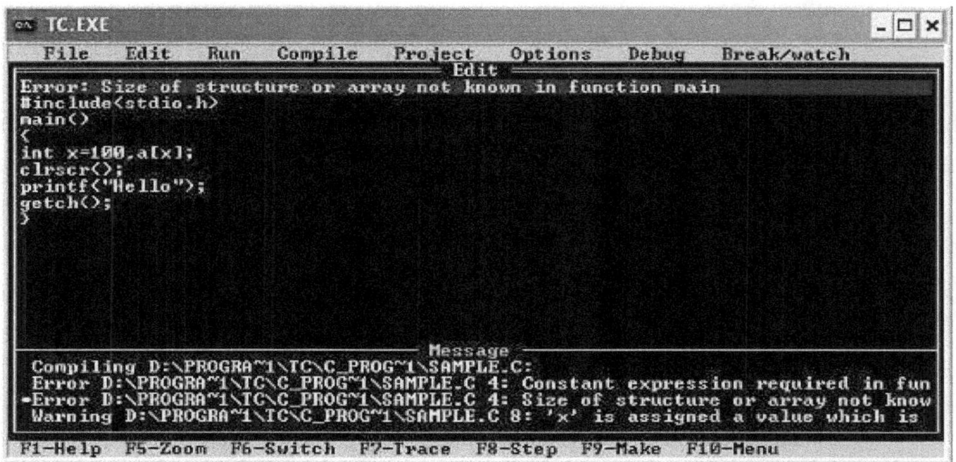

Figure 4.2 Error generated due to improper declaration of array

For example, if we want to access the **fifth** element of the above array, then the corresponding statement would be a[4]. We see that the actual value within the square braces is one less than the actual value we had specified and this is because the array index starts at '0' and not at '1'.

The programmers must process the array elements very carefully. Single operations involving the entire arrays are not permitted in C. For example, if 'a' and 'b' are two similar arrays (that is of same size and data type), the operations like comparing the elements of the arrays, finding the sum of the elements, etc. must be done on an element-by-element basis (that is the operations to be carried out on a piece-meal basis and not all together). The programmer could accomplish this using different loop structures that we had already learnt in the Chapter Two in detail.

Even though the arrays in C are most efficient, there is a slip. In C, there is no way to check whether the array subscript exceeds the array size. This means, there is no **bounds checking** in C. The programmers should take of this issue while programming and see to that the array subscript does not exceed the array size.

Arrays are classified based on two aspects, namely type of elements they store and their dimension. Based on the type of elements they store, they are classified as numerical arrays and character arrays. All those arrays that store elements of type integer, float, double and long double are called as numerical arrays. In our discussion of numerical arrays in the following section, we would consider the integer arrays only and whatever is being discussed equally holds good for the other numerical arrays. The other type of array is the character array, which is capable of storing alphanumeric characters and is otherwise termed as strings.

We had mentioned earlier in this section that if we would like to process the array elements, we must use loop control and in the loop control we would be obviously using a loop counter. The idea of loop counters had been already discussed in Chapter Two. The loop counters are normally incremented or decremented in steps of 'n' ('n' value depends on the nature of the program). When the loop counters are incremented or decremented in terms of '1', then we have an operator used for that purpose and the operator is called as the modification operator. This is a **unary operator** and hence takes only one operand. The two modification operators used in C are ++ (read as Plus Plus) and – – (read as Minus Minus). These operators could be used in two forms namely, pre-increment/pre-decrement (before the operand) and post-increment/post-decrement (after the operand). The programmers could use these operators only when they increment or decrement in steps of '1' and not in any other cases. Also, the programmers shall not be tempted to use an operator symbol '+++' or '– – –' for incrementing or decrementing a loop variable in steps of '2' and it is not logical to do so.

Let us see how these two operators work. For that consider the following program fragment.

```
int a=10;
printf("%d",a++);
```

What do you think the output of the program would be? If your answer is '11', it is wrong. Upon using the increment operator, the programmer should not come to the conclusion that it will simply increment and the output will be generated. The programmer shall see in which context the operator is used. Here in the above program, the operator is used in the function printf() in the post-increment form. Before the function printf() gets executed, the compiler breaks it into two statements as follows:

```
printf("%d",a);
a=a+1;
```

This means, the existing value of 'a' is used in the function printf() and following the execution of the function, the value of 'a' is incremented. So, the existing value of 'a' is used and then it is incremented. This is the real meaning of the **post-increment**. Had we used one more printf() and modified the above program fragment in the following manner, we would get the outputs '10' and '11'.

```
int a=10;
printf("%d",a++);
printf("%d",a);
```

Suppose that we have used the '++' operator in the pre-increment form as in the following program fragment.

```
int a=10;
printf("%d",++a);
```

What do you think about the way the compiler would treat the function printf() here and what would be the result? The function printf() would be treated as follows:

```
a=a+1;
printf("%d",a);
```

Thus the value of 'a' is first incremented and then the function printf() is executed. So, the existing value of 'a', which is 10, is incremented to '11' and passed to the function printf(), which in turn prints the same. So, the operation **pre-increment** means increment the existing value of the variable first and then use it.

Hence, wherever we have used (so far or henceforth) an increment or decrement operation in steps of '1', we could replace the respective expression with the modification operators. That is, if we had used, say, a=a+1 in a piece of coding, we may replace it with the expression 'a++' and if the expression that has been used is a=a-1, then it could be replaced with 'a– –'. (Readers are requested to check with this before they proceed further).

The same kind of explanation holds good for the decrement operator (– –). I would leave it to you for further discussion and conclusion. Let us proceed with the discussion of the arrays.

4.3 NUMERICAL ARRAYS

4.3.1 One-Dimensional Numerical Arrays

Consider the case of a one-dimensional numerical array, specifically an integer array. Let us learn how the integer array works through the programs. In all our programs, we will be using the memory picture of the arrays for a better understanding of their functions. Let us start with a program in which we read an array of 'n' integer elements and display the same on the output. The following is the program code for the process of reading and displaying 'n' integers.

```
/* Program to read and display an array of 'n' integers */
#include<stdio.h>
main()
{
/* Variables declaration */
int a[100],i,n;
clrscr();
/* Getting the value for number of elements in the array */
printf("\n\nEnter how many elements you want to enter : ");
scanf("%d",&n);
/* Reading the values for the individual elements of the array */
printf("\n\nEnter the elements of the array : ");
for(i=0;i<n;i++)
scanf("%d",&a[i]);
/* Displaying the elements read in Same Order */
printf("\n\nThe %d elements of the array are : ",n);
for(i=0;i<n;i++)
printf("%d\t",a[i]);
getch();
}
```

Figure 4.3 shows the output for the above program.

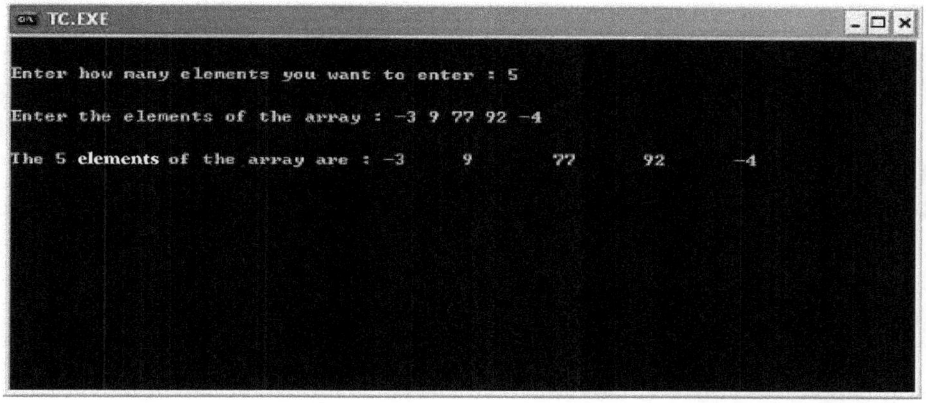

Figure 4.3 Reading and displaying an array of 'n' integers

Let me elaborate on how the program works using the memory picture of the array 'a'. The Figure 4.4 shows the memory picture of 'a', the loop counter 'i' and the variable 'n'.

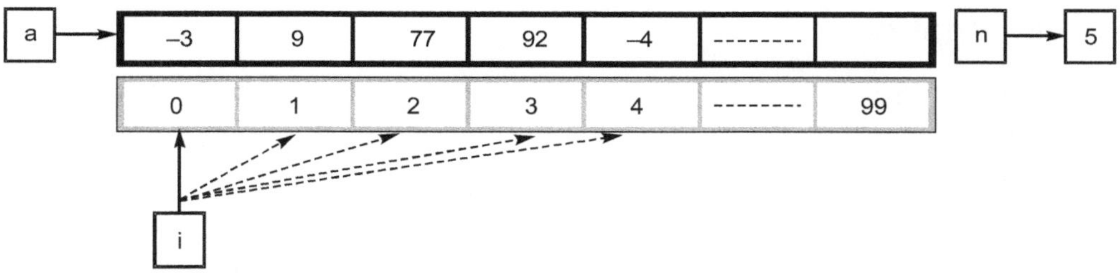

Figure 4.4 Memory picture of the array and the other elements

From the memory picture, we could see that the loop counter 'i' takes the initial value as '0' (the beginning of the array) and as we proceed incrementing the value of it inside the loop, it takes the values according to one less than the maximum limit 'n' (The values that are taken by 'i' during each cycle of the loop is shown as dotted lines). So, the value of the loop counter value would be from '0' to '4' (Even though the array size is 100 and is capable of storing 100 integers, we make use of only 5 of them. But memory will be reserved for all the 100 integers regardless of whether we use it or not.)

In the program, we are getting the value of 'n' first and this is considered to be the upper limit for the loop counter 'i'. Once the value of 'n' is obtained, we get the individual elements of the array using another loop and see that the function scanf() has been coded as **scanf("%d",&a[i])**. This implies the real meaning of the function scanf(), which we had already seen elaborately in the section 2.2.7. The function scanf() here tries to access the address of the individual elements of the array based on the value of 'i' and drops in the values that the user gives in the corresponding addresses.

Once we get the values of the individual elements and store them in the memory, we are ready to print them by retrieving them from the memory. This is done in the next *for* loop which uses a similar loop structure as before, but uses the function printf() (instead of the function scanf()). See that when we use another loop, we reinitialize the value of 'i' since the last value of 'i' in the earlier loop would be '5'. If we want to print the elements of the array in the order of the inputs given, then we have to reinitialize the value of 'i'. You can see from the output shown in the Figure 4.3 that the elements are printed according to the order in which the user has given them as inputs. What would you do in case you want to print the elements in reverse order? That will be our next illustration.

In the above program code, do retain the coding for everything except the place where you print the elements. The following code could be replaced there for the existing code.

```
/* Displaying the elements read in Reverse Order */
printf("\n\nThe %d elements of the array are : " ,n);
for(i=i-1;i>=0;i--)
printf("%d\t",a[i]);
```

Here, we see that the initial value of 'i', the conditional expression inside the loop and the modification part of the loop have changed when compared to the earlier program. This is because, we wanted to print the elements that are read in reverse order. After the array elements have been read, the value of 'i' would be '5' and we would like to print the elements that are read from the last position, which is '4'. Hence we reinitialize the value of 'i' to the last but one position, that is '4' and we decrement it each time when it cycles through the loop and we stop the execution of the loop once it reaches a value below '0'. We see that the last valid value of 'i' would be '0' and if it becomes '–1', the execution stops. Figure 4.5 shows the output of the above logic.

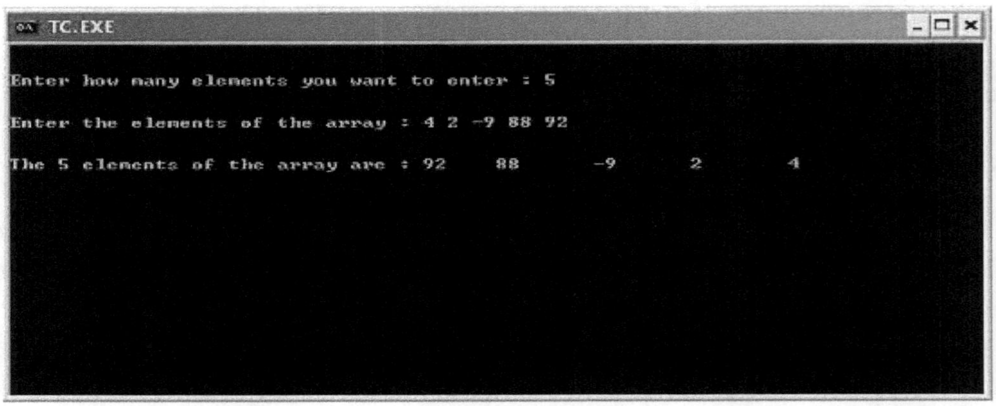

Figure 4.5 Reading and reverse displaying an array of 'n' integers

The memory picture in Figure 4.6 shows the way in which 'i' is reinitialized and the values are printed in reverse order. The dotted arrows show that the value of 'i' is decremented in each cycle and finally reaches '0'. Any further decrement would result in the value to be '–1' and invalidate the condition and terminate the execution of the loop.

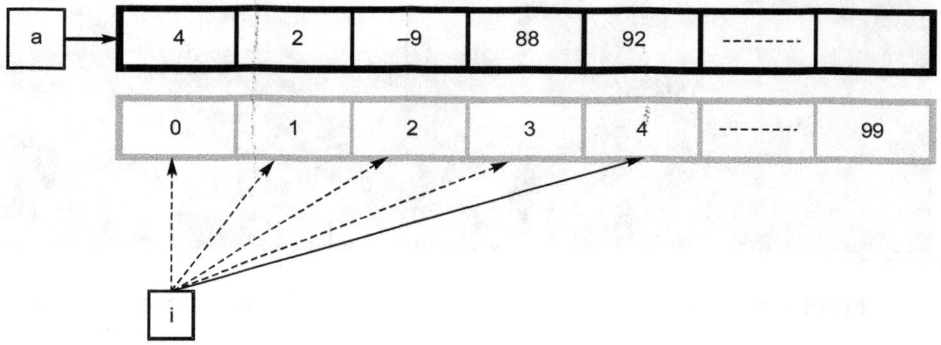

Figure 4.6 Memory picture of the array when the elements are reversed

In the above program coding, we could see that the array values are obtained from the keyboard using the function scanf(). One can initialize the array elements much in the same way as a variable is initialized during the time it is declared. The following is a program that initializes a one-dimensional array and displays the same.

```
#include<stdio.h>
main()
{
int a[5]={23,45,-7,453,89},i;
clrscr();
for(i=0;i<5;i++)
printf("%d\t",a[i]);
getch();
}
```

Figure 4.7 shows the output of the above program code and we could easily understand that array elements could also be initialized just like any other variable.

Figure 4.7 Program in which the array elements are initialized

4.3.2 Two-Dimensional Numerical Arrays

A two-dimensional array is declared in exactly the same way as a one-dimensional array, except that we need to supply an extra square braces as shown below, to imply the second dimension. A 2-D array can be thought of as an **array of arrays**.

```
int a[100][100];
```

In the above declaration we have the size of the first dimension to be 100 and that of the second dimension to be 100. In practice, one can imagine a two-dimensional array as a matrix that consists of rows and columns (any basic book on matrices and determinants would give you a fair idea of what matrices are). It is absolutely not necessary that both the dimensions shall be of same size. One can have a two-dimensional array declared as follows:

```
int a[20][15];
```

The above declaration means that along first dimension, we can store 20 elements and along the second, we can store 15 elements and all together, one store 300 elements (simply multiply the sizes to get the total number of elements that could be stored in the array).

Let us now analyse how the memory picture of a two-dimensional array would look like. For example, consider the following declaration of a two-dimensional array.

```
int a[2][3];
```

The above declaration of the two-dimensional array could store six elements. According to the general assumption of a 2-D array, there could be two rows and three columns in the above array. But when it comes to the storage of the elements in the memory, do you think they will be stored in rows and columns? If your answer is yes, then you are wrong. The individual elements would not be stored in the way that is shown in the Figure 4.8.

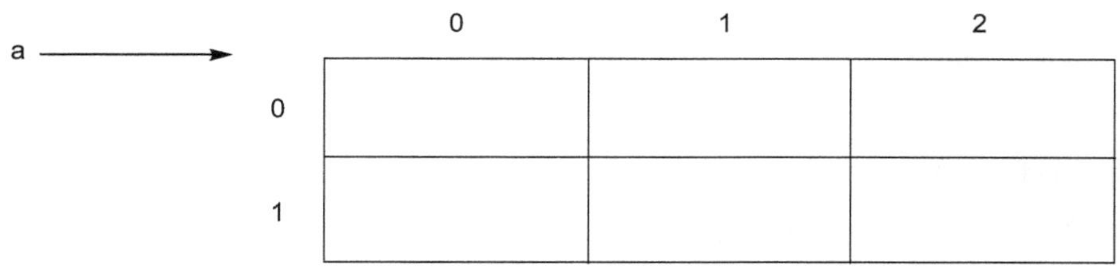

Figure 4.8 Memory picture of a 2-D array of size 2×3

Rather the array would be stored linearly and would be interpreted by the compiler as a combination of two one-dimensional arrays as shown in the Figure 4.9.

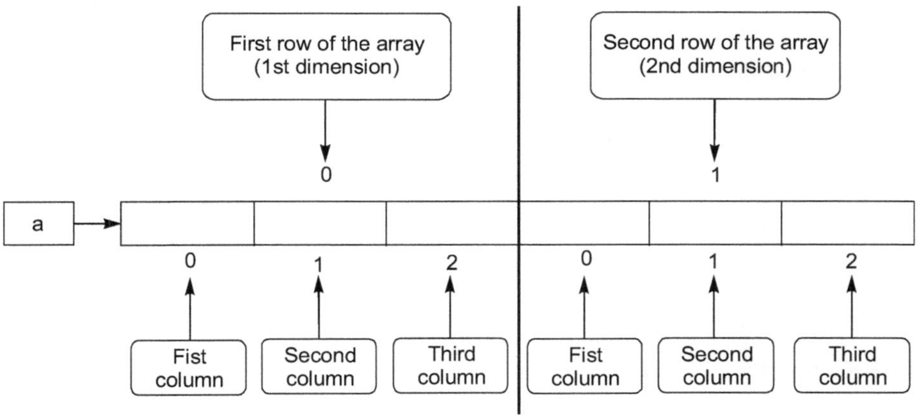

Figure 4.9 Arrangement of rows and colums of a 2-D array

It is easily understandable from Figure 4.8 that regardless of the dimension, the elements are stored in a linear fashion in the memory. The programmer would, for his/her easy understanding and for writing programs, consider the arrangement as a matrix. But in reality, it is not so. Let us write a few programs using the 2-D arrays.

As always, we will consider reading and displaying a 2-D array of size m × n, where 'm' is the size of the first dimension (number of rows in terms of matrices) and 'n' that of the second dimension (number of columns in terms of matrices). The following would be the program for the same.

```
/* Program to read and display a 2-D array of size m × n */
#include<stdio.h>
main()
{
/* Variables declaration */
int a[100][100],i,j,m,n;
clrscr();
/* Getting the size of the first and second dimension (number of rows and columns) */
printf("\n\nEnter the size of the first and second dimension : ");
scanf("%d%d",&m,&n);
/* Reading the values for the individual elements of the array */
printf("\n\nEnter the elements of the array : ");
for(i=0;i<m;i++)
for(j=0;j<n;j++)
scanf("%d",&a[i][j]);
/* Displaying the elements read */
printf("\n\nThe elements of the array are\n");
for(i=0;i<m;i++)
{
for(j=0;j<n;j++)
printf("%d\t",a[i][j]);
printf("\n");
}
getch();
}
```

The output of the above program coding would be like the one shown in the Figure 4.10.

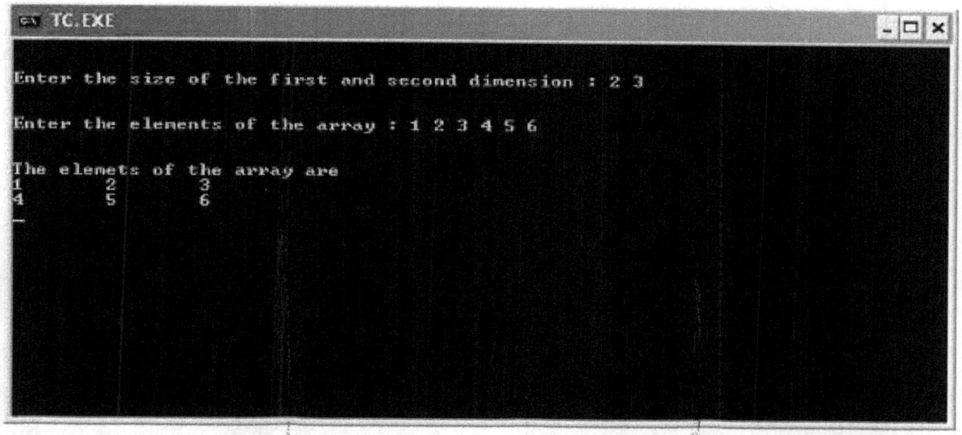

Figure 4.10 Reading and displaying the elements of an array of size 2 × 3

Let us closely analyse the coding written above. As said earlier, the array is declared with two dimensions (with two square braces following the array name). Apart from the array, we have declared four integer variables, two for holding the values of number of rows and columns (m and n) and two for accessing the rows and columns (i and j) while reading and displaying the values.

Initially we get the values of 'm' and 'n', which in turn would be the values corresponding to the number of rows and columns (values corresponding to the first and second dimension) in the array. Following that we get the array elements. See that we use two *for* loops, one after another (the concept of nesting *for* loops is implemented here), one for accessing the row elements and one for accessing the column elements. The outer loop has the access to the first dimension or the row and the inner loop has the access to the second dimension or the column. For each value of the outer loop, the inner loop will execute for three times, since the size of the first dimension is **two** and that of the second dimension is **three**. The outer loop executes **twice** (since the size of the first dimension is two) and hence there will be totally **six loop cycles** in all (two for the outer loop and for each cycle of the outer loop, the inner loop executes thrice).

As a general case, for any 2-D array of size m × n, where 'm' is the number of rows (size of the first dimension) and 'n' is the number of columns (size of the second dimension), there will be 'm' cycles in the outer loop and for each of the cycle, there will be 'n' cycles in the inner loop and hence, in total, there will be 'm × n' cycles and hence there will be 'm × n' elements in the array.

We could see from the above piece of coding that the function scanf() is coded as **scanf("%d",&a[i][j])**. This means, the base address of the entire array is passed to the function scanf(), and it happens to be the location of the first one-dimensional array. So, one can conclude that a two-dimensional array is nothing but a combination of many single-dimensional arrays.

In the example given above, the array is of order 2 × 3 and so there are two one-dimensional arrays and each has three elements. You will know this idea more clearly when we discuss pointers and 2-D arrays in the Chapter Six. When we try to print the addresses of the individual elements of the above 2-D array, we would get the following as the output shown in Figure 4.11.

Figure 4.11 Reading and displaying the elements of an array of size 2 × 3

The following coding is responsible for generating such an output:

```
#include<stdio.h>
main()
{
int a[2][3],m,n,i,j;
clrscr();
printf("Enter the values of m and n : ");
scanf("%d%d",&m,&n);
for(i=0;i<m;i++)
for(j=0;j<n;j++)
printf("%d\t",&a[i][j]);
getch();
}
```

We could easily understand from the output that the base address of the entire array is –64 and it happens to be the base address of the first one-dimensional array and then follows the three elements in the first one-dimensional array. Then we have the address –58, which happens to be the base address of the second one-dimensional array. Likewise, if we have varied sizes for the first and second dimensions, the base addresses of the one-dimensional arrays would also vary accordingly. We also see that each element consumes two bytes of memory, since it is an integer array.

Similar to the one-dimensional arrays, the 2-D arrays could also be initialized when they are declared. This is done as shown below with the help of a program fragment.

```
#include<stdio.h>
main()
{
int a[3][3]={{1,2,3},{4,5,6},{7,8,9}},i,j;
clrscr();
for(i=0;i<3;i++)
{
for(j=0;j<3;j++)
printf("%d\t",a[i][j]);
printf("\n");
}
getch();
}
```

The output of the above program code would be the one shown in Figure 4.12. We see that the elements are printed in the form of a matrix with rows and columns.

Figure 4.12 Output of a 2-D array that is initialized

4.3.3 Multi-Dimensional Numerical Arrays

Multi-dimensional arrays are simply an extension of one- and two-dimensional arrays and whatever we had acquired while learning them is to be logically extended to have a more clear understanding of these arrays. A typical multi-dimensional array is a three-dimensional array declared in the following manner. A multi-dimensional array can be thought of as an array of arrays of arrays.

```
int a[2][3][3];
```

In the above declaration, we have the outer array (the first square bracket) has 2 elements, each of which is a two-dimensional array of 3 rows, each of which has 3 elements. Thus the total number of elements in the array would be $2 \times 3 \times 3 = 18$ elements. This means, while getting the input for the array, we would use three **for loops**. The outermost loop would have **two cycles** and for each cycle second loop would execute **thrice** and for each cycle of the second loop, the innermost (third loop) would execute **thrice**. Thus we would have totally 18 cycles in all and hence 18 elements. Further discussion of the multi-dimensional integer arrays would be beyond the scope of this book.

4.4 CHARACTER ARRAYS OR STRINGS

4.4.1 How Character Arrays Differ from Numerical Arrays?

So far we have been discussing a numerical array and we have taken an integer array as the case. As told earlier, the discussion holds good for any other numerical array, the only difference being the data type used. Character arrays are entirely different in C and also have a different treatment. The following is how one may declare a one-dimensional character array.

```
char a[100];
```

The above declaration tells the compiler that there is a character array by name 'a' and it is capable of storing 100 elements (of type character). Once we declare the array and execute the program, the compiler reserves memory space for the 100 elements (100 bytes in this case, since each character occupies 1 byte of memory). The way the above array would look like in the memory is shown in Figure 4.13.

a[0]	a[1]	a[2]	a[3]	a[4]	a[5]	---------	a[99]
'h'	'e'	'l'	'l'	'0'	'\0'	---------	

Figure 4.13 Character array elements

From the figure, we an easily understand the basic difference between the numerical array and the character array. The end of the character array is terminated with a '\0' (NULL Character), which is the bookmark for the compiler for recognizing the end of the array. In numerical array we know well in advance about the number of values that we would input and hence there is no need for keeping a separate terminating character for such arrays. But in character arrays, we would not be in a position to know the number of characters that we are going to feed in, since we cannot sit and count the input for this. Counting the number of characters for small inputs would be easy, but that would be impossible if the input is too lengthy and too bulky (as is the case with any text input). So, to let the compiler know that the array has terminated and to let the user feel comfortable when the array is read back, the character '\0' (NULL character) is appended with any character array. Null character is the one that differentiates a character array from a numerical array and it is a combination of two characters a backslash ('\') and a zero ('0'). The programmer shall supply the NULL character explicitly when the program is coded with a character array. The ASCII value of the NULL character is ZERO. One more thing the user must keep in mind is that, when a character array of size 'n' (where 'n' is a positive integer) is declared, the array could accommodate only n–1 characters due to the fact that one location is reserved for the NULL character. This is unlike the numerical array where one can store exact number of elements as declared.

4.4.2 One-Dimensional Character Arrays

Let us now start programming with the character arrays straight away to understand it more clearly and in a better way. To start with, we will read and display a string (as we did for an integer array). The following is the program coding for the above task. Remember the fact that the accession of a character array is very much similar to that of a numerical array, except for the NULL character appendage at the end of the array after it has been read.

```
/* Program to read and display a string */
#include<stdio.h>
main()
{
/* Array and other variables used are declared */
char a[100];
int i;
clrscr();

/* The string is read */
printf("\nEnter the string : ");
```

```
i=0;
scanf("%c",&a[i]);
while(a[i]!='$')
{
i++;
scanf("%c",&a[i]);
}
a[i]='\0';

/* The string that is read is displayed */
printf("\n\nThe string you have entered is : ");
i=0;
while(a[i]!='\0')
{
printf("%c",a[i]);
i++;
}
getch();
}
```

Figure 4.14 shows the output for the above program coding.

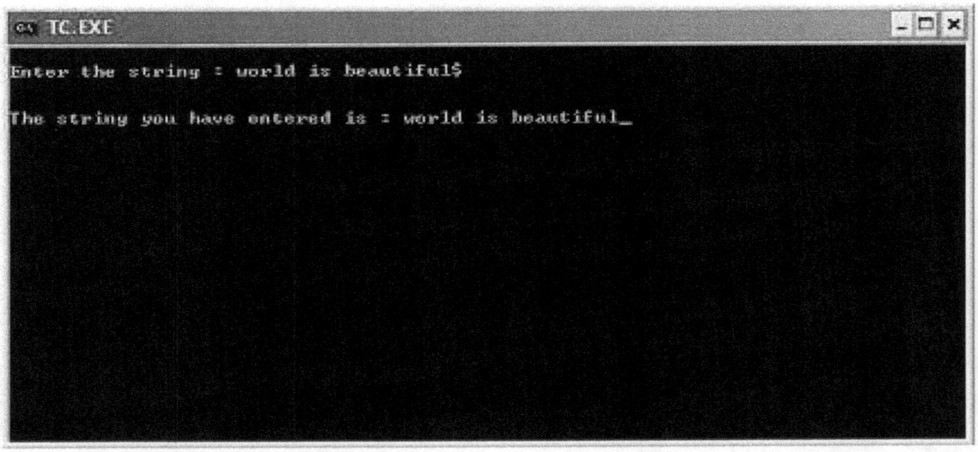

Figure 4.14 Output for the program that reads and displays a string

We can see so many new things from the above program coding. We see that the array is declared as character array, whose size is 100 and it could accommodate only 99 characters given by the user of the program and one space would be used for the NULL character (this point has been discussed earlier).

We also see that the loop counter (array subscript) 'i' is declared as integer. The users of the program shall understand the fact that regardless of the type of array that is used in the program, the loop counter or the array subscript is always a positive integer.

We have used a *while* loop for reading the input string and displaying the read string. We had followed the general syntax of the loop and hence, we had initialized the value of 'i' to be 'zero'. Following this we read the first character of the input string using the function scanf() and see that we have used the format specifier '%c', since the input read is a character. Coming to the conditional expression in *while*, we have used a[i]!='$'. What does this mean? As the user proceeds giving the inputs one after another and at one point exhausted with the inputs, he/she must have a provision to make the compiler know that there are no more inputs. In integer array or any numerical array, we have achieved this by getting the number of inputs for the array ('n') in advance from the user. But in character array, as discussed earlier in this section, there is no way to know the exact number of inputs that would be given by the user. Thus, the user tends to use a character as criteria for stopping the process of giving inputs. The user must be careful in choosing this character and must ensure that it is not an integral part of the array. That is why I have used a special character ('$') instead of an alphabet.

Thus, the conditional expression will always return **true** whenever the character entered is not a $ and when it is a $, further inputs would not be stored (mind you the characters that are given after $ may be accepted while the user gives from the user screen, since the enter key is used as the key for passing the control to the compiler to execute the logic) in the array. Figure 4.15 (a modified version of the Figure 4.14) proves this fact. When the conditional expression returns **true**, the character that is entered from the keyboard gets stored in the array and inside the loop, we first increment the value of 'i', so that the array is readied for accepting the second character and this process continues as long as we do not give a $ from the keyboard. See that the function scanf() is repeated inside the loop, since the first scanf() is considered to be the initialization part of the loop (along with the statement i=0) and hence would be executed only once in its entire life cycle (recall the procedure for executing loops).

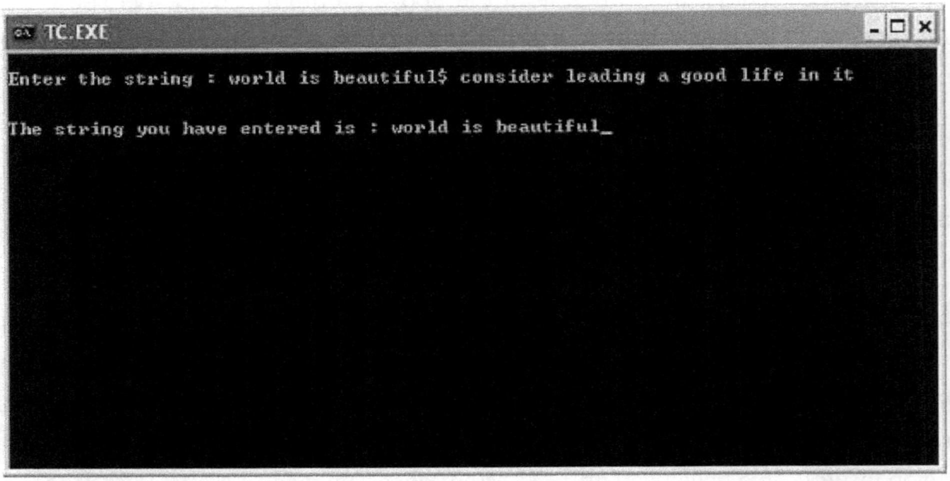

Figure 4.15 Excess input characters after '$' are ignored while display

Once the inputs exhaust, then the user has to store a NULL (\0) character to mark the end of the array. That is what is done at the end of the reading process in the above program coding. After getting the input, we have to display the string that is read. For that, we again use a *while* loop in which we reinitialize the array subscript (i), since it will be pointing to the location of the NULL character. This means we begin to display the string or the content of the array from the beginning and we go ahead by incrementing the loop counter. Figure 4.16 shows the memory picture for the above program.

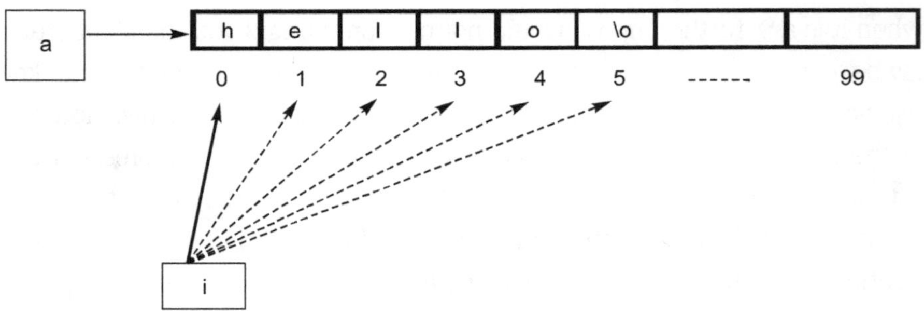

Figure 4.16 Memory picture for reading and displaying a string

We can see that the variable 'i' has been reinitialized to '0' (shown using dark arrow in Figure 4.15) and as we cycle it through the loop, it changes its value (in steps of '1' and shown using dotted arrows in Figure 4.15) and finally after we have given all the inputs, we store the NULL ('\0') character in the array and mark the end of the array. Once we read the array back, we check whether the array has terminated or not, using this NULL character only and that is what is being done in the *while* loop that is used for printing. Each time, the array is checked for the existence of a NULL character and if it is not, then the corresponding character will be printed in the output. Now, what would you do when the above is to be reversed? (like what has been already done with the numerical). The above program code will remain the same, except for the last section, where we print the string that is read. The following shall be the code for printing the reverse of the string that is read.

```
/* Reverse printing the string that is read */
printf("\n\nThe reversed string is : ");
i=i-1;
while(i>=0)
{
printf("%c",a[i]);
i--;
}
```

Figure 4.17 shows the output when the above coding is replaced for the section where the string that is read is displayed, in the program given.

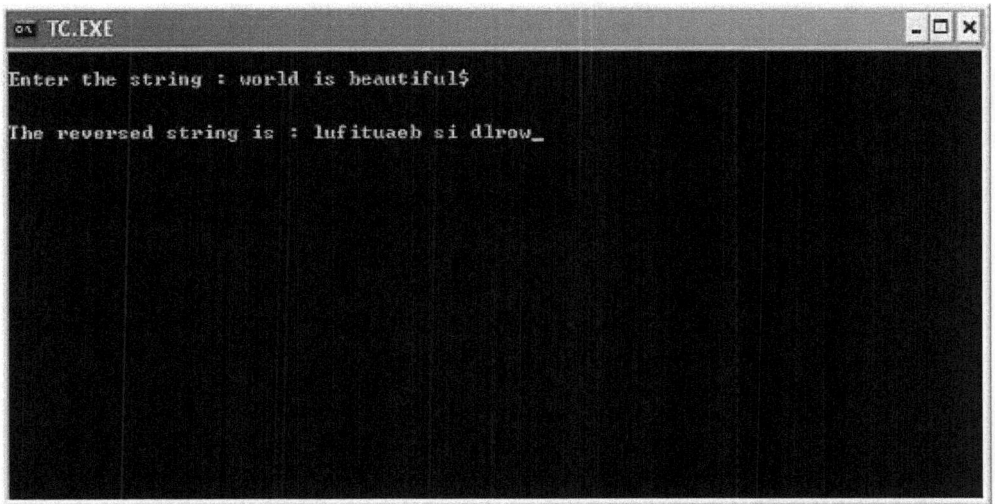

Figure 4.17 Reading and reversing a string

Figure 4.18 is the memory picture for the above logic (see that in the memory picture we have used a different string for reversing, due to convenience and not the one shown in the Figure 4.17.)

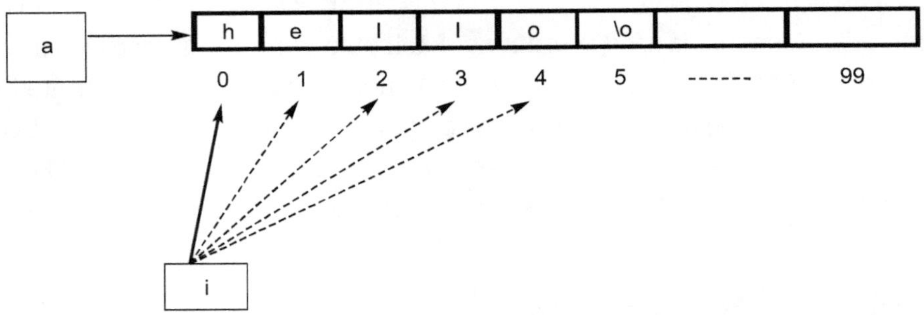

Figure 4.18 Reading and reversing a string

We could see from Figure 4.18 that after the string is read, the variable 'i', which would have been pointing to the NULL character at the position '5' of the above array, is brought to the position just before it, i.e., to '4', since our aim is to print the string that is read from the character 'o' and not from the NULL character (this is shown with a dark arrow) and as we go back, we decrement the value of 'i' and finally reach '0' and as we do so, we print the characters one after another (shown by the dotted arrows). The last character that would be printed is 'h' and after that the variable 'i' would take value '–1', thereby making the conditional expression in the *while* loop false and hence terminating the execution of the loop. We will get the reversed form of the word 'hello' in the output screen.

All such manipulations are possible with the strings and a few of them include finding the length of the string, converting a string in uppercase to lowercase, converting a string in lowercase to uppercase, finding the number of occurrences of a letter in the string and so on. You will be getting all these manipulations in the self-learning exercises section of this Chapter.

The same programs could be done with *for* loop too and the following program shows how it can be done.

```
/* Program to read and display a string using for loop */
#include<stdio.h>
main()
{
/* Variables declaration */
char a[100],i;
clrscr();

/* Reading the characters of the array */
printf("\n\nEnter the string :  ");
```

```
for(i=0,scanf("%c",&a[i]);a[i]!='$';i++,scanf("%c",&a[i]));
a[i]='\0';
/* Displaying the string */
printf("\n\nThe string you have entered is : ");
for(i=0;a[i]!='\0';i++)
printf("%c",a[i]);
getch();
}
```

From the above coding, we can see that the same task of reading and displaying the string could be achieved using *for* loop and we also notice that it is simpler than using the *while* loop.

Don't you think that for loop in the program above seems to be lengthy? Could it be shortened? Yes. C has provision to cut short the *for* loop that is used in the above program. Consider the following modified version of the same *for* loop.

```
for(i=0;(a[i]=getchar())!='$';i++); (a compressed for loop)
a[i]='\0';
```

We could see that a console I/O function, which we had discussed in the earlier Chapter, has been used here. The function getchar(), as you know, will reflect the character that is given from the input device (keyboard) on the screen, unlike the function getch(). It is more or less equivalent to using the function scanf() while the latter is used for getting a character input from the keyboard. Also, we see that a semicolon is supplied at the end of the loop (in both the versions of the *for* loop) and it is much against the basics of *for* loop. Why does a semicolon appear there? The answer is quite simple. If you can recall the general procedure of executing a loop, you can notice that there are four steps or components and you may remember that all the four steps or components are essential for the loop. Just go thru the above loop where we have the initialization of the loop variable (i = 0), conditional-cum-assignment expression ((a[i]=getchar())!='$') and modification of the loop variable (i++). Obviously the component loop statement or loop body is missing. The compiler when compiling the program and executing it would expect the same to be present, as it goes in a logical order only. Also, not providing a proper loop body is a logical error and not a compiler error. So that would not be flashed as an error to you, but when you give inputs and execute the program, you would get an output that would not be anticipated by you. Figure 4.19 shows the output of such a program to justify this argument. See that the inputs are accepted without any problem.

Figure 4.19 O/P of the program in which the *for* loop has no; at the end

On providing the semicolon, we let the compiler know that there is a statement for the loop and it will be considered as a **NULL statement** by the compiler. A NULL statement is that which has no valid C syntax, except for a semicolon. We knew already that any statement is terminated with a semicolon in C programs and the presence of a semicolon itself is an indication that there is a statement. The compiler would treat the coding for the *for* loop given above as the one given below.

```
for(i=0;(a[i]=getchar())!='$';i++);
a[i]='\0';
```

Thus the semicolon given is not for the loop, but for the compiler and it is treated as a statement and hence one of the components of the loop. If one types the semicolon after the loop in a separate line, there won't be any problem, since the compiler itself would treat the semicolon to be available in a separate line. The *for* loop when used with the character arrays can be compressed in the above manner and could be used throughout. It is not possible to do such a compression with the integer arrays.

Another point the programmers who use the arrays should keep in mind is that, when they access the array from the program, the accession is done as a[i]. But the same expression could be written as i[a] too. This point is discussed in detail in the Chapter Six which deals with pointers. This means when you try to access the 4th element in an array, you would type a[3] from your program and it will be equally fine if you type 3[a] from the coding.

Like the way we initialize an integer array, we can initialize a character array too, as shown below:

```
char a[4]={'x','y','z'};
```

Normally when we use a character array and scan the characters from the keyboard, we would be using "%c" as the format specifier. But when we want to print the string (which is a collection of characters) where all the characters are printed in one go, we could use a special format specifier "%s", which does not require a loop at all. For example, in the program, in which we read and displayed a string, we could have used a single line of coding given below for displaying the string, instead of displaying it character-by-character.

```
/* Displaying the string */
printf("\n\nThe string you have entered is : %s",a);
```

If you test the above coding after replacing this in the corresponding program, you may probably know that the result is the same as you got for the earlier version of the program. When you execute the program, you would get the output similar to the one shown in the Figure 4.19. But one should be cautious in not using the %s format specifier with the function scanf(), while getting the input due to technical reasons. Consider the following program coding as an example.

```
/* Program to read and display a text */
#include<stdio.h>
main()
{
char x[100];
clrscr();
printf("Enter the string : ");
scanf("%s",x);
printf("\nThe string you have entered is : %s",x);
getch();
}
```

In the above program, we get a string using the format specifier %s and printing the same. When we give an input text "world is beautiful", the output will have only the word "world" in it and the remaining part of the input will simply be ignored. The Figure 4.20 justifies this fact. This is because of the technicality involved with the format specifier %s. This format specifier will have its default delimiters as space or tab or enter. Thus the programmer when giving the input would give a space between the words (three words in our case and hence two spaces and one enter) and so the format specifier %s would take only one word of the three and stores it in the memory. When we execute the program, we get only the first word of the input and the remaining words are not reflected.

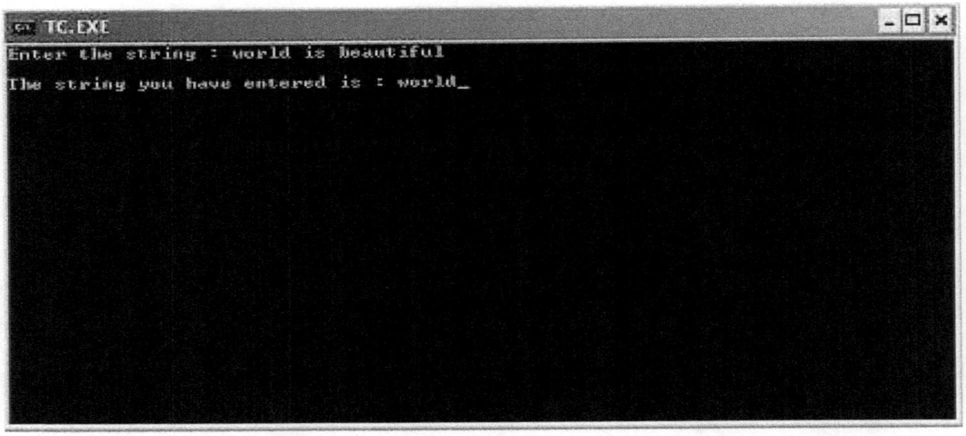

Figure 4.20 O/P of a program when %s is used with scanf()

After doing some good programs from the exercises that I have given at the end of this Chapter, you would surely appreciate the way the *for* loop is useful and would like to do more such activities. We could justify the fact that the *for* loop is the most versatile and handy loop in C language.

4.4.3 Two-Dimensional Character Arrays

A two-dimensional character array is nothing but an array of strings and one can also term it as a multi-dimensional array. Consider the following example program that initializes 5 strings in a two-dimensional character array and prints them.

```
/* Two-dimensional character array or an array of strings */
#include<stdio.h>
main()
{
char strings[5][15] = {"Protein", "Amino Acid", "Nucleotide", "DNA", "RNA"};
int i;
clrscr();
for(i=0;i<5;i++)
printf("%s\n\n",strings[i]);
getch();
}
```

Figure 4.21 shows the output for the above program. From the output we could understand that the initialized strings are stored in more than one dimension and printed using *for* loop.

Figure 4.21 Strings that are stored in a 2-D character array

Figure 4.22 shows the memory representation of how the above strings are stored. One can easily justify the concept behind a 2-D character array from the given memory picture.

	0	1	2	3	4	5	6	7	8	9	10	11	12	13	14
0	P	r	o	t	e	i	n	\0							
1	A	m	i	n	o		A	c	i	d	\0				
2	N	u	c	l	e	o	t	i	d	e	\0				
3	D	N	A	\0											
4	R	N	A	\0											

Figure 4.22 Memory picture of the 2-D character array

In the above program the variable "strings" is an array of strings. It is represented by the size [5] [15]. This representation is shown in the Figure 4.22. The size of the second dimension is given as 15 and it specifies the longest one-dimensional array (including the terminating character \0) and each of the strings that are stored are accessed by the first dimension and with the help of the variable i. One can explore more into a 2-D character array and could easily understand the working of the same. This is possible had they understood two things: one, the way the 2-D integer array works and two, the way a 1-D character array works. These two things have been discussed in length and breadth in this Chapter.

SUMMARY

We have come to the concluding part of this Chapter and now it is time to summarize what we have discussed. We started the discussion with the basic understanding of arrays and their usage in C. Their importance was highlighted and the way they are declared and used in programs was also discussed in detail.

The classification of the arrays based on their dimension and on the type of elements they store was explained. Based on the dimension, we can classify them as single- and multi-dimensional arrays and based on the type of elements, we classify them as numerical arrays and alphanumerical arrays. We went further in studying the details of declaring an array and how elements in an array could be accessed and used.

In this Chapter, we could learn the concepts behind the one-dimensional and two-dimensional integer and character arrays. We have discussed the topics with examples and these examples were supported by proper screen shots and explanations. We had used what is called a **memory picture** in learning and accessing the individual elements in the array. We learnt how to initialize the one- and two-dimensional arrays, instead of getting values for them during the execution of the program.

While we started with the character arrays, we first understood the basic differences between the numerical array and a character array. We also learnt the importance of using a NULL character and how to use it. We discussed a few illustrations in both one- and two-dimensional integer and character arrays and more work is given in the **self-learning exercises** section of this Chapter.

We could also see how the *for* loop could be compressed when it is used with the character arrays and prove the fact that it is the most versatile loop in C language. We had learnt how a NULL statement could be used in programming, especially when we would like to supply all the components of a loop.

We had also discussed the two operators ++ and – –, which are unary and could be used in postfix and prefix forms with the operands. We also came to understand that these operators are very much handy when we use them in loops.

Thus, after studying this Chapter, you will be able to grasp and understand the way the one- and two-dimensional integer and character arrays work and even carry out with better programming with them. If you had thoroughly understood the concepts of these arrays, especially the character arrays, you will fall in love with them. I would suggest you try to work out all the programs that are given at the end of this Chapter to make yourself a master in handling the arrays.

CRITICAL THINKING QUESTIONS

What will be output of the following programs?

1. main()

```
{
int i,a[5]={10,20,30,40,50};
for(i=0;i<5;i++)
printf("%d",i)
}
```

2. main()

```
{
int i,a[5];
0[a]=1;
1[a]=2;
2[a]=3;
a[3]=4;
a[4]=5;
for(i=0;i<5;i++)
printf("%d",a[i]);
}
```

3. int a[10]; what will be the sizeof(a)?

4. main()

```
{
int a[3][9]={ {1,1,1}, {2,2,2}, {3,3,3} };
int i,j;
for(i=0;i<2;++i,printf("\n"))
for(j=0;j<2;++j)
printf("%d",a[i][j]);
}
```

5.
```
main()
{
Char str[]="ABCD";
str[1]='\0';
printf("%s\t%s",str,str+2);
}
```
6.
```
main()
{
char a[4]={'x','y','z'};
printf("%s",a);
}
```
7. `main() { int x=20,y=35; x = y++ + x++; y = ++y + ++x; printf("%d %d\n",x,y); }`
8. `main() { int a=2, b=3; printf("%d", a+++b); }`
9. `int a=-3, b=2, c=0, m; m=++a && ++b || ++c; printf("%d,%d,%d,%d", a,b,c,m);`
10. `int a=2; printf("%d,%d", ++a, ++a);`

SELF-LEARNING EXERCISES

One-dimensional integer array

1. Write a program to read an array of 'n' integers and find out the number of positive and negative numbers and zeroes in it.
2. Write a program to read an array of 'n' integers and find out the number of odd and even numbers and zeroes in it.
3. Write a program to read an array of 'n' integers and find the sum of them.
4. Write a program to read an array of 'n' integers and find the sum of the odd and even numbers separately.
5. Write a program to read an array of 'n' integers and find the sum of positive and negative numbers separately.
6. Write a program to read an array of 'n' integers and find the biggest among them.

7. Write a program to read an array of 'n' integers and find the smallest among them.
8. Write a program to read an array of 'n' integers and find the second biggest among them.
9. Write a program to read an array of 'n' integers and find the second smallest among them.
10. Write a program to read an array of 'n' integers and swap the biggest and smallest among them.
11. Write a program to read an array of 'n' integers and swap the second biggest and second smallest among them.
12. Write a program to read an array of 'n' integers and sort them in ascending order.
13. Write a program to read an array of 'n' integers and sort them in descending order.
14. Write a program to read an array of 'n' integers and sort the first half of the array in ascending order and second half in descending order.
15. Write a program to read an array of 'n' integers and check whether a given integer exists in the array or not.
16. Write a program to read an array of 'n' integers and check the number of occurrences of a given integer in the array.
17. Write a program to read an array of 'n' values and find the average of them and print the same.
18. Write a program to read an array of 'n' values and find out the arithmetic mean, standard deviation and variance.
19. Write a program to find the median of the given 'n' values.

Two-dimensional integer array

20. Write a program to read a matrix of order m×n and find the inverse of the same.
21. Write a program to read a matrix of order m×n and find the determinant value of the same.
22. Write a program to read two matrices of order m×n and find the sum of them (check the compatibility for adding them).
23. Write a program to read two matrices of order m×n and find the difference between them (check the compatibility for adding them).
24. Write a program to read two matrices of order m×n and find the product of them (check the compatibility for multiplying them).
25. Write a program to read a matrix of order m×n and find the transpose of the same.
26. Write a program to read a matrix of order m×n and find whether the upper triangle and lower triangles are equal.

One-dimensional character array

27. Write a program to read a string and convert all the uppercase alphabets in it to lowercase.
28. Write a program to read a string and convert all the lowercase alphabets in it to uppercase.
29. Write a program to read a string and find out the abbreviation of the same.
30. Write a program to read a string and check the number of occurrences of a particular character.
31. Write a program to read a string and encrypt the same.
32. Write a program to read a string and decrypt the same.
33. Write a program to read a string and count the number of uppercase, lowercase alphabets, digits and special characters in it.
34. Write a program to read a string and count the number of words, lines and characters in it.
35. Write a program to read a string and count the number of sentences in it.
36. Write a program to read a string and check whether it is a palindrome or not (a palindrome is one which reads the same way when it is read backwards).
37. Write a program to read a string and replace a comma (wherever appears) with a semicolon.
38. Write a program to read a string and copy it to another array.
39. Write a program to read a string and find the length of it.
40. Write a program to read a string and extract the first 'n' characters from it and store it in another array.
41. Write a program to read a string and extract the last 'n' characters from it and store it in another array.
42. Write a program to read a string and extract the middle 'n' characters from it and store it in another array.
43. Write a program to read a string, which is an assignment statement in Pascal and convert it to a FORTRAN assignment statement.
44. Write a program to read a string, which is a FORTRAN assignment statement and convert it to a Pascal assignment statement.
45. Write a program to read a string and ignore any extra spaces in between the words.
46. Write a program to read a string and check the number of palindromes in it.
47. Write a program to read a string and check the number of occurrences of a particular word in it.

48. Write a program to read a string and skip a particular word from it and copy the remaining part to another array.
49. Write a program to read a string and replace an existing pattern with another one.
50. Write a program to read two strings and concatenate them.
51. Write a program to read two strings and compare them.
52. Write a program to read a word and arrange the alphabets in that in alphabetical order.
53. Write a program to read a nucleotide or protein sequence and print the sequence in uppercase letters and also find the length of it.
54. Write a program to compute the percentage of AT and GC content in a given DNA sequence.

Two-dimensional character array

55. Write a program to arrange the given 'n' names in alphabetical order.
56. Write a program to compare the given string with the existing list of strings and store the positions where they occur in an integer array.
57. Implement the programs from 46 through 49 using 2-D character array.

5 STRUCTURES AND UNIONS

5.1 STRUCTURES

A structure is a data structure in which the individual elements can differ in type. A single structure may contain simple type data like integer, character and float elements. It also includes pointers, arrays and other structures. The individual elements of the structure are called as its members. A structure is a heterogeneous data type whereas the array is a homogeneous one. The main use of structure is to lump together collections of disparate variable types, so they can conveniently be treated as a unit.

The concept of structures is so important that when we learn C++, we would study what is called a class for which we require all the ideas behind a structure. Classes in C++ inherit many properties of the structure in C.

5.1.1 Why Structures?—Is it Advantageous Over Arrays?

We had been discussing arrays in the earlier Chapter and we knew how significant they are in programming with elements of similar data type. Suppose we want to store elements of different data types in a single array. Is it possible to do so? The answer is no. One can store elements of similar type in an array and clubbing different types is not possible. What if we write a program to store the bio-data of a person? The bio-data would contain attributes like name, age, sex, address, salary (if the person is employed) and similar ones. We could see that each attribute is of different type and if we use the attributes in our program by declaring them as usual, then we would be able to write a program but the program would be technically weak, even though it works fine. For example, the following would be the coding for storing the bio-data of a person.

```
/* Program to store the bio-data of a person */
#include<stdio.h>
main()
{
/* Variables declaration */
char name[100],sex,add[100];
int age,i;
float sal;
clrscr();
/*  Getting the name of the person */
printf("\nEnter the name : ");
for(i=0;(name[i]=getchar())!='$';i++);
name[i]='\0';
/* Getting the age of the person */
printf("\n\nEnter the age of the person : ");
scanf("%d",&age);
fflush(stdin);
/* Getting the gender of the person */
printf("\n\nEnter the sex : ");
scanf("%c",&sex);
fflush(stdin);
/* Getting the salary of the person */
printf("\n\nEnter the salary : ");
scanf("%f",&sal);
fflush(stdin);
/* Getting the address of the person */
printf("\n\nEnter the address of the person : ");
for(i=0;(add[i]=getchar())!='$';i++);
add[i]='\0';
/* Printing the bio-data of the person */
```

```
printf("\n\nName : %s",name);
printf("\n\nAge : %d",age);
printf("\n\nSex : %c",sex);
printf("\n\nSalary : %f",sal);
printf("\n\nAddress : %s",add);
getch();
}
```

In the program, we see that the variables pertaining to the bio-data are declared accordingly and used. But will they be grouped together in a particular memory chunk? No, they won't. Instead they would be scattered here and there and the overheads involved in the execution of the program would be high. Also, if we would like to store bio-data of more than one person, we have to declare as many set of variables as the number of persons for whom we are storing the bio-data. Thus, we must have some alternative where all the variables related to the bio-data are grouped together and stored in a single variable, so that they are stored one after another and not scattered in the memory. Also, if we can do like this, we would be able to reduce the overheads of the compiler and the execution time for the program would be comparatively less. C language provides a special aspect to deal with this idea. They are nothing but structures. It is a user-defined data type. Once we define a structure, we can declare for it and use it very much like a variable declared for built-in types. In the forthcoming sections, we would see more structures. Thus structures are far more advantageous over the arrays.

5.1.2 The Concept of Structures

The concept of the structures is simple and in fact easy to understand. The structure is defined in a program using the keyword **struct**. Any structure would have its name or tag and following that, we would declare the members of the structure. The members of the structure are nothing but the individual attributes that we would have used, had the structure not been defined. The following is the general syntax of defining a structure.

```
struct <name of the structure>
{                                    ——> Structure name/tag
<data type> member-1;  \
<data type> member-2;   } ——> Structure members
___________________     |
<data type> member-n;  /
}<variable(s)>;
```

If we want to declare the structure for the above program, then the following code would be the corresponding coding.

```
/* Program to store the bio-data of a person using structure */
#include<stdio.h>
/* Structure definition */
struct bio
{
char name[100],sex,add[100];
int age;
float sal;
};
main()
{
/* Structure variable declaration */
struct bio A;
int i;
clrscr();
/* Getting the bio-data of the person using the structure variable */
printf("\nEnter the name : ");
for(i=0;(A.name[i]=getchar())!= '$';i++);
A.name[i]='\0';
printf("\n\nEnter the age of the person : ");
scanf("%d",&A.age);
fflush(stdin);
printf("\n\nEnter the sex : ");
scanf("%c",&A.sex);
fflush(stdin);
printf("\n\nEnter the salary : ");
scanf("%f",&A.sal);
fflush(stdin);
printf("\n\nEnter the address of the person : ");
for(i=0;(A.add[i]=getchar())!='$';i++);
A.add[i]='\0';
```

```
/* Printing the bio-data of the person using structure variable */
printf("\n\nName : %s",A.name);
printf("\n\nAge : %d",A.age);
printf("\n\nSex : %c",A.sex);
printf("\n\nSalary : %f",A.sal);
printf("\n\nAddress : %s",A.add);
getch();
}
```

We see from the above program that a structure by name bio is defined in the beginning before the function main() starts. It has the variables (members) pertaining to the bio-data of a person and we see that all are grouped in one container and that container is nothing but the structure. The name of the structure (bio) is a new data type for which we can declare variables and access the members of the structure. Since the structure is defined outside the function main(), we can say that the structure is global in scope and can be used by any other function that may be defined by the user in the same program. We would see more about global scope in the chapter related to Functions.

The variable for the structure is declared inside the function main(). It could be declared along with the syntax too, in which case it would also be having a global scope as the structure. Here, the variables that are declared inside the function main(), are local to main() and would not be visible outside it.

Once we declare the structure variable, we are ready to access the members of the structure.

After the declaration, we get the values of name, age, sex, salary and address of a person (A) and finally print them. If you closely observe the program, you may notice that while accessing a member of the structure, we have used a dot operator (.) and this is one of the operators used in structures. Thus for reading in the name and other details of the person A, we have used the dot operator. See that the dot operator is used for both reading and displaying the details of the person.

We can declare as many variables for the structure bio as we wish. If we need more variables corresponding to the structure defined, then we would go in for an array of structure. Figure 5.1 shows the memory map for the structure defined and used in the above program. We could see from the figure that the members of the structure bio are distributed continuously in the memory, unlike the earlier version of the same program, where the same variables would have been distributed randomly. This is the primary advantage of using structures in our programs.

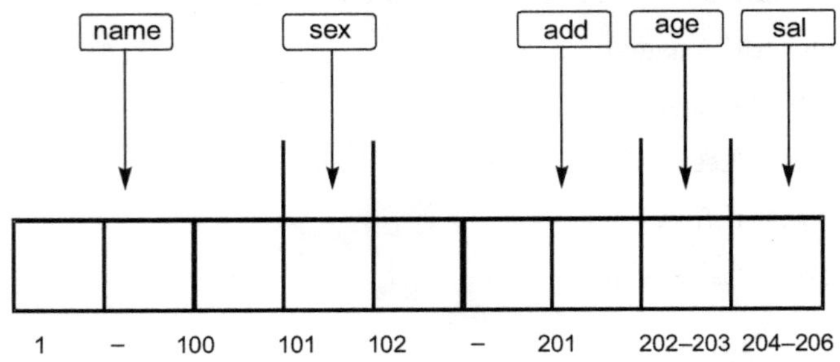

Figure 5.1 Memory map of the structure bio

5.1.3 Structure Operators

As we had seen in the earlier section, **dot operator** is one of the operators used in structures. Apart from the dot operator, we have one more structure operator denoted by the symbol –>, called arrow operator. This operator is used when we use pointer for a structure and hence will be discussed in the Chapter Six and we will defer the discussion of this operator here. The dot operator is the fundamental operator needed for accessing any member of the structure from the functions including main(). This concept has laid the foundation for the object-oriented programming and the concept of structure was extended to class later and the accessibility of class members was made possible by using the dot operator.

5.1.4 Forms of Structure

Even though there is general syntax available for a structure (as given in the section 5.1.2), there are three forms of a structure as discussed below.

Structure without tag (anonymous structure) The following is an example of a structure without a proper tag and is called technically as an anonymous structure.

```
struct
        {
        char name[100];
        int a,b,c;
    } A, B;
```

Normally an anonymous is not preferably used by the professional C programmers. One shall be aware of the fact that a declaration like the one above is possible in C. The accessibility rule for an anonymous structure is the same as for any other structure.

Structure with tag The following is an example for a structure that has a proper tag.

```
struct bio
        {
        char name[100];
        int a,b,c,;
        } A, B;
```

We have already discussed about such a structure in the section 5.1.2 with an example program.

Structure as a new data type Consider the following definition of a structure.

```
struct book
        {
        char bname[100], aname[100], pname[100];
        float price;
        int no_of_pages;
        };
struct book A;
typedef struct book BOOK;
BOOK B;
```

In the above definition BOOK is a user-defined structure type. It is just an alias for the structure book and one can declare variables for the structure either using the tag book or using the tag BOOK. This is a provision available in C language and this is achieved by using the keyword typedef. Even for built-in types, one can use this typedef and create alias. For example, the following program is perfectly valid.

```
#include<stdio.h>
main()
{
typedef int I;
I a;
clrscr();
scanf("%d",&a);
printf("%d",a);
getch();
}
```

Here we first create an alias I for the type int and then use the alias to declare variables for the type I.

5.1.5 Initializing Structure Variables

Like any ordinary variable, a structure variable could be initialized and used. The following program code illustrates how it is done.

```
/* Program to initialize a structure and store the bio-data */
#include<stdio.h>
/* Structure definition */
struct bio
{
char name[100], sex, add[100];
int age;
float sal;
};
main()
{
/* Structure variable declaration */
struct bio A,B={"johnson",'m',"coimbatore",25,23000};
int i;
float sal;
};
main()
{
/* Structure variable declaration */
struct bio A,B={"johnson",'m', "coimbatore",25,23000};
int i;
clrscr();
/* Getting the bio-data of the person using the structure variable */
printf("\nEnter the name : ");
for(i=0;(A.name[i]=getchar())!='$';i++);
A.name[i]='\0';
```

```
printf("\n\nEnter the age of the person : ");
scanf("%d",&A.age);
fflush(stdin);
printf("\n\nEnter the sex : ");
scanf("%c",&A.sex);
fflush(stdin);
printf("\n\nEnter the salary : ");
scanf("%f",&A.sal);
fflush(stdin);
printf("\n\nEnter the address of the person : ");
for(i=0;(A.add[i]=getchar())!='$';i++);
A.add[i]='\0';
/* Printing the bio-data of A using structure variable */
printf("\n\nName : %s",A.name);
printf("\n\nAge : %d",A.age);
printf("\n\nSex : %c",A.sex);
printf("\n\nSalary : %f",A.sal);
printf("\n\nAddress : %s",A.add);
/* Printing the bio-data of B using structure variable */
printf("\n\nName : %s",B.name);
printf("\n\nAge : %d",B.age);
printf("\n\nSex : %c",B.sex);
printf("\n\nSalary : %f",B.sal);
printf("\n\nAddress : %s",B.add);
getch();
}
```

We see that the structure bio is defined and two variables are declared for it namely A and B. The attributes for A are obtained from the keyboard and displayed, whereas the attributes for B are initialized during the program coding time itself. This is the way to initialize a structure. See that we have followed the same order as we have defined the members of the structure, when we initialize the attributes for B. It is compulsory that we have to follow the same order. Even in the memory map in Figure 5.1, you can see that the order of the variables is very much the same as they are declared inside the structure.

What if we want to initialize one structure variable and would like to assign the same values to another one, i.e., copy one structure value to another? When we want to achieve this for built-in types, we use the assignment operator (=). Will it be applicable for the structures, since they are user-defined types? You would just like to know whether the following code is accepted in the above program.

```
struct bio A,B={"johnson",'m',"coimbatore",25,23000};
A=B;
```

It is acceptable to use such a code. This is another aspect, which we would say as the highlighting feature of any object oriented programming languages, namely operator overloading, is implicitly available in the structures of C language. But the concept is not fully supported in C. Again I would justify how it is not fully utilized in C. Consider the following fragment of a program, for example.

```
/* Program to justify that operator overloading is not supported fully in C structures */
#include<stdio.h>
struct one
    {
    int a;
    }A;
struct two
    {
    int a;
    }B;
struct three
    {
    int a;
    }C;
main()
{
A.a=4, B.a=5; /* initialize the variables A and B */
clrscr();
C=A+B; /* perform the addition of the variables A and B */
printf("%d",C.a);
getch();
}
```

In the above coding, A, B and C are the variables pertaining to the structures one, two and three respectively. See that we are initializing the variables A and B with the values 4 and 5 respectively. In the next line of coding, we add them and try to store the result in C. This kind of addition is performed having a presumption that the values will be added exactly like those of built-in types. But this will flash an error; since the operator '+' is designed only to add the values pertaining to built in types and direct usage of it on the structure variables is not allowed. But when the same expression is used in C++, it would be correct, since the operator '+' would be considered as overloaded (provided overloading function is written). The above line of coding C=A+B shall be modified in the following way to make the program work fine.

```
C.a=A.a+B.a;
```

The error flashed by the above program is shown in the Figure 5.2.

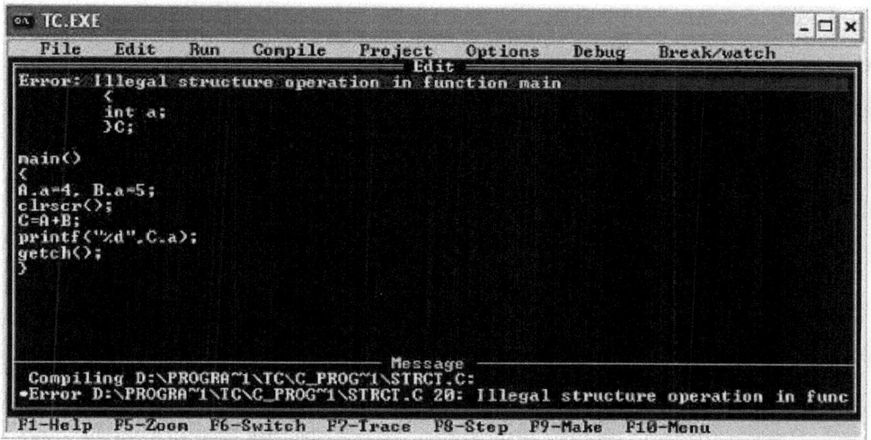

Figure 5.2 Error due to illegal usage of '+'

Once you make the correction that is mentioned above in the error-prone line, the program will work fine and get you the output.

5.1.6 Structures within Structures

One can declare a structure within another structure and thereby nest structures to any level. The following program code illustrates how it is done.

```
/* Structure within another structure */
#include<stdio.h>
struct bio1 /* PARENT STRUCTURE */
```

```
    {
    struct bio2 /* CHILD STRUCTURE */
        {
        int age;
        }A;
    float sal;
    }B;
main()
{
B.A.age=22;
B.sal=25000;
clrscr();
printf("\n\nAge : %d",B.A.age);
printf("\n\nSalary : %f", B.sal);
getch();
}
```

We could see that the structure bio2 is declared within the structure bio1 and we have the corresponding variables A and B declared for them respectively. When we try to access the member of the structure bio2, then we have to use the variable A, but since that structure is within the scope of bio1, the variable corresponding to that structure must be primarily used. That is why the protocol is B.A.age. The member sal is in the parent structure and hence it could be directly accessed using its own variable. This way, we could nest the structures to any level and provide the correct variable to access the individual members of the structure.

5.1.7 Mutually Referential Structures

Two or more structures that are defined in each other are called as mutually referential structures. For example, if we have two structures A and B, if we define a variable for the structure B in the definition of the structure A and if we define a variable for the structure A in the definition of structure B, then we would say that they are mutually defined. This concept was later used in C++ for the classes and is given the name friend classes. Here we can term one structure as a friend of another. This kind of using a structure is more prevalent when structures are used with pointers. The following program fragment shows how it is done.

```
/* Mutually Referential Structures */
#include<stdio.h>
struct B; /* Forward Declaration */
struct A
    {
    int a;
/* Declaring a pointer for the structure B in the structure A */
    struct B *bptr;
    };
struct B
    {
    int b;
/* Declaring a pointer for the structure A in the structure B */
    struct A *aptr;
    };
```

We shall discuss more on this when we deal with linked lists later in the chapter 6. We could see from the above fragment that a pointer for structure B is declared in the definition of structure A. So, the compiler should know that there would be a structure by name B later in the program. This is the reason why we have used what is called as **forward declaration** and by doing so, we let the compiler know that the definition of structure B would follow later in the program and even before that, there is a pointer or a related variable that would be declared in another structure. In the absence of such a forward declaration, the compiler would be confused about the status of the variable *bptr that is declared in the structure A and hence would flash an error while compiling the program. The real usage of such structures would be in linked lists which we will discuss later in this book.

5.1.8 Self-referential Structures

A self-referential structure contains a pointer member that points to a structure of the same structure type. Consider the following definition of a structure.

```
struct node {
  int data;
  struct node *nextPtr;
};
```

The coding above denotes that a pointer variable of type node is created within its own definition and hence we say that the structure refers to itself. This is something like a recursive function where the function calls itself from within its own code. This type of structure is useful when we write programs for implementing linked lists and binary trees, which we will discuss in the forthcoming chapter.

5.1.9 Structures with Bit Fields

A bit field is an additional declaration in a structure that can be used to specify the number of bits. Its length is set off from the declarator for the field name by a colon. A bit field is interpreted as an integral type. This means a bit field is used in expressions in exactly the same way as a variable of the same base type would be used, regardless of how many bits are in the bit field. If the bit fields are <=1 byte, then it takes one byte. Similarly if bit fields are >1 and <=2, then it takes two bytes. If it is >2, it will lead to error.

Consider the following program.

```
/* Bit Fields */
#include<stdio.h>
struct with_bits
    {
    unsigned first : 5;
    unsigned second : 9;
    };
main()
{
union bits
    {
    struct with_bits b;
    int i;
    }a;
    a.i=0;
    a.b.first=9;
clrscr();
printf("%d,%u",a.i,a.b.first);
getch();
}
```

We see that the bit fields are treated as unsigned integers. When we execute the above program, the output would be 9, 9. The identifier with_bits is a structure with two members, **first** and **second**. The member **first** is an unsigned integer with 5 bits and **second** is also an unsigned integer with 9 bits. Both the numbers can be stored in a 16-bit entity now rather than two separate integers. Combined with unions, such structures can be used with great effectiveness, as in the above example. We will not delve into the details of bit fields as it is beyond the scope of this book.

5.1.10 Array of Structures

One can form an array of structures exactly in the same way an array of integers or float is formed. The array thus formed will have the individual structures as its elements. Similar to declaring structure variables, an array of structures can be declared in two ways, either following the structure definition or in the function where we would like to use it. Consider the following example.

```
/* Array of Structures - Processing Employee Records */
#include<stdio.h>
struct date
    {
    int day, month, year;
    };
struct person
    {
    char name[20];
    struct date bday;
    float salary;
    };
main()
{
struct person emprec[10];
int i,n;
clrscr();
printf("\n\nEnter how many records you want to store : ");
scanf("%d",&n);
printf("\n\nPress any key to continue....");
getch();
```

```
clrscr();
for(i=0;i<n;i++)
{
printf("\n\nYou are now going to enter the details of the %d employee\n",i+1)
printf("\n\nPress any key to continue....");
getch();
printf("\n\nEnter the name of the person : ");
scanf("%s",emprec[i].name);
printf("\n\nEnter the date of birth : ");
scanf("%d",&emprec[i].bday.day);
printf("\n\nEnter the month of birth : ");
scanf("%d",&emprec[i].bday.month);
printf("\n\nEnter the year of birth : ");
scanf("%d",&emprec[i].bday.year);
printf("\n\nEnter the salary of the person : ");
scanf("%f",&emprec[i].salary);
printf("\n\nPress any key to continue....");
getch();
clrscr();
}
for(i=0;i<n;i++)
{
printf("\n\nYou are now going to view the details of the %d employee\n",i+1);
printf("\n\nPress any key to continue....");
getch();
printf("\n\nName of the person : %s",emprec[i].name);
printf("\n\nDate of birth : %d\\%d\\%d",emprec[i].bday.day,
emprec[i].bday.month,emprec[i].bday.year);
printf("\n\nSalary of the person : 0.2%f",emprec[i].salary);
printf("\n\nPress any key to continue....");
getch();
clrscr();
}
```

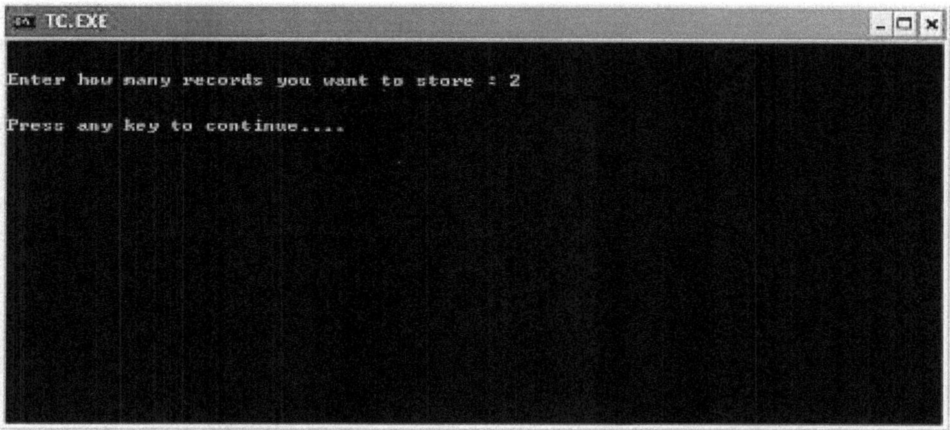

Figure 5.3 Getting the input for the number of employee records

In the above program, we define two structures namely **person** and **date**. The variable for the structure **date** is declared within the structure **person** (recall the concept of the structure within another structure). Thus we have the definition of structure **date** followed by that of **person**. Then we declare an array variable **emprec[10]** for the structure **person**, so that it can take the details of the employees according to the value of 'n' that we give from the keyboard (number of records of employees). We get the name of the person directly using the variable declared for the structure **person** and amidst getting the values for the other members, we get the date of birth (date, month and year) of the person through another structure **date**. The program has been written interactively so that the user knows what happens at each point. The user could easily understand the program since the program uses *for* loop for getting the inputs and displaying the records. Figures 5.3–5.8 show the various stages of the execution of the above program. The program is executed for getting the details of two employees and hence would have two records stored in the given structure variable **emprec[10]**.

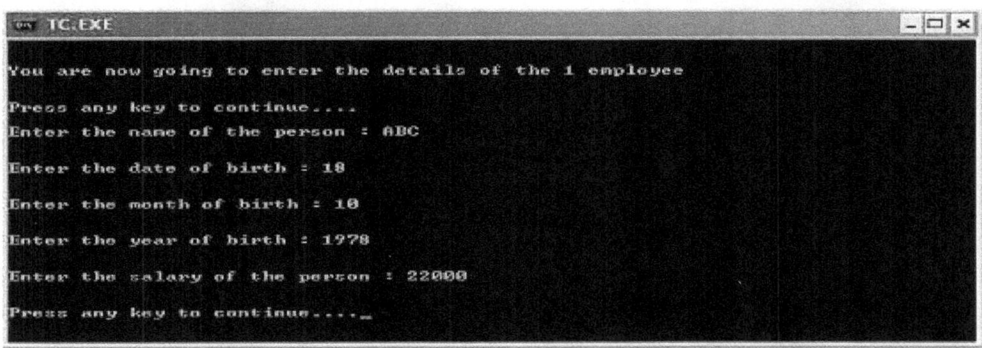

Figure 5.4 Getting the input for first employee record

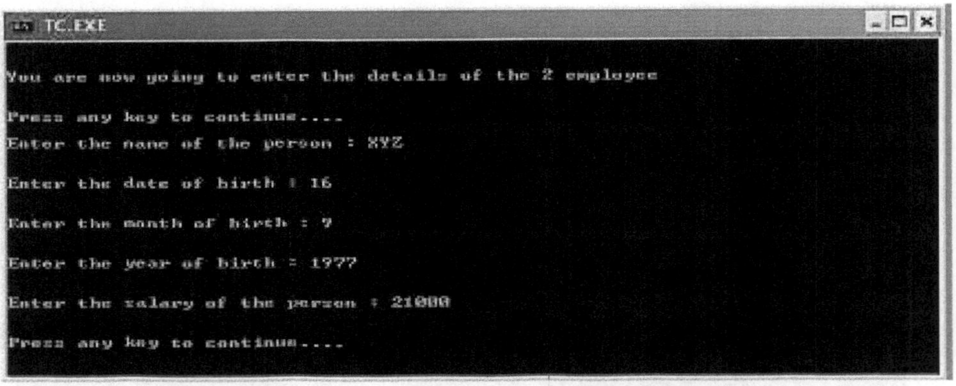

Figure 5.5 Getting the input for second employee record

Figure 5.6 Displaying the first employee record

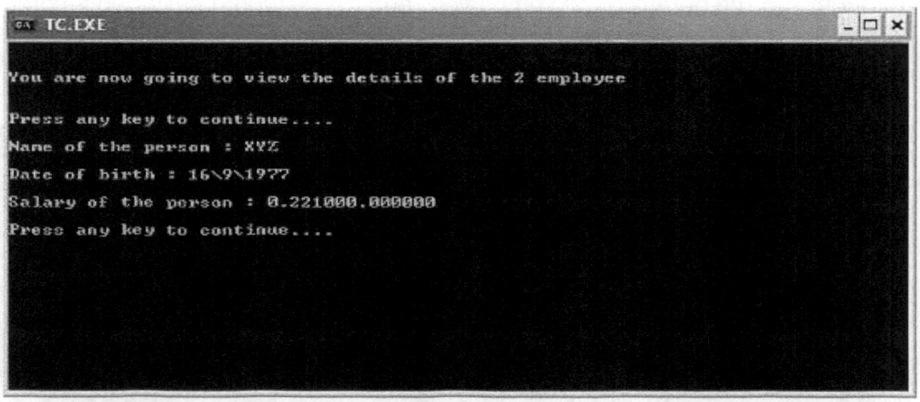

Figure 5.7 Displaying the second employee record

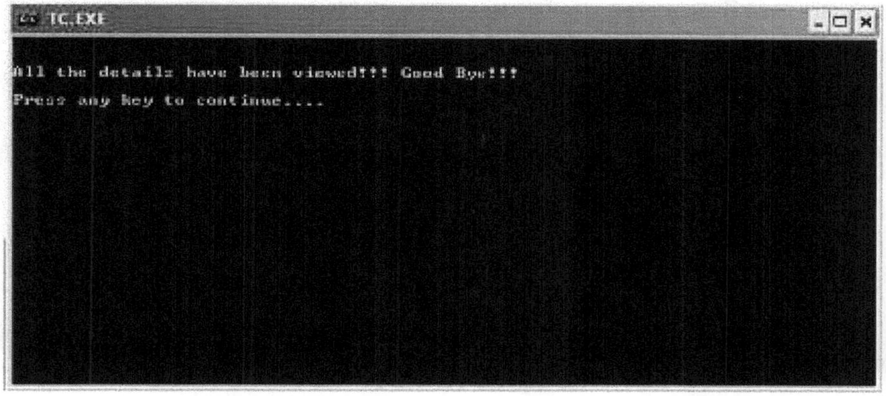

Figure 5.8 Winding up the program

5.2 UNIONS

Unions are similar to structures and they are declared using the keyword **union**. This means, a union would also contain members of different types. A union declaration specifies a set of variable values and optionally a tag naming the union. As said earlier, the same set of rules govern the declaration and usage of both the structure and union members. Like structures, unions could also have bit fields. A union type declaration is also a template like that of a structure and memory would not be reserved until the variable is declared for it and used.

The primary and the only difference between a structure type and union type is the memory space used by their members. When we declare a structure and declare a variable for it, the memory space used by that variable would be equal to the total of size of the individual members of that structure template. Consider the following declarations of a structure and a union.

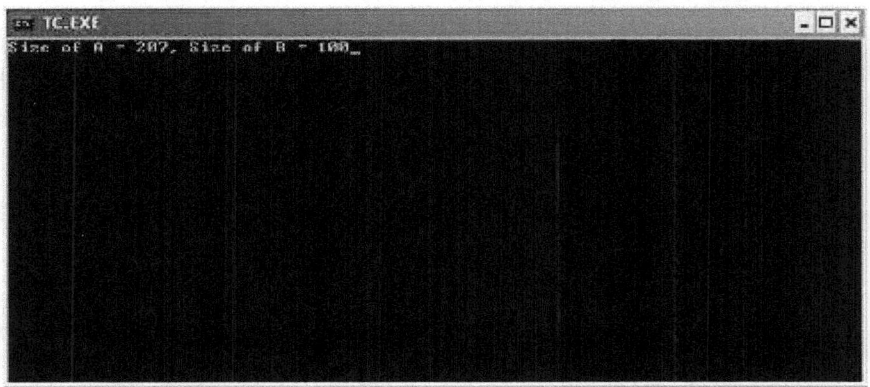

Figure 5.9 Size of the structure variable A and union variable B

```
/* Structures and Unions - Primary Difference */
#include<stdio.h>
struct bio_1
     {
     char name[100],sex,add[100];
     int age;
     float sal;
     }A;
union bio_2
     {
     char name[50],sex,add[100];
     int age;
     float sal;
     }B;
main()
{
clrscr();
printf("Size of A = %d, Size of B = %d", sizeof(A), sizeof(B));
getch();
}
```

The output for the above program would be 207 and 100 as shown in Figure 5.9. Note that we have used a new operator called as sizeof, which will find the size of the variable pertaining to a data type and returns that value. Thus it is obvious that the storage associated with a union variable is the storage required for the largest member of the union. When a smaller member is stored, the union variable can contain unused memory space. All the members are stored in the same memory space and start at the same address. The stored value is overwritten each time a value is assigned to a different member of the union.

Consider the following program for getting the information regarding the storage of members of the union type.

```
/* Union - Storage of members */
#include<stdio.h>
main()
{
```

```
union test
    {
    int i;
    char ch[2];
    };
    union test t;
    t.i=64;
    clrscr();
    printf("\n\ntest.i = %d",t.i);
    printf("\n\ntest.ch[0]=%d",t.ch[0]);
    printf("\n\ntest.ch[1]=%d",t.ch[1]);
    getch();
}
```

This data type would occupy only 2 bytes in the memory, whereas the same would occupy 4 bytes with the **struct** definition. Here in this program, we are initializing the value of 'i' for the structure variable't' using a statement t.i and the value is 64. The same memory locations used for t.i are also being used by t.ch[0] and t.ch[1]. This means the memory locations used by t.i can also be accessed using t.ch[0] and t.ch[1]. Figure 5.10 shows the output for the above program when the value of t.i is 64.

Figure 5.10 Output of the program when the value of t.i is 64

We know that 64 is an integer and hence requires 2 bytes of memory. Figure 5.11 is the memory representation of the number 64 (see that the number is converted to its binary equivalent and stored in memory).

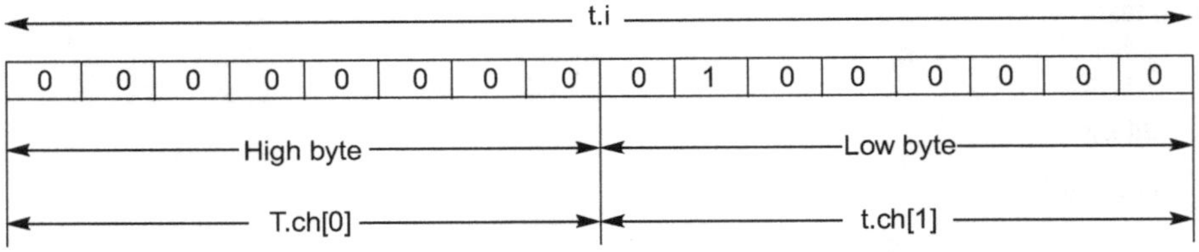

Figure 5.11 Memory representation of 64

The member 'i' is an integer and hence is represented in two bytes in Figure 5.11, namely a low byte and a high byte. We can see that the low byte is occupied before the high byte. Thus, the output in Figure 5.10 has the value of t.i as 64, t.ch[0] as 0 and t.ch[1] as 64. Considering the value of t.i to be 256 in the program above, we will have the memory picture as in Figure 5.12. In this case, we have the output for t.i as 256, for t.ch[0] as 0 and ch[1] as 1 as shown in the Figure 5.13.

You may try out with other values for t.i and check the values of t.ch[0] and t.ch[1]. This obviously means that the union type allows its members to share the memory space. In unions the variables will have only one member at any time and hence any reference to the wrong type of data will produce meaningless results. Unions are used when all the members need not exist in the memory at the same time. All the union members are stored in the same memory space, with the same starting address, but not at the same time.

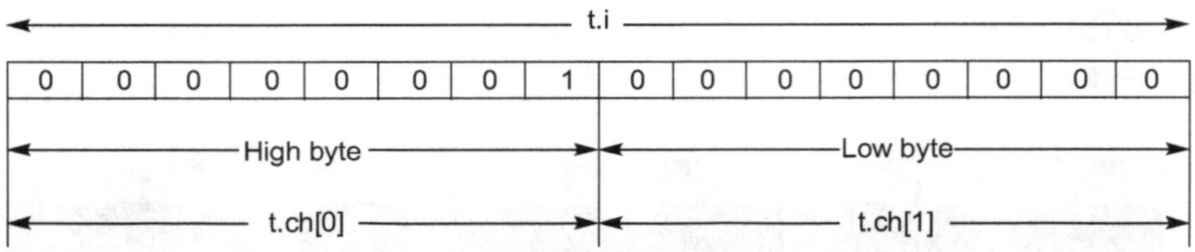

Figure 5.12 Memory representation of 256

Figure 5.13 Output of the program when the value of t.i is 256

SUMMARY

This chapter has dealt in detail with two of the user-defined data types namely structures and unions. Even though they are similar in many aspects, technically they are different and we saw that memory occupied by the union variables is comparatively lesser than that occupied by the structure variables. Both of them are synonymous in almost all ways. We had seen this important significant difference in detail and I am sure that the figures and the screenshots would have been of great help to you in learning this technical difference very clearly.

When we started off with the introduction to structures, we discussed the importance and advantage of structures over arrays. We had also discussed how the concept of structures was useful in extending the same to classes later in the programming language C++. The discussion of structures also dealt with a few of the object oriented concepts that are supported by it and its limitations have been explained in a lucid manner.

We then discussed the different forms of structure. We also learned how to initialize a structure variable and what are the technical details involved in it. We learnt the concept of using and accessing a structure within another structure. We then discussed different types of structures like mutually referential structures and self-referential structures. We also understood that the ideas of mutually referential structures are extended in C++ as friend classes. We then discussed the usage of bit fields in structures and how useful it is while programming. We concluded the discussion of structures with a program and explanation of how to declare an array of structures. We processed the employee records and had a few figures explaining how the array of structure is accessed. The learners of this book shall try out in getting the memory picture of an array of structure and you may use the Figure 5.1 in getting the same.

We concluded the discussion with a very brief note on the unions, and the idea of memory sharing by the members of the union was explained using proper pictures and surely it could have provided the readers with a very clear understanding of how advantageous a union is in most complicated programming applications, where the user intends to use a structure.

The readers have to keep in mind that this chapter has given an overall view of using structures and unions and I have tried to give most of the technical issues related to them. The concept of structure is further extended in the Chapters 6 and 7, when I discussed pointers and functions. The structure is a very big ocean in C. In fact, C ends in structures, where C++ begins. When one starts learning the idea of class in C++, he/she must have the background knowledge of structure and this is where the migration begins. So, I would advise the readers to do more programs in structures (they may try out all those at the end of this Chapter and the other Chapters that follow as well) and get practical exposure to them. Once they have a thorough knowledge on these, I am sure they would not have any difficulty while migrating from C to C++.

CRITICAL THINKING QUESTIONS

1. What is wrong in the following structure declarations?

```
struct s
{
int i;
char s[10];
};
main()
{
________
}
```

2.

```
struct s1 { float f2; };
struct s2 { char c; };
main()
{
struct s1 v1,v2;
struct s2 v3,v4;
clrscr();
printf("%d,%d,%d,%d",sizeof(v1),sizeof(v2),sizeof(v3),sizeof(v4));
getch();
}
```

What is the output of the above program?

3.
```
struct { int a,b; } aa={3,4,5};
main()
{
printf("%d",sizeof(aa));
}
```
What is the output of the above program?

4.
```
struct aa
{
    int i:2;
    int j:4;
    int k:2;
    };
union
    {
    struct aa a;
    unsigned char c;
    }bb;
main()
{
bb.c=199;
clrscr();
printf("%d,%d,%d",bb.a.i,bb.a.j,bb.a.k);
getch();
}
```
What is the output of the above program?

5.
```
union c
    {
    int i;
    char c;
```

```
    }cc.c=' ';
main()
{
clrscr();
printf("%d",cc.i);
getch();
}
```

What is the output of the above program?

SELF-LEARNING EXERCISES

1. Write a program to process the students' records by getting the inputs for marks of five subjects and find out the total, average and grade.
2. Write a program to do the basic arithmetic operations with two given complex numbers.
3. Write a program to create a database of books and authors with other details such as publisher's name, year of publication, number of pages and price of the book.
4. Write a program to create a small database of a library.
5. Write a program to create a database of a cricket team.
6. Write a program to create a database of employees in a particular industry.
7. Repeat the above programs using unions.
8. Write a program to create a small database of protein sequences you have downloaded from a standard website.
9. Write a program to create a small database of nucleotide sequences you have downloaded from a standard website.
10. Write a program to store the PDB ids and the corresponding protein names that you would have collected from Protein Data Bank.

POINTERS

6.1 INTRODUCTION

If you ask a professional C programmer as to which concept he/she found very difficult to understand when they learned C, the answer will be very simple and obvious. Yes, the answer would be **pointers**. The significance of pointers in C is the flexibility it offers in programming. It is C's clever use of pointers that has made it an excellent and an efficient language. As a professional programmer, you can fiddle with the pointers long enough until you finally mess it.

The basic problem with the C programmers in understanding the pointers would be in not programming with them, but in understanding the concept and the terminology involved in it. So, the point lies in understanding the terminology very clearly.

Simply stated, a pointer is nothing but an address. Instead of being a variable, it is a pointer to a variable stored somewhere in the address space of the program. For the C programmers to understand a pointer more clearly, they must understand the organization of memory in computers. We discuss the organization of the memory in the following paragraph.

Memory is organized as a sequence of byte-sized (1 byte = 8 bits) locations. These bytes are numbered beginning with a zero. The number associated with a byte is known as its address or memory location. A pointer, therefore, is an entity that contains a memory address. Thus, a pointer is a number which specifies a location in memory.

Thus pointer variables are defined as variables that contain the memory addresses of data or executable code. Like any other variable, the pointer variable must also be defined, so that it will work only with data of a given type. Just as an integer variable can hold only integer values, character variable can hold only character values and so forth, a pointer variable can point only to a specific type of data.

6.1.1 Why Pointers?

Although a lot of programming can be done without using the pointers (as we have done in the earlier Chapters), their usage would enhance the capability of the program to manipulate data very easily and efficiently. In C, we have a concept called **Dynamic Memory Allocation** where the application and usage of pointers becomes indispensable. For example, consider that we are reading marks of a set of students and processing them. For the process, we have to declare an array as the one given below:

```
float marks[100];
```

But, the obvious limitation of this declaration is that the maximum number of students is limited to 100. If during run time, we need one more student's marks to be entered, it may not be possible, since the size of the array is fixed. On the other hand, using the dynamic memory allocation, we can design the program such that the limit for the maximum number of students is restricted only by the amount of available memory and the number could be decided by the user during the execution of the program.

By using the pointer effectively, variables can be created and defined when the program is being executed. Memory can be obtained from the system, if available, and allocated for the creation of a new variable or a data structure. This is called dynamic memory allocation. Dynamic variables so created can only be accessed by the use of pointers.

Pointers offer tremendous flexibility in the creation of dynamic variables, accessing and manipulating the contents of memory locations, and releasing the memory occupied by the dynamic variables which are no longer needed. Some of the uses of pointers in C are:

- Accessing the array elements
- Passing arguments to functions by reference
- Passing arrays and strings to functions
- Creating data structures such as linked lists, trees, graphs and so on
- Obtaining memory from the system dynamically

6.1.2 Pointer Operators—Storing Addresses using Pointers

In pointers and related discussions, we will be using two operators designated by & (ampersand) and * (asterisk). These two symbols are already familiar to you. But their meaning is different in pointers and it is not the same as that discussed in the earlier chapters. The symbol '&' was used to imply the bitwise logical AND operator and '*' was used as multiplication operator. Here in pointers, they have entirely a different name and operation.

The symbol '&' is called as **address of** operator. We have already used this operator in the syntax of the function scanf() and I had explained clearly about this operator's function in the Chapter 2, when we were discussing I/O instructions. In the function scanf(), it is provided to fetch the addresses of the arguments. That is why we call this operator as 'address of' operator. Consider the following piece of program. What do you think will be the output of the printf() statement in this?

```
int a=10;
printf("%d",&a);
```

So far, we have been using '&' only in the function scanf() and in the above piece of coding, we have used this in the function printf(). Will it flash any error? No. The above coding is perfectly valid and would give an output. It would print the address of the variable 'a'. As we have used '&a' as the argument to the function printf(), it is printing the address, since we knew that '&' will fetch the address of the corresponding variable that is used after it.

Consider the following declaration in a program.

```
int a=10;
```

What exactly happens when this line of program is executed by the C compiler? Three things happen at the compiler level. They are:

1. Allocate space in memory to hold an integer value
2. Associate the name 'a' with that location
3. Store a value 10 in that location

The following memory picture would imply the above three steps very clearly.

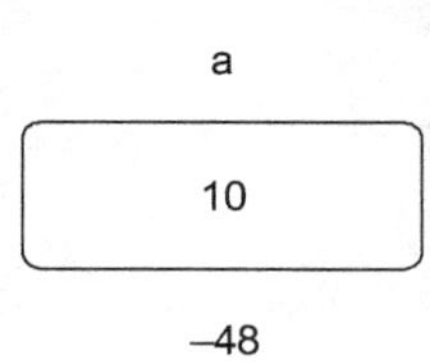

Figure 6.1 Memory map of the variable 'a'

We can see from Figure 6.1 that the variable 'a' is allotted with a memory space (in terms of bytes) and given an initial value (as given in the piece of coding). When no value is given for 'a', it would assume any garbage value within the range for an integer. We further see that the address of the variable 'a' is –48 and it is the one that is selected by the C compiler and not by the programmer or the user. When you declare 'a' and use it, the compiler may select a different location for 'a'. The point worth noting here is that the address of 'a' in memory is a number. As discussed in the earlier paragraph, this address can be printed using the function printf().

The symbol '*' is called as 'value at address' and it will fetch the 'value' from the corresponding address as instructed. It is otherwise called as 'indirection' operator. Consider the following piece of coding.

```
/* Program for demonstrating the function of '*' */
#include<stdio.h>
main()
{
int a=10;
clrscr();
printf("\n\nValue of a = %d",a);
printf("\n\nAddress of a = %d",&a);
printf("\n\nValue of a = %d",*(&a));
getch();
}
```

When we execute the above program, we can get the output as shown in the Figure 6.2.

Figure 6.2 Output for the program—'&' and '*' in action

Thus, we see that printing the value of *(&a) is the same as printing the value of 'a'. We conclude that the operators '*' and '&' complement each other.

We had defined earlier a pointer as a variable that stores the address of another variable. So, let us see how they are declared and used in programs. Is it possible to store the address of a variable in another? Yes. We can do it by using the assignment operator (=). The following program just does that.

```
/* Program to demonstrate the function of '&' and '*' */
#include<stdio.h>
main()
{
int a=10,*b;
b=&a;
clrscr();
printf("\n\nValue of a = %d",a);
printf("\n\nValue of a = %d",*b);
printf("\n\nValue of a = %d",*(&a));
printf("\n\nValue of b = %d",b);
printf("\n\nAddress of a = %d",&a);
printf("\n\nAddress of a = %d",b);
printf("\n\nAddress of b = %d",&b);
getch();
}
```

Figure 6.3 shows the output of the above program.

Figure 6.3 '&' and '*' in action

The following memory picture in Figure 6.4 would provide more idea on how the above output is generated.

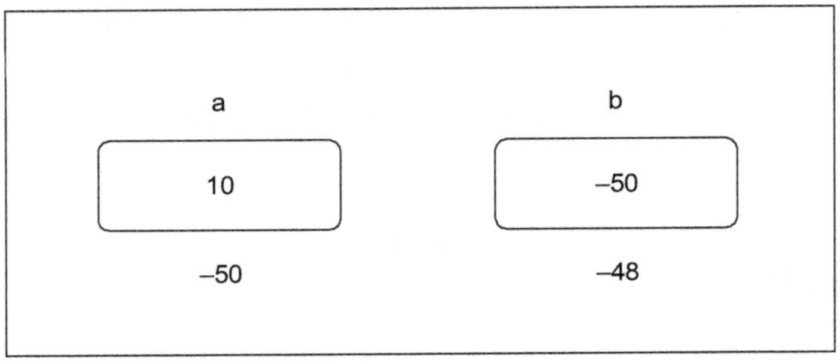

Figure 6.4 Memory map of 'a' and 'b' of the above program

We can understand from Figure 6.4 that a's value is 10 and b's value is a's address. According to the convention in C, no variable could be used without declaring it. If you closely observe the program, we have defined both 'a' and 'b'. But, 'a' is an ordinary variable and 'b' is a pointer variable and it is declared as **int *b**. This declaration tells the compiler that 'b' would be storing the address of an integer value and in other words, 'b' would point to an integer. Can you justify the usage of the operator '*' in the declaration of 'b'?

For understanding this, let us recall what actually is meant by '*'. It is 'value at address' operator. Thus the declaration **int *b** would mean, the value at the address contained in 'b' is an **integer**. Go through the program above and correlate the same with the Figures 6.3 and 6.4 and you will easily get to know the concept of pointers.

To put in a nutshell, we would say that pointers are variables which contain addresses and the addresses are always whole numbers. Hence pointers would always contain whole numbers. For example, if we have a declaration **float *b** in a program, it does not mean that 'b' is going to contain a float value and all that it means is 'b' is going to contain the address of a float variable or it is going to be a pointer for a float variable.

The idea of pointers could be extended to any level and let us see with an example how it is extended to the next level. Consider the following program.

```
/* Pointer to Pointer */
#include<stdio.h>
main()
{
int a=10,*b,**c;
```

```
b=&a;
c=&b;
clrscr();
printf("Value of a = %d",a);
printf("\n\nValue of a = %d",*b);
printf("\n\nValue of a = %d",*(&a));
printf("\n\nValue of a = %d",**c);
printf("\n\nValue of b = %d",b);
printf("\n\nValue of c=%d",c);
printf("\n\nAddress of a = %d",&a);
printf("\n\nAddress of a = %d",b);
printf("\n\nAddress of a = %d",*c);
printf("\n\nAddress of b = %d",&b);
printf("\n\nAddress of b = %d",c);
printf("\n\nAddress of c = %d",&c);
getch();
}
```

Figure 6.5 Output for the program illustrating pointer to pointer

In the program, we see that three variables are declared and used, namely 'a', 'b' and 'c', out of which 'a' is an ordinary variable and 'b' and 'c' are pointer variables. 'b' is a pointer to 'a' (from the statement b=&a), since it stores the address of 'a', and 'c' is a pointer to 'b' (from the statement c=&b),

since it stores the address of 'b'. Since, 'c' points to not just an ordinary variable, but to a pointer, 'c' is called a pointer to pointer. We can go on extending this idea to any level and the most comprehensive version is a pointer to pointer. The output of the above program is shown in Figure 6.5. The memory map in the Figure 6.6 will help you better understand the way the variables and their addresses are stored in the memory.

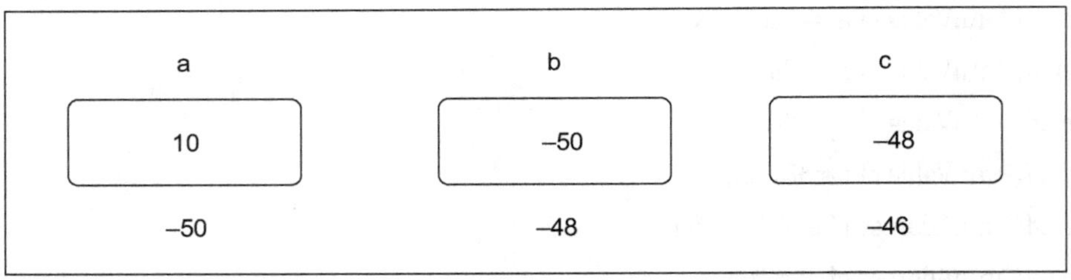

Figure 6.6 Memory Map showing the status of variables a, b and c

6.2 BASIC PROGRAMMING USING POINTERS

Now that we know how a pointer variable stores the addresses and how they could be accessed, we are ready to do some basic programs using the pointer notations and operators. Let us begin with a very simple example of adding two integers using pointers and display the result. The following is the coding.

```
/* Adding two integers and displaying the result */
#include<stdio.h>
main()
{
int a,b,c,*pa,*pb,*pc;
pa=&a;
pb=&b;
pc=&c;
clrscr();
printf("\n\tEnter the values of a and b : ");
scanf("%d%d",pa,pb);
*pc=*pa+*pb;
printf("\n\tThe sum of %d and %d is %d",*pa,*pb,*pc);
getch();
}
```

The output of the above coding is shown in Figure 6.7. From the program, we can understand that a, b and c are the integer variables and *pa, *pb and *pc are pointers and they point to a, b and c respectively. It is a convention while using pointers that the letter 'p' is used as prefix and it denotes that the variable is a pointer and it is followed by the name of the variable to which the pointer is actually pointing. Thus, 'pa' points to 'a', 'pb' points to 'b' and 'pc' points to 'c'. Figure 6.8 shows the memory map of the variables.

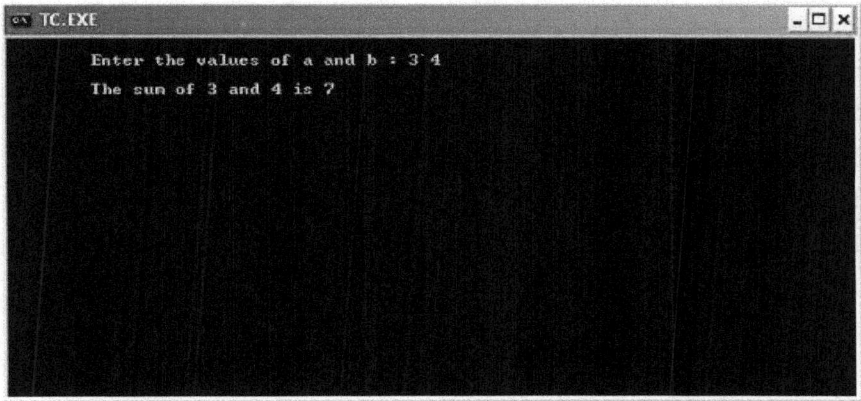

Figure 6.7 Output of adding two integers

To make yourself familiar with pointers you may try using pointers in some of the basic programs which I have given as self-learning exercises. This practice is essential since as we go further in learning pointers, we will be using these kind of jargon in a much advanced manner and as a programmer one shall not feel the difficulty in either writing programs or understanding them.

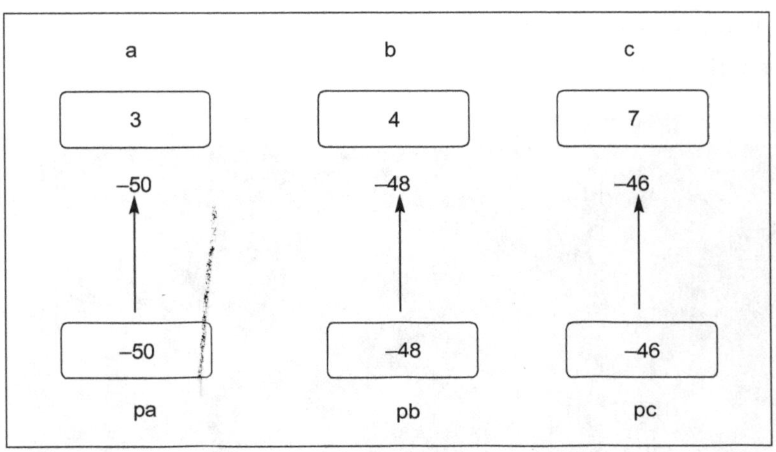

Figure 6.8 Memory picture for the program adding two integers

Let us do one more program here to make the concept clearer. Consider the program in which we find the temperature in Celsius for the given input temperature in Fahrenheit. The following would be the program coding for the same.

```
/* Program to convert temperature in Fahrenheit to Celsius */
#include<stdio.h>
main()
{
float cel,fa,*pcel=&cel,*pfa=&fa;
clrscr();
printf("\n\tEnter the temperature in Fahrenheit : ");
scanf("%f",pfa);
*pcel=(*pfa-32)*5/9;
printf("\n\tThe temperature in Celsius is %0.1f for %0.1f in Fahrenheit",*pcel, *pfa);
getch();
}
```

Figure 6.9 shows the output of the above program. We could see from the program that the output is formatted corresponding to one place after the decimal point. This is possible while we print the output using the corresponding format specifier (%f in this case) and this is possible with only float and its associated data types like double and long double. Also from the program, we notice that when the pointers are declared, they are assigned with the addresses of the respective variables. This is unlike the earlier program, where we declare the pointers in one step and assign the addresses in the subsequent step. This type of initializing the pointers when declaring them is allowed in C. This is very much similar to initialization of a variable during its declaration, which is possible while writing programs in C.

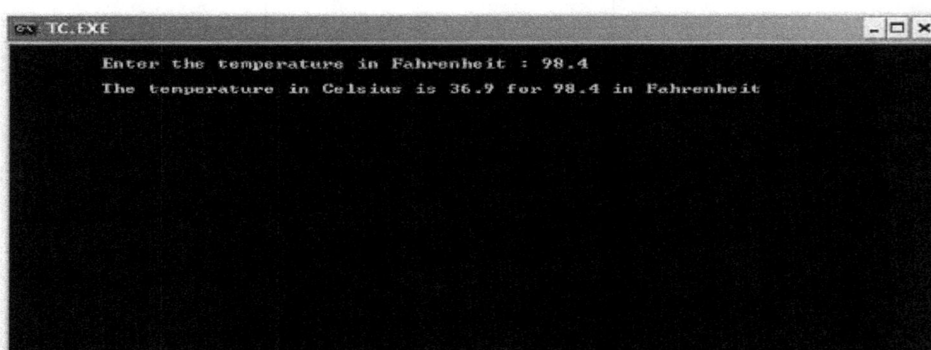

Figure 6.9 Output for the temperature conversion using pointers

Figure 6.10 shows the memory picture of the variables declared and their corresponding pointers in the above program.

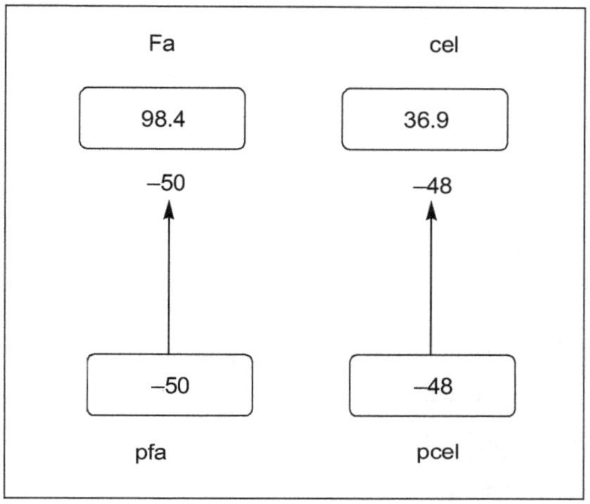

Figure 6.10 Memory picture of variables in temperature conversion program

6.3 POINTER ARITHMETIC

Pointer arithmetic is important for a C programmer to understand how a pointer works with an array, since pointers are most powerful when they are combined with arrays. Moreover, pointers and arrays are like couples and they are inseparable. For knowing about the arithmetic involved with the pointers, consider the following program.

```
/* Pointer Arithmetic */
#include<stdio.h>
main()
{
int a,*pa=&a;
float b,*pb=&b;
char c,*pc=&c;
double d,*pd=&d;
clrscr();
printf("\n\nBefore incrementation");
printf("\n********************");
printf("\n\nAddress of a = %u",pa);
```

```
printf("\n\nAddress of b = %u",pb);
printf("\n\nAddress of c = %u",pc);
printf("\n\nAddress of d = %u",pd);
pa++;
pb++;
pc++;
pd++;
printf("\n\nAfter incrementation");
printf("\n*******************");
printf("\n\nAddress of a = %u",pa);
printf("\n\nAddress of b = %u",pb);
printf("\n\nAddress of c = %u",pc);
printf("\n\nAddress of d = %u",pd);
getch();
}
```

What do you guess about the output of the above program, assuming that the addresses of the variables a, b, c, and d are 65468, 65470, 65475 and 65478 respectively? See that we have used the format specifier %u for printing the addresses, since we would sometimes like to have the addresses in terms of positive integers.

We have used two similar sets of function printf() in the program. For the first set of the function printf(), we would get the output where the corresponding addresses of the variables would be printed. For the next set of the function printf(), we would get the output as 65470, 65474, 65476 and 65486 respectively for a, b, c and d. This is because of the arithmetic involved with the pointers and due to the four increment statements (pa++, pb++, pc++ and pd++) following the first set of function printf().

Since 'pa' is a pointer to an integer and an integer occupies two bytes in memory, when we increment 'pa', it would increment and point to the next integer that would be available two bytes after the present position, i.e., the variable 'a' which is pointed to by 'pa' is presently occupying the bytes 65468 and 65469 and when we increment the pointer by one, it would increment by two bytes and hence point to the next location of the integer, which is 65470. Similar justifications could be given for the other three variables b, c and d. Thus, when 'b' occupies the addresses 65470, 65471, 65472, 65473 and it is pointed to by 'pb' and when 'pb' is incremented by one, it would point to the next location, which is four bytes after the present location and hence 65474. Since a character variable occupies only one byte, when we increment it we get the result as 65476. A double takes

8 bytes and hence the value is incremented from 65478 to 65486 (variable 'd' occupies the bytes 65478, 65479, 65480, 65481, 65482, 65483. 65484, and 65485).

Figure 6.11 shows the output for the above program code. The main point that the programmers are expected to understand and assimilate from the above program and its output is that when we increment/decrement a pointer, it would increment/decrement according to the data type it is pointing to. Upon incrementing it by '1', when it is pointing to an integer, it would increment by 2 bytes, when pointing to a float, it would increment by 4 bytes and so on. This point is to be remembered from here, as we will be using pointers with the numerical and character arrays and there we would apply this concept.

Figure 6.11 Output of the pointer arithmetic program

Apart from this, a C programmer shall also know what operations are possible with pointers. The following list would be helpful to you when you use pointers in your programs. Keep in mind the points, since any misuse of the pointers will ultimately result in malfunctioning of the program and in the worst case may lead to the program or system crash. Figure 6.12 shows the memory picture of the above program.

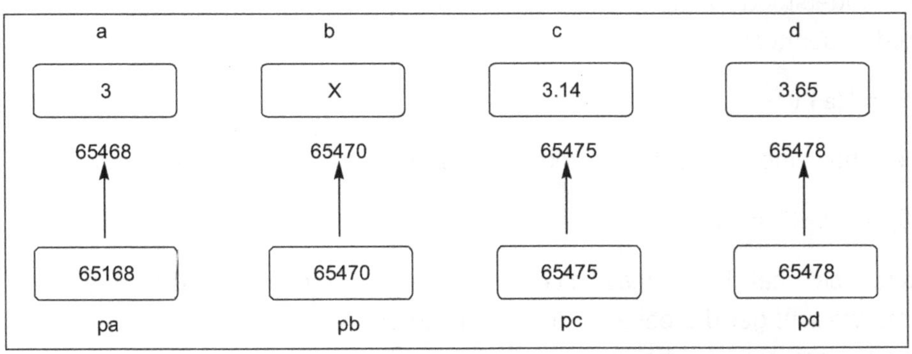

Figure 6.12 Memory picture of the pointer arithmetic program

Do's and Don'ts with the pointers

1. Do not add, subtract, multiply or divide two pointers (as it is equivalent to performing arithmetic operations between two memory addresses which will not yield anything fruitful). Consider pa, pb, and pc which are pointing to integers a, b, and c respectively. We are not allowed to carry out the following arithmetic operations with these pointers.

   pc=pa+pb, pc=pa-pb, pc=pa*pb and pc=pa/pb

2. You can add a constant with a pointer or subtract a constant from a pointer. But you cannot multiply a constant with a pointer or divide a pointer by a constant. Consider for example, 'n' is a constant. We can perform the following operations between the pointer 'pa' and the constant 'n'.

   pa=pa+n and pa=pa-n.

   We cannot perform the following operations between 'pa' and 'n'.

   pa=pa*n and pa=pa/n

6.4 POINTERS AND ARRAYS

For a C programmer to master the concept of pointers, he/she must be efficient in handling the pointers not just with ordinary variables, but with the arrays. This is because, arrays are the structured data types in C language and they implicitly use pointers to access the individual elements of the array. If you recall the discussion on arrays, we would have used what is called as a **subscript**. An array subscript is essential for a programmer to identify the individual elements of the array and normally alphabets like i, j, k are used for designating them. When we want to access the i'th element of an array 'a', then we say a[i] from our program coding and give various values for the variable 'i' by allowing it to go through the loop. But when we say a[i] from the program, it is implicitly converted to pointer notation by the C compiler and it never uses the notation that is used and understood by the programmer. Let us see how this is done. The expression would be converted to pointer notation as shown below.

```
a[i] = *(a+i)
```

When the value of 'i' is '0' the above expression would be like the one below.

```
a[0] = *(a+0) = *a
```

Thus we see that the expression is reduced to *a. When we refer the array name anywhere, in a program we will get the base address (starting address) of the array . This is a very important fact to be remembered by the programmer whenever he/she uses any type of array. Thus, in the above expression, we get the base address of the array 'a' and assume it to be the '1000th' byte of memory, then we would get the result of *(a+0) as *(1000+0) which is equal to *1000 and you know what the operator '*' does in the pointers. Yes, you are right. It will fetch the value at the address '1000'. Thus, on increasing the value of 'i', we would get the corresponding values at the subsequent addresses. If the base address that is quoted above is that of an integer array 'a', then upon giving a value of '1' for 'i' we will get the expression as *(1000+1), which is equal to 1002. The reason behind this will be easily understood from the earlier section, where we had been discussing the pointer arithmetic. Since the array 'a' is an integer array and its base address is '1000' and when we say a[1], it is converted to an equivalent expression *(a+1) and when we have the name of the array, we will get the base address of the array and hence the above expression gets further reduced to *(1000+1) and since an integer takes 2 bytes in memory, the result of the above expression is 1002 and not 1001.

We have mentioned earlier in this section that the expression a[i] would be converted internally by the compiler into *(a+i). Now the question is, can you express a[i] as i[a]? Yes. Obviously, you can. Let me justify how this could be done.

Consider the expression a[i], which could be written according to the compiler as *(a+i) and *(a+i) is the same as *(i+a), due to the commutative property of addition. If we can express a[i] as *(a+i), then why not *(i+a) could be expressed as i[a]? As we had discussed earlier, the compiler will not bother about whether it is a[i] or i[a], since it converts either of these expressions to their pointer equivalents. I would advise you that, you check the programs that you have done in the Chapter Four by rewriting the expression a[i] as i[a] to understand this concept more clearly. It is just a simple mathematics, which is summarized below.

```
a[i]=*(a+i)=*(i+a)=i[a].
```

I would suggest you go through the earlier sections till this point once again so that you make yourself to get prepared for learning pointers and arrays.

6.4.1 Pointers and Numerical Arrays

Pointers, as mentioned earlier, work more efficiently with arrays and invariably pointers and arrays are interlinked. In the previous section, we discussed how the C compiler converts the notation a[i] to the pointer notation internally and whether the user likes or not, the conversion happens internally and the C compiler does not bother about the array subscript that is being used by the programmer. Let us see how pointers work with numerical array with the help of a few programs. We will consider the integer array for our discussion, which will hold good for any type of numerical array. Consider the following program code, where we read and display 'n' integers using pointer to the integer array declared in the program.

```
/* Read and display an array of 'n' integers using pointers */
#include<stdio.h>
main()
{
/* Array and a pointer and 'n' are declared */
int a[100],*pa,n;
clrscr();
/* Getting the value of 'n' */
printf("\n\nHow many elements you want to enter : ");
scanf("%d",&n);
/* Validating the value of 'n' */
if(n<0 || n>100)
{
printf("\n\nInvalid value entered for n");
getch();
return;
}
/* Getting the values of the individual elements of the array */
printf("\n\nEnter the elements one after another : ");
for(pa=a;pa<(a+n);pa++)
scanf("%d",pa);
/* Displaying the values read one after another */
printf("\n\nThe values entered are :  ");
for(pa=a;pa<(a+n);pa++)
printf("%d\t",*pa);
getch();
}
```

We see that the program is more or less the same as we had written earlier without using pointers. The variable declaration part has the array declared and a variable 'pa', which is used as a pointer to the array. We also have a variable 'n' which is used to get and store the number of elements in the array. Once we get the value for 'n', we check whether it exceeds the limit of the array or takes an invalid value. If so, we flash an error message and return. This has to be done explicitly in any program using any kind of array since, in C there is no check to see if the array subscript exceeds the array size or not. If the 'n' value is acceptable, we proceed getting the elements of the array. This is done using *for* loop.

We can see that in the loop, we initialize the pointer with the base (starting) address of the array. This is achieved with the statement pa=a. We know that when we refer the name of the array at any point of the program, we would get its base address. This fact is used in writing such an expression in the loop. Since the variable 'pa' is a pointer, it stores the base address of the array 'a'. Following the initialization, we check the condition using the expression pa<(a+n). This expression, again, appears quite strange. When we use the array name, we would get the base address and we see that we add the value of 'n' with it in the above condition. We are allowed to do like this because we have learnt earlier in pointer arithmetic that a constant value could be added to a pointer. Thus the right hand side of the above expression becomes a constant and the value of 'pa' changes in each cycle of the loop. Assuming the base address of the array to be 1000 and the value of 'n' to be '5', the initialized value of 'pa' would be 1000 and the RHS of the expression would be 1000+5=1010. This is because, when we add five integer values with the base address of the array, it is equivalent to adding 10 bytes to it, since an integer occupies two bytes in memory. On incrementing the value of 'pa' in steps of '1', it would point to the next memory location that an integer would occupy. Thus upon incrementing 'pa' for the first time, it would take the value 1002, since that will be the location of the next integer in the array 'a'.

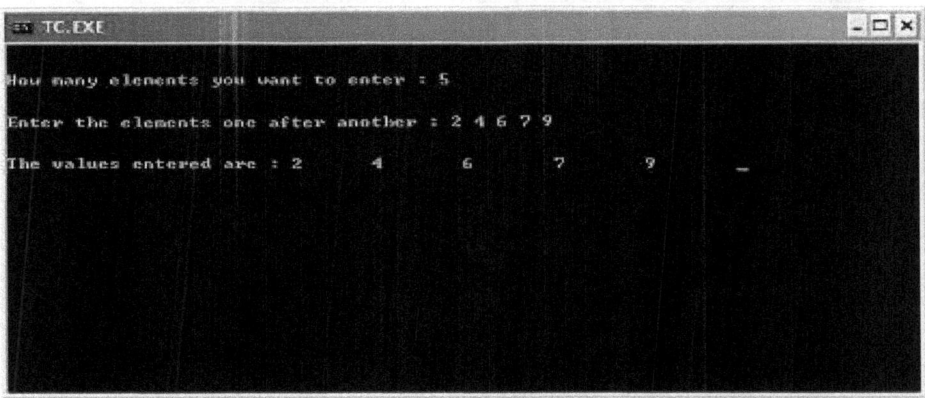

Figure 6.13 Output of the program that reads and displays 'n' integers

As we get the n'th value in the array, the value of 'pa' would increment from 1008 to 1010, thereby making the expression to return a false and hence that will cause the loop to terminate. Once we get all the required inputs, we would like to display them and hence use another *for* loop to achieve this. Since at the end of the reading process, the pointer would be at 1010, we have to bring it back to the initial position or the base address of the array and hence we reinitialize it by writing pa=a. Then we proceed with the process of printing the array elements one by one, in the order they have been read. This is very much done like the earlier loop. Except for the function scanf(), we are using the function printf() in that part of the loop. Also, see that we use the escape sequence '\t' in order to leave a tab spacing in between the elements when they get printed. This will increase the clarity of the output of the program. Figure 6.13 is a sample output of the above program.

Figure 6.14 shows an output of the same program when we give a value for 'n' which is not acceptable and hence an error flashed. Figure 6.15 shows the memory picture of the array 'a' and the pointer 'pa' used in the above program. The memory picture shall make the things clearer and better.

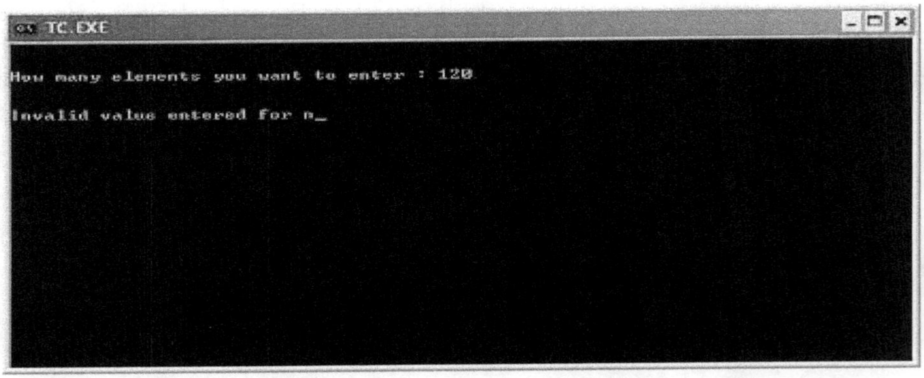

Figure 6.14 Output of the program when 'n' value is invalid

How will you modify the above program in case you are reversing the array? You would find the task easier, if you have clearly understood the way the pointer is accessing the array elements using their memory addresses. The only place where you would be modifying the above program would be the section where you display the array elements. Instead of displaying them in the same manner as you have read, you would reverse-display them. This is achieved by the following loop (that you would have replaced in the appropriate position of the above program).

```
printf("\n\nThe values in reverse order are : ");
for(pa=pa-1;pa>=a;pa--)
printf("%d\t",*pa);
```

We see that the initial value of 'pa' is 'pa-1'. This is because, once we have read the array, 'pa' would be at 1010 and we have to bring it one position less than that it is actually pointing to. We know that we can subtract a constant from the pointer and hence this kind of initialization is allowed. We see the condition is written as pa>=a and the modification part is written as pa– –. As we decrement the value of 'pa' in steps of '1', it keeps pointing to the preceding locations and at one point of time would reach the base address. That is the reason why we have used the expression pa>=a, since we know very well that mentioning the name of the array would give its base address. Figure 6.16 would help you in understanding things in a better way.

In Figure 6.16, we have given the memory map for the above program that is drawn in the interest of the learners of this book and you try to figure out the ideas from the memory map, on your own.

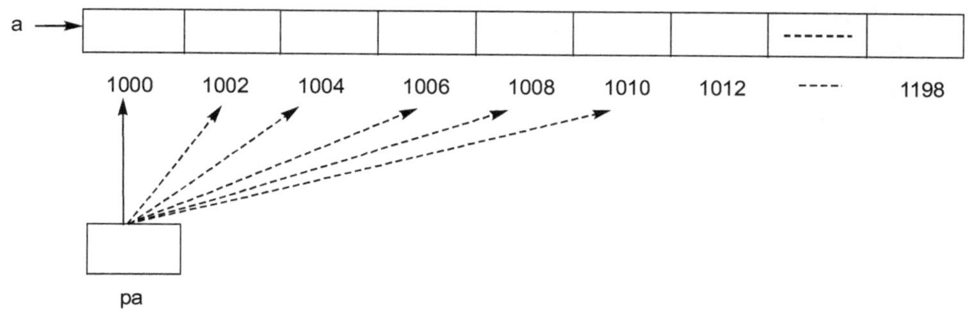

Figure 6.15 Memory picture—program to read 'n' integers

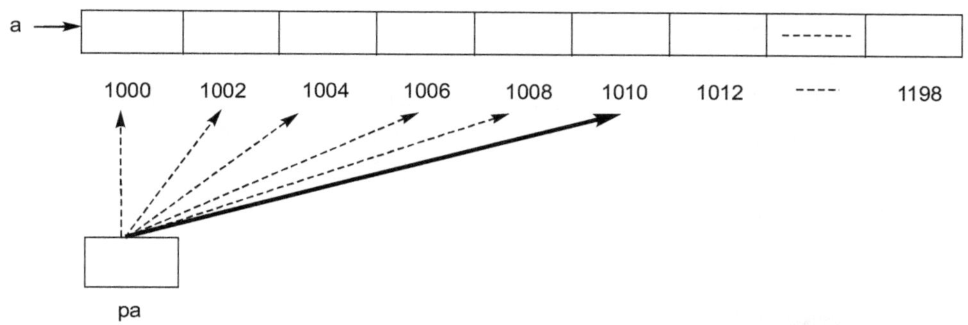

Figure 6.16 Memory picture—program to read and reverse 'n' integers

I am sure that you would have understood how the pointers work with the arrays. I would suggest that you work with all the programs that are given in the self-learning exercise in the Chapter 4, which will make you better understand the pointers and you will ultimately master them.

6.4.2 Pointers and Character Arrays

Pointers work with character arrays in the same way as with a numerical array. Since we knew the way the characters are stored in an array and how the character array is terminated, we will straight away take a program to illustrate the way the pointer is used with the character array. Consider the program where we read and reverse a string or a text. The following is the program code for the same.

```
/* Program to read and reverse a string using pointer */
#include<stdio.h>
main()
{
char a[100], *pa;
clrscr();
printf("\n\nEnter the string : ");
for(pa=a;(*pa=getchar())!= '$';pa++);
*pa='\0';
printf("\n\nThe string you have entered is : %s",a);
printf("\n\nThe reversed string is : ");
for(pa=pa-1;pa>=a;pa--)
putchar(*pa);
getch();
}
```

We can see that the pointer variable 'pa' is used in exactly the same way as it was used earlier with the numerical array. Figure 6.17 shows the output of the above program and Figure 6.18 shows the memory picture of the array and pointer variables that are used in the program. The memory picture will give you a clearer explanation of the program and hence the way the pointer works with the character array.

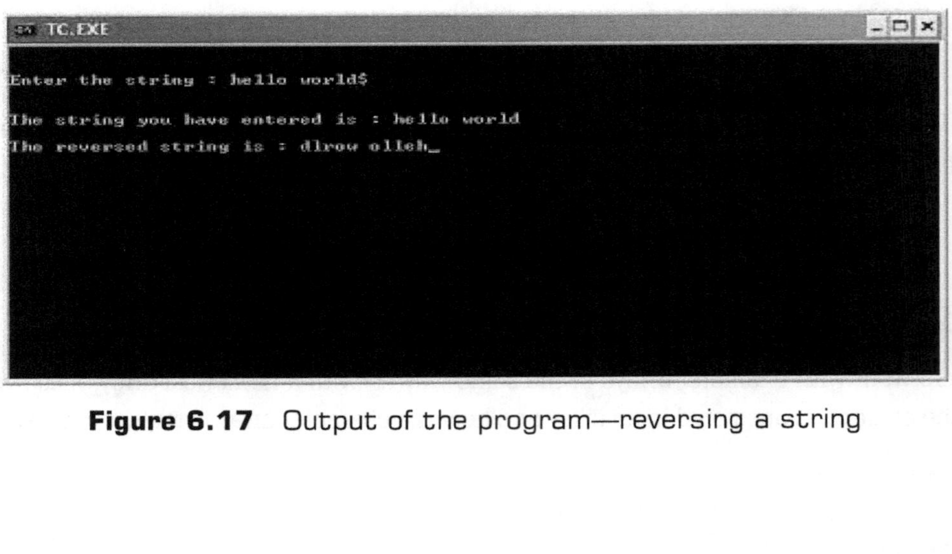

Figure 6.17 Output of the program—reversing a string

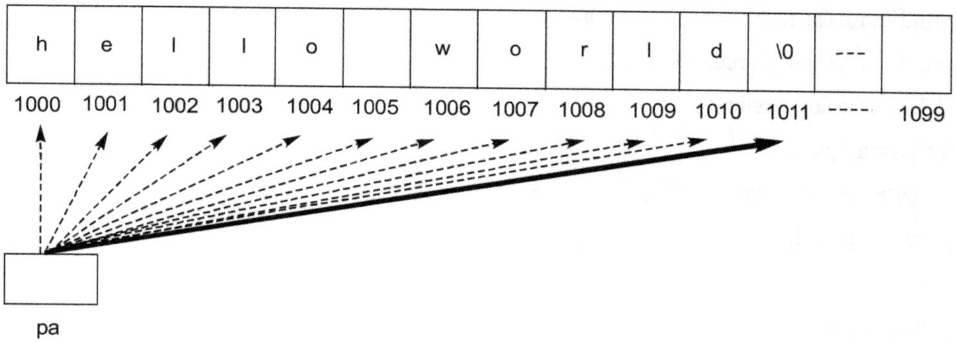

Figure 6.18 Memory picture—reversing a string

We have assumed that the array 'a' has the base address 1000 and hence the end address would be 1099 (since the array is of size 100). We get the input text for the array using *for* loop and once we finish getting all the input characters, we terminate the array using a NULL character, which is essential for any character array. Since in the above program we reverse the string that is read, we reinitialize the pointer with the value pa-1 in the second *for* loop. This is because, once we read the array elements one after another, the pointer would be having the address 1011, when we store the NULL character. When we want to reverse the array, we must store the address 1010 in the pointer and hence we deduct '1' from the present value of 'pa' and hence we use the expression pa-1. I am sure that this explanation would help you in understanding the way the pointers work with character array. See also that we have used the function putchar() (works in exactly the opposite manner to the function getchar()), which will display one character at a time and the advantage of using this is that we need not use a format specifier for the character type values we are displaying (unlike the function printf() which essentially needs format specifier for any type of values used in it).

Let us do another program, in which we will convert the lowercase alphabets of the given array to their equivalent uppercase and display. The following is the program code for the same.

```
/* Program to convert the lowercase alphabets to uppercase using pointer */
#include<stdio.h>
main()
{
char a[100], *pa;
clrscr();
printf("\n\nEnter the string : ");
for(pa=a;(*pa=getchar())!='$';pa++);
*pa='\0';
```

```
printf("\n\nThe string you have entered is : %s",a);
printf("\n\nThe uppercase string is : ");
/* Logic for Conversion */
for(pa=a;*pa;pa++)
if(*pa>=97 && *pa<=122)
putchar(*pa-32);
else
putchar(*pa);
getch();
}
```

Please observe the segment of the program where we convert the lowercase string to uppercase. There we have used a condition, if(*pa>=97 && *pa<=122), which means the input when given as mixed (both uppercase and lowercase), only that part of the input which has lowercase character(s) has to be converted with this logic, otherwise the character has to be displayed as such. Figure 6.19 shows the output of the program.

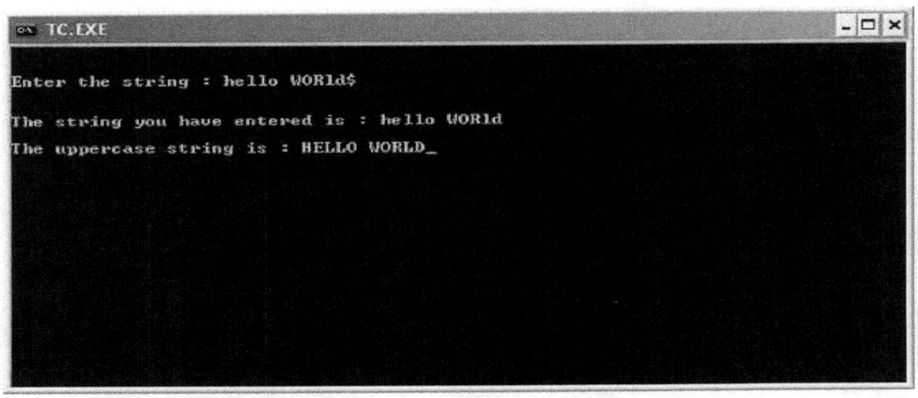

Figure 6.19 Output of the program—lowercase to uppercase

You may also observe the logic used for the conversion process, wherein we subtract a value 32 from the existing alphabet. This is because, the standard difference between the ASCII values of uppercase and lowercase alphabets is 32. Thus, subtracting 32 from an alphabet will yield its corresponding uppercase equivalent. Similarly when we want to convert an uppercase alphabet to the equivalent lowercase, we have to add 32 to the existing alphabet.

I hope the above programs would clearly imply to you, how efficiently we can work with the arrays clubbed with pointers. Try to rewrite all the programs corresponding to the character array that are given as exercise to you in the Chapter 4 using pointers and it will surely make you to use pointers more with arrays more efficiently.

6.5 POINTERS AND STRUCTURES

We have dealt with the concept of structures already in Chapter 5. This section will let you know how to declare a pointer to a structure and access the same. Let us declare a structure called **book** where we store the details of a book such as book name, author name, publisher's name, year of publication, number of pages and price. Of these attributes, the first three are character arrays, fourth and fifth attributes are integers and the last one is a float type. Since a structure allows different types of data to be clubbed together, the above declaration of the **book** structure is acceptable. The following is the complete program which reads details of a book and displays the same using pointer to the declared structure.

```
/* Declaring and using a pointer to a structure */
#include<stdio.h>
/* Inclusion of Floating Point Emulator */
void LinkFloat(void)
     {
     float a=0,*b=&a;
     a=*b;
     }
/* Declaration of the book structure */
struct BOOK
     {
     char bname[100],aname[100],pname[100];
     int nop,yop;
     float price;
     };
main()
{
int i;
/* Declaration of variable and pointers of type of BOOK */
struct BOOK A,*PA;
/* Assigning the addresses of A to PA */
```

```
PA=&A;
clrscr();
/* Getting the details of the BOOK A using the pointer PA */
printf("\nYou are now ready to enter the details of A");
getch();
printf("\n\nEnter the name of the book : ");
for(i=0;(PA->bname[i]=getchar())!='$';i++);
PA->bname[i]='\0';
fflush(stdin);
printf("\n\nEnter the name of the author : ");
for(i=0;(PA->aname[i]=getchar())!='$';i++);
PA->aname[i]='\0';
fflush(stdin);
printf("\n\nEnter the name of the publisher : ");
for(i=0;(PA->pname[i]=getchar())!='$';i++);
PA->pname[i]='\0';
fflush(stdin);
printf("\n\nEnter the number of pages in the book : ");
scanf("%d",&PA->nop);
printf("\n\nEnter the year of publication : ");
scanf("%d",&PA->yop);
printf("\n\nEnter the price of the book : ");
scanf("%f",&PA->price);
/* Displaying the details of the BOOK A */
printf("\n\nBook Name : %s", PA->bname);
printf("\n\nAuthor Name : %s", PA->aname);
printf("\n\nPublisher Name : %s", PA->pname);
printf("\n\nNumber of Pages : %d",PA->nop);
printf("\n\nYear of Publication : %d",PA->yop);
printf("\n\nPrice : %0.2f",PA->price);
getch();
}
```

We see from the program that many new things have been included while the coding is written. The first and most important of them is the inclusion of a segment called **Floating Point Emulator**. This is a simple fragment of coding, which will enable us to use floating point data type in the structure flexibly. When we do not use this, we may get an error saying **floating point formats not linked**, especially when we use the floating point data types in a structure. The inclusion of the above coding would enable the compiler to link the floating point formats and values properly.

The next important aspect of the program is the usage of a pointer to the structure declared, unlike the normal usage of the structure variable to access its members. Once we declare a pointer to the structure and assign the base address of the structure to it, further accession of the members of the structure could be easily done using it.

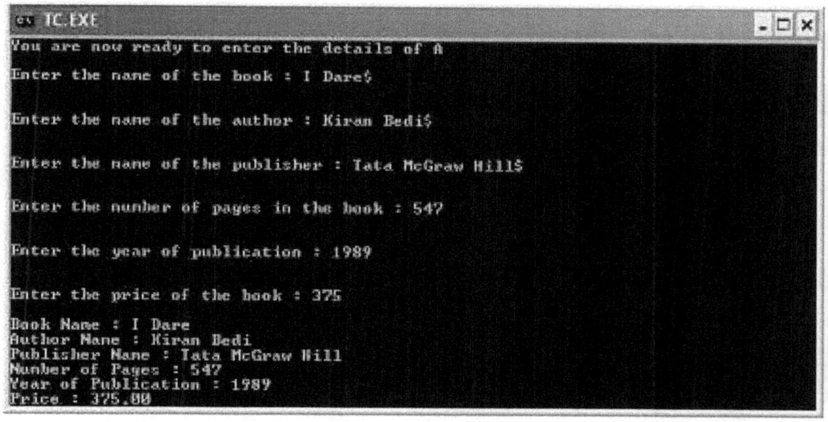

Figure 6.20 Using pointer to a structure

Closely observe the program. You will get the next significant aspect. We have used a new operator (–>) called as arrow operator. This is again an operator in C and could be used for the unions too. Figure 6.20 shows the output of the above program. The memory picture of the structure and the members of the structure are shown in Figure 6.21.

See from the memory picture that the base address of the structure BOOK is the same as that of the first member of the structure. In our case the first member of the structure is the character array **bname** and the base address of **bname** would be the base address of the entire structure.

From the discussion and the memory picture with a sample program, the concept of using pointers with structures would have been clear to you. You may try pointers with array of structures and also unions, which would be very much easier, since the concept of array of structures and unions are already known to you. Read this section of the chapter, which will facilitate you in understanding the concept of linked lists.

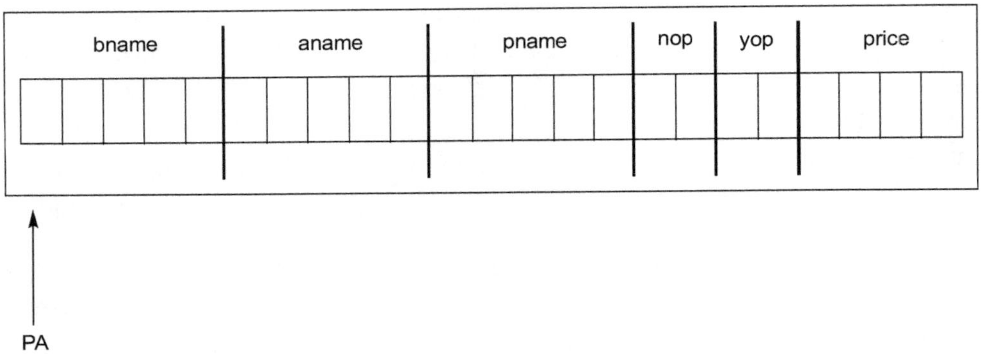

Figure 6.21 Memory picture of the structure BOOK

6.6 LINKED LISTS—INTRODUCTION

The term **list** refers to linear collections of data items. Data processing frequently involves storing and processing data organized into lists. One way to store data elements is by means of arrays. Another way of storing a list in memory is to have each element in the list to contain a field called a link or pointer, which contains the address of the next element in the list.

A linked list or one-way list is a linear collection of data elements called nodes, where the linear order is given by means of pointers, which may vary, in number. A simple way to represent alinear list is to be expanding each node to contain a link to the next node. This representation is called one-way chain or singly linked linear list as shown in the Figure 6.22.

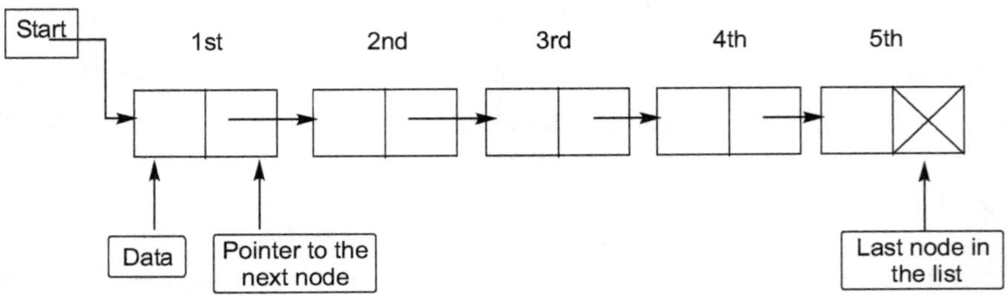

Figure 6.22 Representation of a singly linked list

Each node is divided into two parts. The first part represents the information of the element and the second part contains the address of the next node. The last node does not have a successor node and no actual address is stored in the pointer field and it is left empty. Usually a NULL value is stored as the address in the last node. The operations that are possible with the linked list are insertion, deletion, searching, traversing and modification.

For writing a program for linked lists and the operations that are possible with them, we use self-referential structures, which we had seen at the introductory level in the Chapter 5. Before we proceed to write a program to explain the operations on a singly linked list, let us have an idea about advantages and disadvantages of linked lists.

Advantages

- Overflow can never occur unless the memory is actually full.
- Insertions and deletions are easier than for contiguous lists.
- With large records, moving pointers is easier and faster than moving the items themselves.
- These lists are dynamic.

Disadvantages

- The pointers used in the lists require extra space.
- Linked lists do not allow random access.
- Time must be spent traversing and changing the pointers.
- Programming is typically trickier with pointers.

Linked lists often confuse the beginners even though they are good in using the pointers and structures. Those who have gained expertise in both are exceptional to this.

6.7 IMPLEMENTATION OF SINGLY LINKED LIST

The following is the program for the various operations on a singly linked list which uses pointers and structures.

```
/* Operations on Singly Linked List */
#include<stdio.h>
#include<alloc.h>
/* Structure Declaration */
struct node
        {
        int data;
        struct node *next; /* Self-referential structure */
        };
/* Declaration of pointers to the structure node */
```

```
struct node *first=0,*last=0;
main()
{
int ch;
clrscr();
do
{
printf("\n\n1.Append");
printf("\n\n2.Deletion");
printf("\n\n3.Display");
printf("\n\n4.Search");
printf("\n\n5.Insertion");
printf("\n\n6.Quit");
printf("\n\nEnter your choice : ");
scanf("%d",&ch);
switch(ch)
        {
        case 1:
                append();
                break;
        case 2:
                delete();
                break;
        case 3:
                display();
                break;
        case 4:
                search();
                break;
        case 5:
                insert();
                break;
        case 6:
```

```
                printf("\n\nGood Bye");
                getch();
                exit();
        default:
                printf("\n\nEnter a valid option");
        }
}while(ch!=6);
}
/* Insert Operation */
insert()
{
struct node *temp,*pp,*cp;
int inse,before;
temp=(struct node*)malloc(sizeof(struct node));
printf("\n\nEnter the element to insert : ");
scanf("%d",&temp->data);
printf("\n\nEnter before which element you want to insert : ");
scanf("%d",&before);
if(first->data==before)
        {
        temp->next=first;
        first=temp;
        }
        else
        {
        for(pp=cp=first;cp!=NULL;pp=cp,cp=cp->next)
                {
                if(cp->data==before)
                        {
                        pp->next=temp;
                        temp->next=cp;
                        }
```

```
                }
        }
}
/* Append Operation */
append()
        {
        struct node *temp;
        temp=(struct node*)malloc(sizeof(struct node));
        printf("\n\nEnter data to insert : ");
        scanf("%d",&temp->data);
        if(first==0)
                first=last=temp;
        else
                last->next=temp;
        last=temp;
        last->next=0;
        }
/* Delete Operation */
delete()
        {
        struct node *pp,*cp;
        int dee;
        printf("\n\nEnter the element to delete : ");
        scanf("%d",&dee);
        if(first->data==dee)
                first=first->next;
        for(pp=cp=first;cp!=NULL;pp=cp,cp=cp->next)
        {
                if(cp->data==dee)
                pp->next=cp->next;
        }
        if(cp==last)
```

```
        {
        last=pp;
        last->next=0;
        }
        }
/* Display Operation */
display()
        {
        struct node *temp;
        printf("\n\nThe singly linked list is :  ");
        if(first==0)
        printf("\n\nThe list is empty");
        for(temp=first;temp!=NULL;temp=temp->next)
        printf("%d->",temp->data);
        }
/* Search Operation */
search()
        {
        struct node *temp;
        int sear;
        printf("\n\nEnter the element to search :  ");
        scanf("%d",&sear);
        for(temp=first;temp!=NULL;temp=temp->next)
                {
                if(temp->data==sear)
                        {
                        printf("\n\nElement Found....");
                        break;
                        }
                }
                if(temp==NULL)
                printf("\n\nElement Not Found....");
        }
```

If you closely observe the program, we have not only used pointers and structures, but also user-defined functions, which is an additional and added feature of C language. Functions are discussed in the chapter 7 and once you read the chapter 7 and understand how user-defined functions are declared, defined and used in C programming, it would be definitely easy to understand the execution of the above program. The concepts of structures and using a pointer with it have been already discussed. The following is the sample execution of the above program coding and it clearly explains the variety of operations that are possible with a singly linked list.

Sample execution

1. Append
2. Deletion
3. Display
4. Search
5. Insertion
6. Quit

   Enter your choice : 1

   Enter data to insert :10

1. Append
2. Deletion
3. Display
4. Search
5. Insertion
6. Quit

   Enter your choice : 1

   Enter data to insert : 20

1. Append
2. Deletion
3. Display
4. Search
5. Insertion
6. Quit

Enter your choice : 1

Enter data to insert : 30

1. Append
2. Deletion
3. Display
4. Search
5. Insertion
6. Quit

Enter your choice : 1

Enter data to insert : 40

1. Append
2. Deletion
3. Display
4. Search
5. Insertion
6. Quit

Enter your choice : 5

Enter the element to insert : 33

Enter before which element you want to insert : 40

1. Append
2. Deletion
3. Display
4. Search
5. Insertion
6. Quit

Enter your choice : 3

The singly linked list is : 10->20->30->33->40->

1. Append
2. Deletion

3. Display
4. Search
5. Insertion
6. Quit

Enter your choice : 4

Enter the element to search : 23

Element Not Found....

1. Append
2. Deletion
3. Display
4. Search
5. Insertion
6. Quit

Enter your choice : 4

Enter the element to search : 33

Element Found....

1. Append
2. Deletion
3. Display
4. Search
5. Insertion
6. Quit

Enter your choice : 5

Enter the element to insert : 29

Enter before which element you want to insert : 30

1. Append
2. Deletion
3. Display
4. Search

5. Insertion
6. Quit

Enter your choice : 3

The singly linked list is : 10->20->29->30->33->40->

1. Append
2. Deletion
3. Display
4. Search
5. Insertion
6. Quit

Enter your choice : 6

Good Bye

Figure 6.23 shows a part of the output screen where we execute one of the existing options related to a singly linked list. The core idea behind using a linked list is that the data are stored somewhere in the memory, but still bound together. This fact makes a list much different from arrays. The elements of the array are stored in contiguous memory locations and those of the linked list are dispersed, still connected through the links.

This section of the chapter discusses only singly linked lists and their implementation using a program. The learners are advised to explore the variety of linked lists like doubly linked lists, circular lists, etc. and their implementation procedures. More important, try to attempt all the questions and programs at the end of this chapter under the head self-learning exercises. That will surely make you the best in using pointers with the structures.

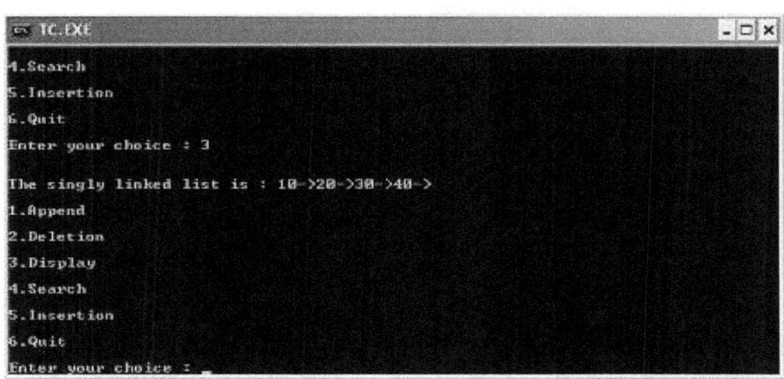

Figure 6.23 O/p of singly linked list—one of the six options in action

SUMMARY

We started off this chapter with an introduction to pointers. Why a programming language shall use pointers was discussed and it was also highlighted that C language efficiently uses pointers. We then followed with the explanation of the operators used in pointers and how these operators are used in storing and retrieving the values and data.

We then did a few basic programs using pointers and used the memory pictures and screenshots of the outputs to understand the way the pointers function in C. We then learnt a very important aspect in pointers, namely, the Pointer Arithmetic and in that, we discussed how a pointer would increment or decrement and how this concept is useful in arrays, where pointers are used much.

The discussion of pointer arithmetic followed that of the usage of pointers with numerical and character arrays. We did a few programs on these two topics and as usual we had the screenshots of the output and memory pictures for more clarity in understanding the concept.

We continued the chapter with a very important discussion on pointers and structures, which was the ultimate concept that would be used in implementing the linked lists. We concluded the chapter with a brief introduction to linked lists and the implementation of singly linked lists and the variety of operations on it.

CRITICAL THINKING QUESTIONS

What will be the output of the following programs?

1.
```
main()
{
int a, *b, **c, ***d, ****e;
a=10;
b=&a;
c=&b;
d=&c;
e=&d;
```

```
printf("a=%d,b=%u,c=%u,d=%u,e=%u",a,b,c,d,e);
printf("%d,%d,%d",a, a+ *b, **c + ***d + ****e);
}
```

2.

```
main()
{
char c, *pc;
int a;
long b;
float d;
c='X';
a=20;
b=67888;
d=3.145;
pc=&c;
printf("\n\nc=%c, pc=%u",*pc,pc);
pc=&a;
printf(("\n\na=%d,pc=%u",*pc,pc);
cc=&b;
printf(("\n\na=%ld,pc=%u",*pc,pc);
cc=&d;
printf(("\n\na=%f,pc=%u",*pc,pc); }
```

3.

```
main()
{
char *c;
int *i;
long *l;
float *f;
double *d;
printf("\n\nc=%u,c+1=%u",c,c+1);
```

```
printf("\n\ni=%u,i+1=%u",i,i+1);
printf("\n\nl=%u,l+1=%u",l,l+1);
printf("\n\nf=%u,f+1=%u",f,f+1);
printf("\n\nd=%u,d+1=%u",d,d+1);
}
```

4.
```
main()
{
int i=10,j=20,diff;
diff=&j-&i;
printf("\n\nAddress of i=%u\tAddress of j=%u",&i,&j);
printf("\n\nDifference between addresses of i and j is : %d",diff);
}
```

5.
```
main()
{
int *i,*j;
j=i*2;
printf("\n\nj=%u",j);
}
```

6.
```
main()
{
int i=10;
printf("\n\nValue of i=%d\tAddress of i=%u",i,&i);
&i=7200;
printf("\n\nNew value of i=%d\tNew address of i=%u",i,&i);
}
```

7.
```
main()
{
int a[]={10,20,30,40,50};
int j;
```

```
for(j=0;j<5;j++)
{
printf("%d",*a);
a++;
}
}
```

8.
```
main()
{
float a[]={13.24,1.5,1.6,5.4,3.5};
float *j,*k;
j=a;
k=a+4;
j*=2;
k/=2;
printf("%f\t%f",*j,*k);
}
```

9.
```
main()
{
int a[25];
a[0]=100;
a[24]=200;
printf("\n%d\t%d",*a,*(a+24),*(a+0));
}
```

10.
```
main()
{
int a[]={10,20,30,40,50};
int i,*k;
k=&a[4]-4;
for(i=0;i<=4;i++)
```

```
{
printf("%d\t",*k);
k++;
}
}
```

11. main()

```
{
int a[]={2,4,6,8,10};
int i;
for(i=0;i<=4;i++)
{
*(a+i)=a[i]+i[a];
printf("%d\t",*(i+a));
}
}
```

12. main()

```
{
int a[]={2,4,6,8,10};
int *ptr;
for(ptr=&a[0];ptr<=&a[4];ptr++)
printf("%d\t",*ptr);
}
```

13. main()

```
{
int a[]={2,4,6,8,10};
int *ptr,i;
for(ptr=a+4,i=0;i<=4;i++)
printf("%d\t",ptr[-i]);
}
```

14. main()

```
{
int a[]={2,4,6,8,10};
int *pa[]={a,a+2,a+1,a+4,a+3};
int **ptr;
ptr=pa;
**++ptr;
printf("\n\n%d\t%d\t%d",**ptr,ptr-pa,*ptr-a);
}
```

15. main()

```
{
char s[]="C Language";
int i;
for(i=0;s[i];i++)
printf("\n%c\t%c\t%c\t%c",s[i],*(s+i),i[s],*(i+s));
}
```

16. main()

```
{
char s[]="C Language";
char *ps;
ps=&s[6]-6;
while(*ps)
printf("%c",*ps++);
}
```

17. main()

```
{
char s[]="Malayalam";
char *ps;
ps=s+8;
while(ps>=s)
```

```
{
printf("%c",*ps);
ps--;
}
}
```

18.
```
main()
{
char a[]="Able was i ere i saw elba";
char *pa,*pb,*pc;
clrscr();
pb=a;
pc=a+strlen(a)-1;
pa=pc;
while(pb!=pc)
{
printf("%c",*pb);
pb++;
printf("%c",*pa);
pa--;
}
}
```

19.
```
main()
{
int a[12];
printf("%d",sizeof(a));
}
```

20.
```
main()
{
char a[]="C Language";
```

```
int i;
printf("%c\n",*a);
a++;
printf("%c",*a);
}
```

SELF-LEARNING EXERCISES

1. Rewrite all the basic programs that you would have done in the earlier chapters with pointers (this includes the programs that you would have done using loops, conditions, numerical and character arrays, structures, unions, etc.). Do all the programs with the memory pictures.
2. Figure out the memory map for the singly linked list program.
3. Write a C coding to implement the doubly linked list. Do go through the basics of a doubly linked lists before you start writing a program.
4. Give suitable points appreciating the efficiency of linked lists in storage and retrieval of data. Explore how this would be useful in Bioinformatics.

FUNCTIONS

7.1 WHY WE NEED FUNCTIONS?

The efficiency of any language is measured from the way it does the task by splitting it into different modules. Divide and Rule is the fundamental fact behind any successful programming language. C has the power of dividing the given task into different modules and accomplishing the same. The gravity of the programmers using the C language could be known from two of its most striking features. The first one is of course, the pointers and the second one is the functions.

In C, a function is a self-contained program segment that carries out some specific, well-defined task. We have invariably used many functions so far such as printf(), scanf(), getchar(), clrscr(), getch(), etc. Above all we have been using the function main() in all our programs and as said earlier, it is the essential one for any program to compile and execute; in simple words, it is the heart throb of programs written in C.

C language is a compiled language and it contains many functions—all of them are built-in, which means they come along with the language and the language designers have given to the users and the programmers of the language. They are categorized under different heads according to the context in which they are used. We have discussed earlier about header files and their usage in programming. Many of the functions are categorized as header files and there are as many as 30 header files in the language. All these headers were made available for the users when the language was compiled. But as the language was used across the programming community and started growing, these built-in functions were not sufficient and hence the language was modified in such a way that one can define his/her own function, what we call as user-defined function. This aspect has made C language most popular among the programming group.

Functions are thus considered an essential part of the programs. The user is at his/her liberty to create as many functions as he/she wants in a program. One more beauty is that a function that is created is not just restricted to the program where it is declared and defined, but could be used across different programs, just like the headers.

This aspect gave C one more face, namely, the idea of code reusability. The code was defined just once and used any number of times at any point of the particular program or across different programs. This feature was inherited later by C++. Thus C++ became popular not just because of its object oriented nature but also because of the procedure oriented features inherited from C. This idea of code reusability gave much power to the language and the users had a feeling that they were given freedom in using the language syntax.

7.2 TYPES OF FUNCTIONS

C provides us with two types of functions, as mentioned in the earlier section, namely, built-in/library functions and user-defined functions. The header files contain the details of the library functions. We, in this chapter, are going to discuss the way the user-defined functions are declared, defined and used. As said earlier, the key advantage of user-defined functions is the code reusability.

7.3 FUNCTION HEADER AND FUNCTION PROTOTYPING

Any user-defined function must be properly declared and defined, before used, just like the variables. We call the declaration of functions as function prototyping, which is the essential part of any C program (even though some versions of C compilers do not expect a function to be prototyped). Prototyping is not an issue in the case of built-in functions, since their declarations are available in the standard header files and we would anyway include the corresponding headers, using the #include macro once we want to use a particular built-in function.

The C function has two parts, namely, the function header and body of the function. The function header typically has the following style.

<return type> <function name>([<type-1> <argument[s]>...<type-n> <argument[s]>])

The above style of using the function is the classical notation and initially when Dennis Ritchie and Brian Kernighan defined functions, the header was quite different and the following was their style and called as K&R notation and most of the modern compilers still accept this style too.

```
<return type> <function name>([argument[s]])
[<type-1> <argument[s]>...<type-n> <argument[s]>];
```

From both the styles, we can conclude that the usage of arguments inside the functions is purely optional and if they are used, we have to clearly specify their types. Also, we would be following the classical notation of declaring the function headers in our discussion.

A function basically accepts zero, one or more inputs or values, processes them and returns (or does not return anything at all) the result. Based on this aspect we can say a function:

- Accepts one or more inputs, but does not return anything back
- Accepts one or more inputs, processes them and returns one or more results (remember an ordinary function cannot return more than one result, where a function that uses a pointer could do so)
- Does not accept anything, but may return some result
- Neither accepts nor returns anything

When we want a user-defined function to work, we would call them from some other function of the same program or a different program. In either case, the user-defined function is said to be **called** by the other function. Thus we can introduce two terms here, namely **caller** and **callee** (similar to an employer and an employee). A function which calls another function is called the caller and the function that is being called is the callee. Remember the fact that a function can be called any number of times, depending upon its requirement. Also remember that, in C language we can just call one function from another, but we cannot define the logic of one function in another (this is why C is technically not a block structured language). One can make use of any function in C language only by calling it. The compiler will recognize the call and compile the programs to success, provided the corresponding header is included in the case of built-in functions and the function is properly defined in the case of user-defined functions.

Now, I will put a question to you, for which the answer will be simple and I would give the answer in the later part of this section (Better do not read the later part of this section immediately after getting the question. I would advise you to give a serious thinking before you look for the answer). I had just told you that any function could be used in C only by calling it. Thus we make use of the functions printf(), scanf() and all such functions by calling them from within the function main(). Thus the function main() is the **caller** for these functions and they are the **callee** functions. But who is the caller for the function main()? Think seriously the answer for this question. The reason why I had raised this question is that, I had told you any function in C could be used only by calling it and the function main() is no exception for this rule.

The answer for the above question is, the **Operating System**. Yes, the operating system under which the C compiler works is the boss for the function main() and hence it will be the caller and the function main() is the callee. If we use the C complier under the DOS platform, then DOS is the caller for the function main(). The caller would change according to the platform that we use for programming with this language. I had used DOS as the platform for all the programs in the book and hence the caller for all of them is invariably the DOS.

The functions in C could be categorized also according to the way they are called. You can call them with the values as the arguments, you call this as **call by value** and you can call them with the addresses as the arguments, where you call this aspect as **call by reference**. The latter method is more efficient and advantageous in many ways when compared with the former. The significant advantage is that when we use call by reference, we can implicitly return more than one value from the function to the calling function, whereas, when we use call by value method, we can return either nothing from the called function or a maximum of one value from it. There are many other technical advantages of using call by reference method, which you would understand as we do programming later in this chapter.

I am sure you are comfortable with the function header and prototyping concepts discussed so far. This would be the base for you to program using functions, from basic ones to the complicated ones.

7.4 WRITING SIMPLE FUNCTIONS

Let us write simple functions in C and validate them. To begin with, let us write a function that adds two integers and returns the result to the calling function. The coding for the same is as follows. We validate the function using a C program.

```
/* Program to add two integers using functions */
#include<stdio.h>
main()
{
/* Declaration of variables */
int a,b,c;
/* Function prototyping */
int add(int,int);
clrscr();
/* Getting the values of a and b */
printf("\n\nEnter the values of a and b : ");
```

```
scanf("%d%d",&a,&b);
/* Calling the function add() */
c=add(a,b);
/* Printing the result of addition */
printf("\n\nThe sun of %d and %d is %d",a,b,c);
getch();
}
/* Definition of the function add() */
int add(int x, int y)
{
int z;
/* Adding the values that are passed from the calling function */
z=x+y;
/* Returning the result of addition to the calling function */
return z;
}
```

In the above program, we have used a very simple user-defined function, by name add() and it is used for adding two integers and returns the result. Here the calling function is main() and the called function is add(). In other words, the function main() is the caller and the function add() is the callee. We see that the function add() is properly prototyped, just like the variables used in the program and we see that the prototyping follows the declaration of variables. The prototyping could also be done outside the function main() and in doing so, we would use it globally, again, just like a variable. The prototyping has the details like the return type, function name and the type of values the function accepts and this is obviously essential for any function that is prototyped. When a function does not accept or return anything, we use the type **void** to imply that. A typical function could be prototyped like the one below.

```
void add(int,int);
```

The above prototype means the function does not return anything, but accepts two integers. Another typical prototype could be like the one below.

```
void display(void);
```

Here, the function neither accepts nor returns anything. The keyword **void** could be optionally used within the parentheses following the function name.

The prototyping of functions itself is optional in C language, as its importance was not realized initially. Later when the concept of functions was inherited by C++, prototyping was made essential in all the C++ programs, thereby removing the anomalies created by the C functions.

Following the prototyping, we get the values of a and b, as usual. Following this, comes the function call c=add(a, b). Earlier to the call, we have let the compiler know that there would be a function defined later in the program with the name add() and it would accept two integer type values and return one integer type value. Now, we call the function add() by passing the two values (a and b) that we have obtained using the function scanf(). This is exactly the call by value method of using the functions. These variables that are in the calling function (main()) are called as the actual arguments. When these values are passed on to the function add(), there must be two variables exactly similar to a and b in that function, which must be compatible to receive and store these values. See that we are using two more variables x and y in the function add() and these variables are called as formal ones or dummy arguments. They are just copies of a and b and their lifetime will be as long as the control is executing in the function add() and once the control goes back to the function main() or the calling function, these formal arguments vanish and that is why we call them as dummy arguments.

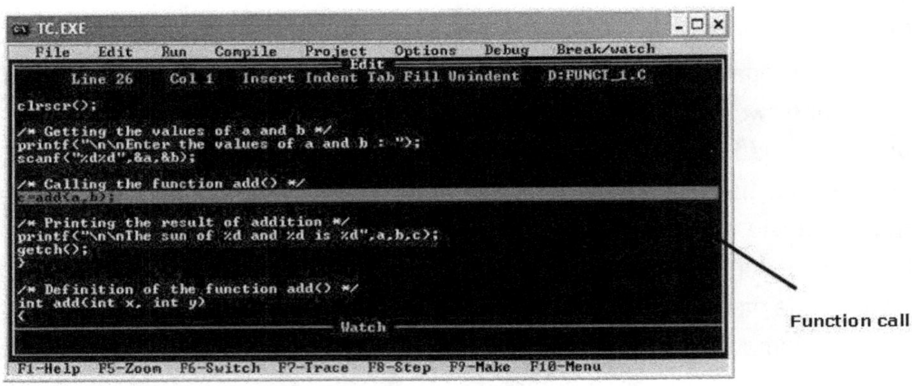

Figure 7.1 Function add() getting called during the trace of the program

When the function add() is called, the execution in the function main() is temporarily suspended and the control gets into the function add() and till it encounters the *return* statement in it, it remains there. The *return* statement explicitly takes the control back to the calling function and the execution in the calling function will be resumed from the place where the control last left. For that matter, in no C function, the control will remain in the called function. It must always come back to the calling function and terminate the program execution from there. I would suggest you check the above program code using the keystroke F7 (tracing the program execution). Figures 7.1 and 7.2 show the trace of the program execution just when the control goes from the calling function (main()) to the called function (add()). Figure 7.3 shows the trace of the program execution when the control resumes action in the calling function after coming back from the called function.

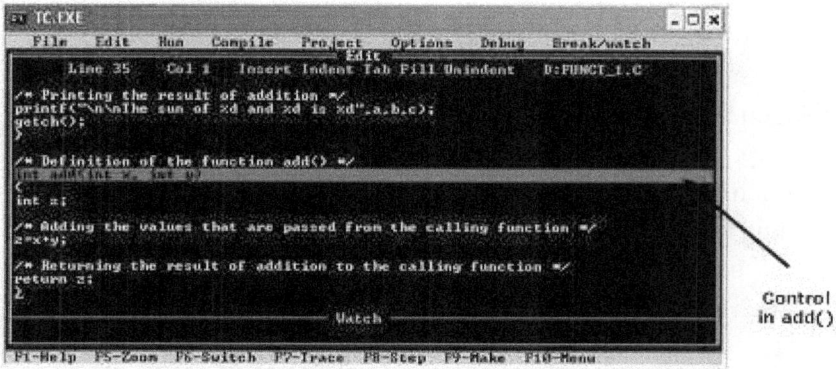

Figure 7.2 Control getting into the function add() from main()

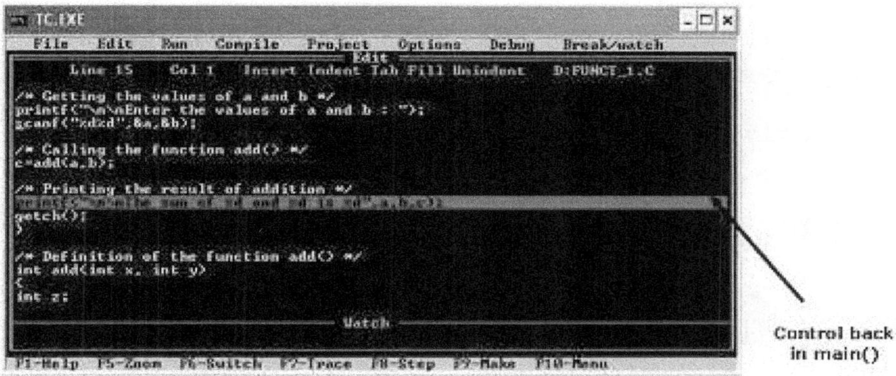

Figure 7.3 Control back in the function main()

I have given the memory picture of the above program in the Figure 7.4 for a better understanding of how the control flows throughout the program.

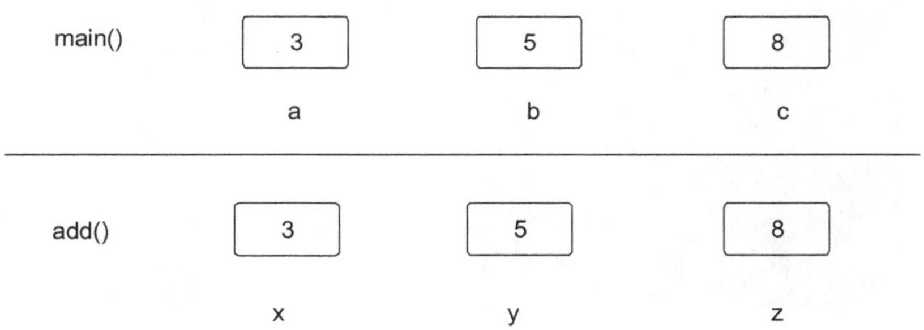

Figure 7.4 Memory picture—adding two integers

From Figure 7.4, we can observe that the values of a and b in the function main() are copied on to the variables x and y respectively in the function add(). The variables a and b are the actual parameters or variables and the variables x and y are formal or dummy parameters. The values of x and y are added and the result that is stored in 'z' is returned to the calling function (main()) using the *return* statement and it is collected in the variable 'c' in main() and then displayed. We see that there is no relation between the variables a and b and x and y, except that x and y are dummies for a and b respectively. There will be memory allocated for the dummy variables when the control is inside the function add() and their memory will be automatically released once the control gets back to the calling function. Another important fact is that the function add() will also be allocated with a memory space, since it is just like another variable. The memory allocated for the function will remain as such as long as the program is in the execution stage. Also, know that it is absolutely not necessary to name the variables in the function add() as x and y and we could have named them also as a and b, since there is no relation between the variables declared in the calling function and called function and hence there would not be any name collision.

In the function add() we could have avoided the declaration of the variable 'z' and we could have written the same coding as follows.

```
int add(int x, int y)
{
return (x+y);
}
```

This is very simple compared to the earlier version of the function add(). Figure 7.5 shows a sample execution of the above program.

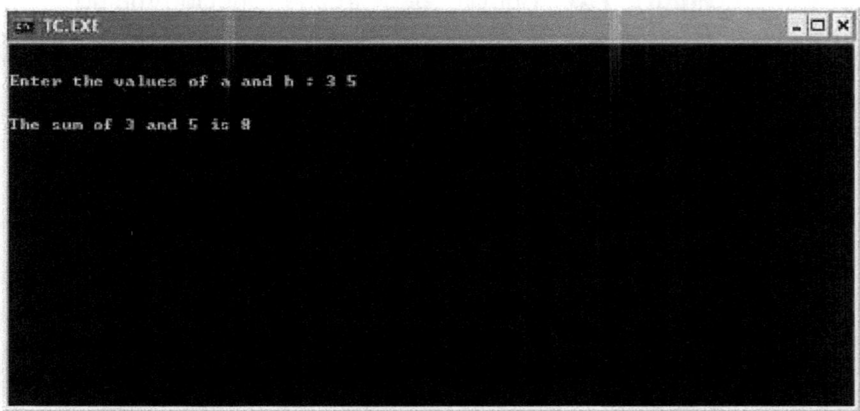

Figure 7.5 Sample run—adding two integers

Assume that we are doing a similar program to add two float values. The following coding for the same.

```
/* Program to add two floats using functions */
#include<stdio.h>
main()
{
/* Declaration of variables */
flaot a,b,c;
/* Function prototyping */
float add(float,float);
clrscr();
/* Getting the values of a and b */
printf("\n\nEnter the values of a and b : ");
scanf("%f%f",&a,&b);
/* Calling the function add() */
c=add(a,b);
/* Printing the result of addition */
printf("\n\nThe sum of %f and %f is %f",a,b,c);
getch();
}
/* Definition of the function add() */
float add(float x, float y)
{
return (x+y);
}
```

Suppose that we do not use the prototyping for the function add() and we do not specify the return type of the function add(). Guess what would happen and what would be the result of adding two float values. The compiler will compile the program to success, but the result will be the one that is not anticipated at all by the programmers and some compilers truncate the result to an integer and return, thereby making the functions default return type as integer. The concluding fact is that all the functions must be properly prototyped to get correct results.

Now that we have a basic understanding of how to use functions, let us proceed in understanding how to use conditions and loops in the functions and how to return the values once the process is done in the function. Consider that we are writing a program to check the odd or even nature of a number. The following is the program code. Observe the program closely to get to know the new ideas that we have used in it.

```
/* Program to check the odd or even nature of a number */
#include<stdio.h>
main()
{
int a,flag;
int check(int);
clrscr();
printf("\n\nEnter a number to check : ");
scanf("%d",&a);
/* Function check() is called */
flag=check(a);
if(flag)
printf("\n\nThe number %d is even",a);
else
printf("\n\nThe number %d is odd",a);
getch();
}
/* Definition of function check() */
int check(int x)
{
return(!(x%2)?1:0);
}
```

We could observe that two variables 'a' and 'flag' are used in the function main(). We get the value of 'a' using the function scanf() and pass the same to the function check() and there we process the value of 'a' that is copied in 'x' and return the result, which is a Boolean true or false.

Thus we have used the variable 'flag' in the function main() as a Boolean variable (since there is no Boolean data type in C, we have declared this as an integer and it is left to the programmer to make sure that this variable stores only a 1 or 0).

In the function check(), we are using the ternary expression inside the *return* statement and this is perfectly allowed. You may also observe a very important aspect both in the function main() and check(). In the function main() we have used a conditional expression if (flag). What is the meaning of saying like this? This has an implicit meaning and called technically as implicit checking of the conditions. When we write the code as if(<variable>), it means that if(<variable> !=0). Instead of using the inequality operator and a zero explicitly, C allows the users to make their presence understood through the implicit checking as given above. We have another kind of implicit checking, which is given as if(!<variable>). When we write a code like this, it means if(<variable>==0). This again reduces the usage of equality operator and the zero explicitly. Note that implicit checking in conditions is possible only when we find the equality or inequality of a variable with zero and no other values.

In the function check(), we have used the second form of implicit checking and return either a '1' or a '0' based on the input number. Figure 7.6 shows the sample run of the above program.

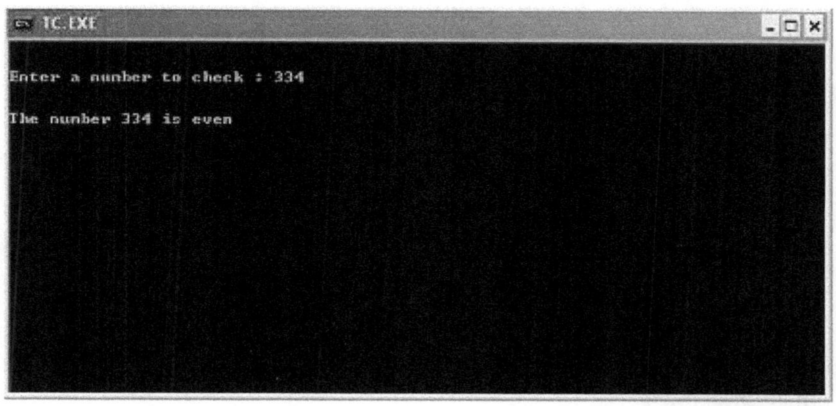

Figure 7.6 Sample run—odd or even nature of a number

Figure 7.7 shows the memory picture for the above program.

Let us see one more program using loop and at the end of this section, you must be in a position to convert all the programs that are found in chapter 2 using the functions, so that you will be ready to work with arrays and functions in the forthcoming sections. Consider that we are finding whether a given number is prime or composite using *for* loop. The following would be a program coding for the same.

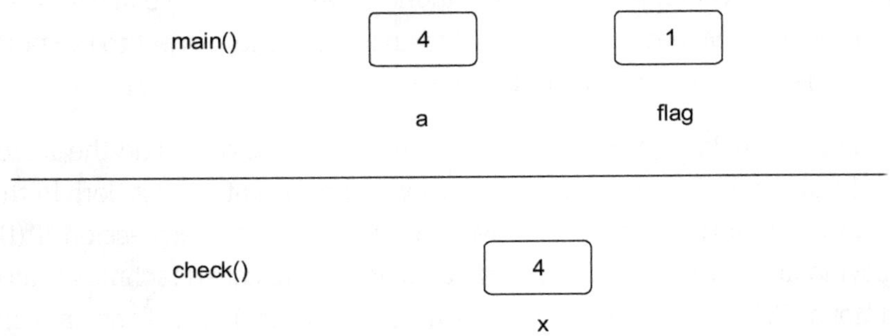

Figure 7.7 Memory picture—odd or even nature of a number

```
/* Program to find whether the number is prime or composite */
#include<stdio.h>
main()
{
int n,flag;
int prime_check(int);
clrscr();
printf("\n\nEnter the number to check : ");
scanf("%d",&n);
if(n==1)
{
printf("\n\n%d is neither prime nor composite",n);
getch();
return;
}
flag=prime_check(n);
if(flag)
printf("\n\n%d is a prime number",n);
else
printf("\n\n%d is a composite number",n);
getch();
}
int prime_check(int x)
```

```
{
int flag=1,i;
for(i=2;i<=x/2;i++)
if(x%i==0)
{
flag=0;
break;
}
return flag;
}
```

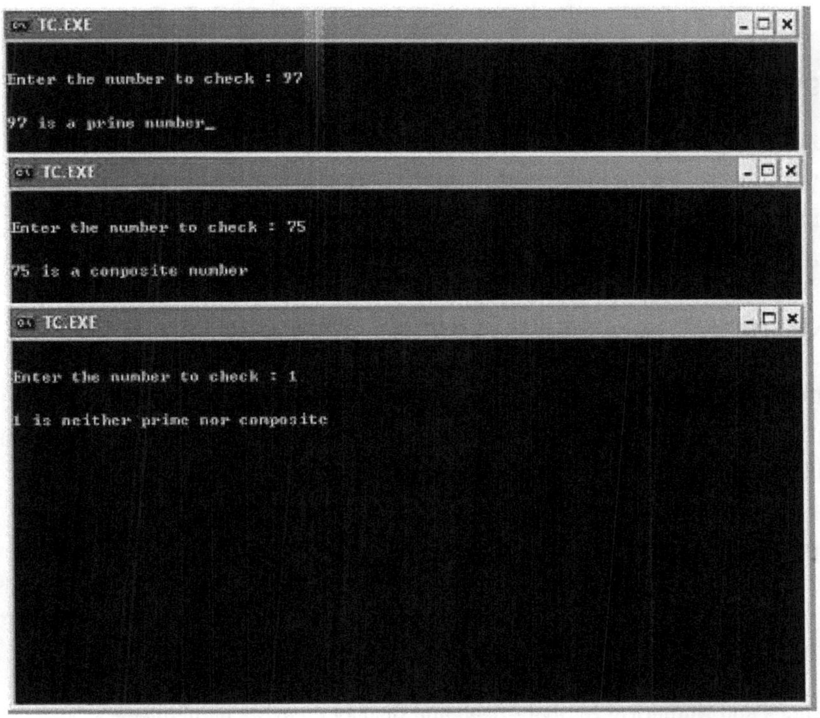

Figure 7.8 Sample run—prime or composite nature of a number

We get the value to check in the variable 'n' and immediately check it for unity. If it is '1' then we flash the corresponding message. If not, we call the function prime_check() and pass the value of 'n' to the dummy variable 'x' and allow it inside a loop and accordingly we set the value of the 'flag' variable (remember that we had used 'flag' variable in our earlier program also and it is used as a Boolean variable and will store a '1' or '0'). Finally we return the value of 'flag' and based on it, we print the result as to whether the given number is prime or composite. Figure 7.8 is a sample run of the above program (it shows the three different executions—prime, composite and neither prime not composite).

Figure 7.9 shows the memory picture for the above program for a prime number.

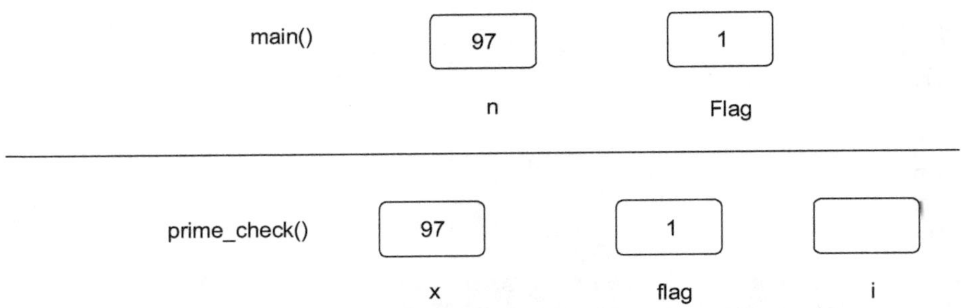

Figure 7.9 Memory picture—prime or composite nature of a number

I am sure that you would definitely be able to work out all the programs at the end of Chapter 2 without any difficulty. It will just mould you up in declaring the function prototypes and define the functions. Later when we discuss the functions and arrays and functions, arrays and pointers, you will find that a child's play. Let us proceed with the discussion on functions and numerical arrays.

7.5 FUNCTIONS AND NUMERICAL ARRAYS

We can write functions by passing not just variables as we have seen in the earlier section. We can also pass arrays also as arguments and in this section we will discuss how functions work with single- and two-dimensional numerical arrays.

7.5.1 Functions and Single-Dimensional Numerical Arrays

Let us consider the program in which we read and display an array of 'n' integers. The following would be the program code for the program.

```
/* Functions and Numerical Arrays - Read and Display an array of 'n' integers */
#include<stdio.h>
main()
{
int a[100],n;
void read_arr(int[],int);
void print_arr(int[],int);
```

```
clrscr();
printf("\n\nEnter how many elements you want to enter : ");
scanf("%d",&n);
printf("\n\nEnter the elements :  ");
read_arr(a,n);
printf("\n\nThe elements you have entered are :  ");
print_arr(a,n);
getch();
}
/* Reading the elements of the array */
void read_arr(int x[], int n)
{
int i;
for(i=0;i<n;i++)
scanf("%d",&x[i]);
}
/* Printing the elements of the array */
void print_arr(int x[], int n)
{
int i;
for(i=0;i<n;i++)
printf("%d\t",x[i]);
}
```

We see from the program that we are using two functions, unlike the earlier programs. One function read_arr() is used for reading in the elements of the array and the other function print_arr() is used for printing the elements that are read. We also see that when we send the array to the functions, we send it by referring its name (just like the other variables) and once we refer the name of the array name in the program, we know that its base address would be accessed (from the discussion of pointers and arrays). Thus it is immaterial to pass the size of the array to the functions, since the address is passed. The functions read_arr() and print_arr() hence could understand the size of the array very easily using the addresses passed. Also note the prototypes used for the functions read_arr() and print_arr(). They contain an array variable (denoted by square braces) and an integer variable. Figure 7.10 shows the sample run for the above program.

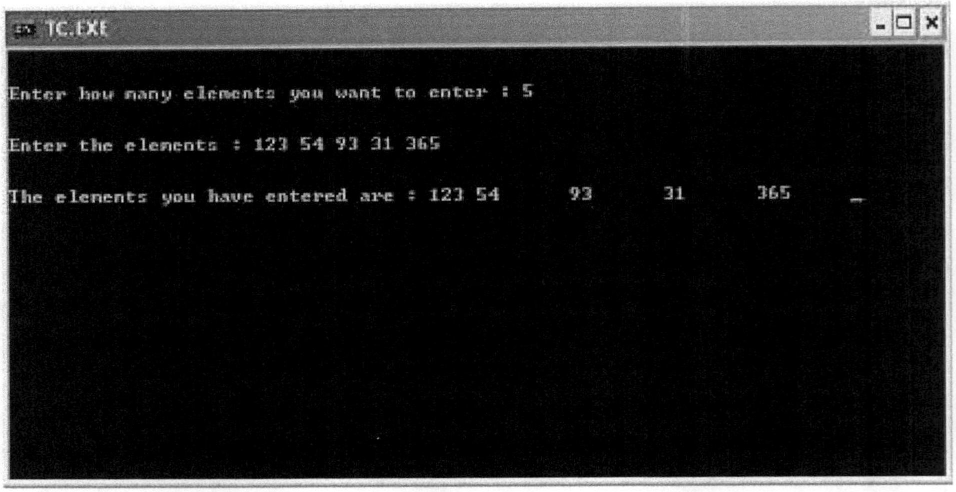

Figure 7.10 Sample run—read and display an array of 'n' integers

The memory picture is shown in the Figure 7.11.

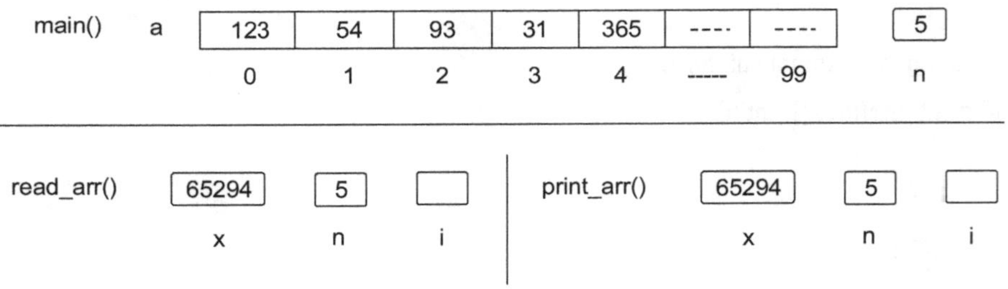

Figure 7.11 Memory picture—read and display an array of 'n' integers

We can see from the memory picture that when we call the functions read_arr() and print_arr(), we pass the address of the array 'a' and the value of 'n' and we get the input for the array and display the same using the respective functions. Note that when we pass the address of the array from the function main(), the dummy variable 'x' receives that and from there on it could access the individual locations of the array 'a'. This is possible for the variable 'x' since it has the base address of the array 'a'. Thus, even though it is not a pointer, it acts as an implicit pointer to the array 'a'. In the memory picture above, 65294 happens to be the base address of the array 'a' and you can see that the variable 'x' has that address when it receives 'a' from the function main(), when the functions read_arr() and print_arr() are called. You may check this by adding the following printf() statement in your program in any one of the two user-defined functions read_arr() and print_arr().

```
printf("%u",x);
```

The above printf() statement would result in printing the value in 'x', which is nothing but the base address of the array 'a'.

Let us see another program for a better understanding of how numerical arrays work with functions. Consider the program which reads an array of 'n' integers and sorts the array in ascending order. The following would be the program for the same.

```
/* Functions and Numerical Arrays – Sorting an array of 'n' integers in ascending order */
#include<stdio.h>
main()
{
int a[100],n;
void read_arr(int[],int);
void print_arr(int[],int);
void ascend(int[],int);
clrscr();
printf("\n\nEnter how many elements you want to enter :  ");
scanf("%d",&n);
printf("\n\nEnter the elements :   ");
read_arr(a,n);
printf("\n\nThe elements you have entered are :  ");
print_arr(a,n); /* First time print_arr() gets called */
ascend(a,n);
printf("\n\nThe array in ascending order is :   ");
print_arr(a,n); /* Second time print_arr() gets called */
getch();
}
void read_arr(int x[], int n)
{
int i;
for(i=0;i<n;i++)
scanf("%d",&x[i]);
}
void print_arr(int x[], int n)
```

```
{
int i;
for(i=0;i<n;i++)
printf("%d\t",x[i]);
}
void ascend(int a[],int n)
{
int i,j,temp;
for(i=0;i<n;i++)
for(j=i+1;j<n-1;j++)
if(a[i]>a[j])
{
temp=a[i];
a[i]=a[j];
a[j]=temp;
}
}
```

We see that we have one more function ascend(), whose prototype is very much similar to the other two functions read_arr() and print_arr(). This function performs the sorting process and once this function gets called from the function main(), the array would have been sorted in ascending order and when we call the function print_arr() for the second time, the sorted array gets printed. This can be observed from the Figure 7.12.

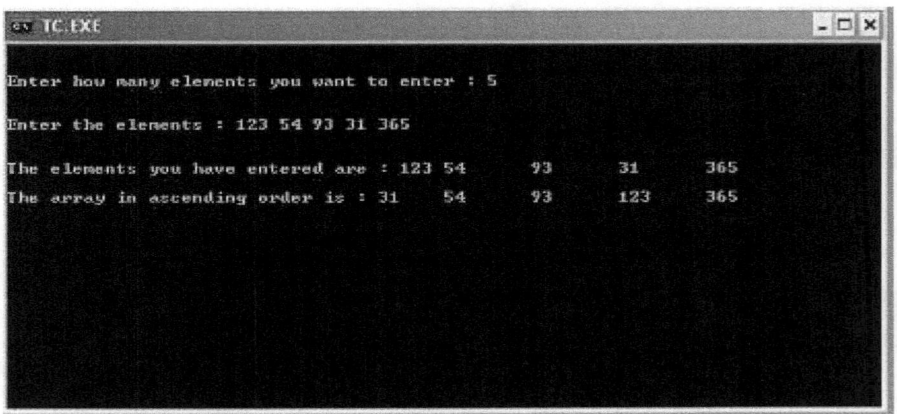

Figure 7.12 Sorting an array in ascending order

The memory picture of the above program is shown in the Figure 7.13. There is just one addition to the memory map shown in the Figure 7.11, the function ascend().

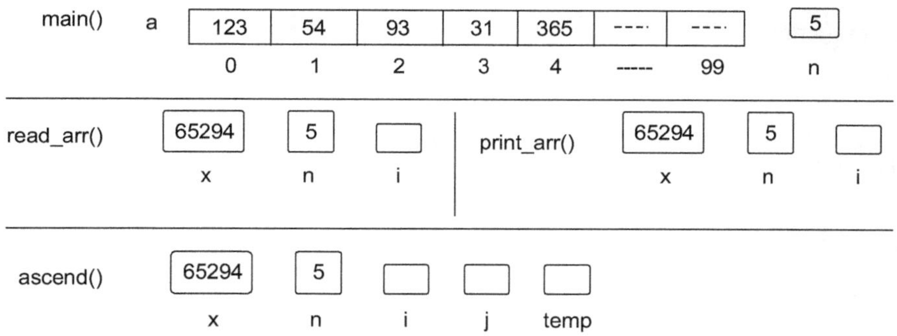

Figure 7.13 Memory picture—sorting an array in ascending order

With this idea of working with single-dimensional array, we proceed to discuss on how the two-dimensional arrays work with functions in the following section.

7.5.2 Functions and Two-Dimensional Numerical Arrays

Two-dimensional numerical arrays work in a similar manner with the functions as the single-dimensional array. Let us consider the program to read and find the transpose of an m×n matrix. The following is the program coding.

```
/* Reading and finding the transpose of an m×n matrix */
#include<stdio.h>
main()
{
int a[10][10],m,n;
void read_arr(int[][],int,int);
void print_arr(int[][],int,int);
void transpose(int[][],int,int);
clrscr();
printf("\n\nEnter the order of the matrix :  ");
scanf("%d%d",&m,&n);
printf("\n\nEnter the array elements :  ");
read_arr(a,m,n);
```

```
printf("\n\nThe array you have entered is\n");
print_arr(a,m,n);
printf("\n\nThe transpose of the matrix is\n");
transpose(a,m,n);
getch();
}
void read_arr(int x[][10],int m,int n)
{
int i,j;
for(i=0;i<m;i++)
for(j=0;j<n;j++)
scanf("%d",&x[i][j]);
}
void print_arr(int x[][10],int m,int n)
{
int i,j;
for(i=0;i<m;i++)
{
for(j=0;j<n;j++)
printf("%d\t",x[i][j]);
printf("\n");
}
}
void transpose(int x[][10],int m,int n)
{
int i,j;
for(j=0;j<n;j++)
{
for(i=0;i<m;i++)
printf("%d\t",x[i][j]);
printf("\n");
}
}
```

We have three functions namely read_arr(), print_arr() and transpose(), to read, display and print the transpose of an m × n matrix. We also see that when we call the function we call it with the base address of the 2-D array 'a' and with the values of 'm' and 'n'. When we define the function, we observe that the size of the second dimension alone is collected in the header and that of the first dimension is left out. This is because once the base address of the entire 2-D array is known when the control comes to the function and once the second dimension's size is also known, the memory will be allocated accordingly in the function. I would suggest you check this fact by using the following statement in the program. This would print the base address of the 2-D array.

```
printf("%u\n",x);
```

Also, try to figure out yourself the memory map of the above program. Figure 7.14 shows a sample execution of the above program.

7.6 FUNCTIONS AND CHARACTER ARRAYS

Character arrays, as discussed in Chapter 4, differ from the numerical arrays only with respect to the NULL character that is used to terminate the array. Apart from this, there is no difference in accessing and manipulating the individual elements of the array.

Let us straight away discuss a program with the character array. Consider the following program in which we reverse the string that is read. We use four functions, readfn(), writefn(), lenfn() and revfn() for reading, displaying, finding length and reversing the given string. The length function is important since the function revfn() must know the length before it reverses the input string.

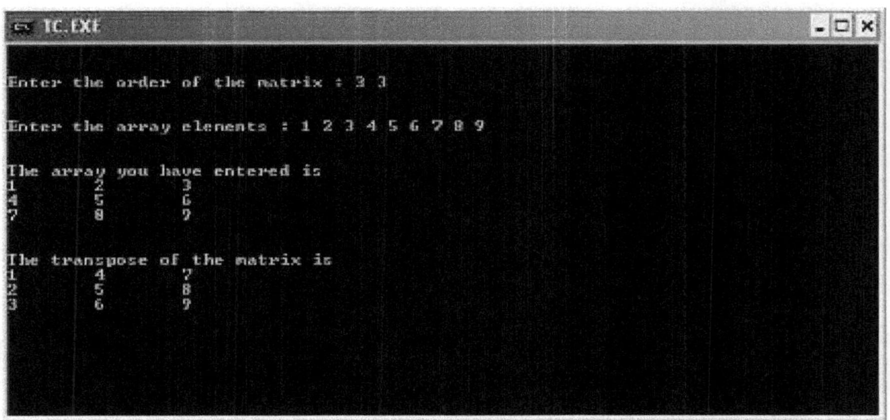

Figure 7.14 Transpose of the matrix of order m × n

```
/* Read and reverse a string using functions */
#include<stdio.h>
main()
{
char a[100];
int l;
void readfn(char[]);
void writefn(char[]);
int lenfn(char[]);
void revfn(char[],int);
clrscr();
printf("\n\n\tEnter the string :  ");
readfn(a);
printf("\n\n\tThe string entered is :  ");
writefn(a);
l=lenfn(a);
printf("\n\n\tThe length of the string is : %d",l);
printf("\n\n\n\tThe reversed string is :  ");
revfn(a,l);
getch();
}
void readfn(char x[])
{
int i;
for(i=0;(x[i]=getchar())!='$';i++);
x[i]='\0';
}
void writefn(char x[])
{
puts(x);
}
```

```
int lenfn(char x[])
{
int i;
for(i=0;x[i];i++);
return i;
}
void revfn(char x[],int l)
{
int i;
for(i=l;i>=0;i--)
putchar(x[i]);
}
```

Figure 7.15 shows the output of the above program and Figure 7.16 shows the memory picture of the program. If we analyse the program, we can understand that when the functions readfn(), writefn() and lenfn() are called, the base address of the array 'a' is passed from the function main(). When the function revfn() is called, along with the base address we pass the value of the length of the string. For more clarity, go through the memory picture of the program. The base address of the array is found to be 65386 in my execution of the program and it may differ when you execute the same program. This address could be obtained by using the following printf() any one of the user-defined functions used in the program.

```
printf("\n%u",x);
```

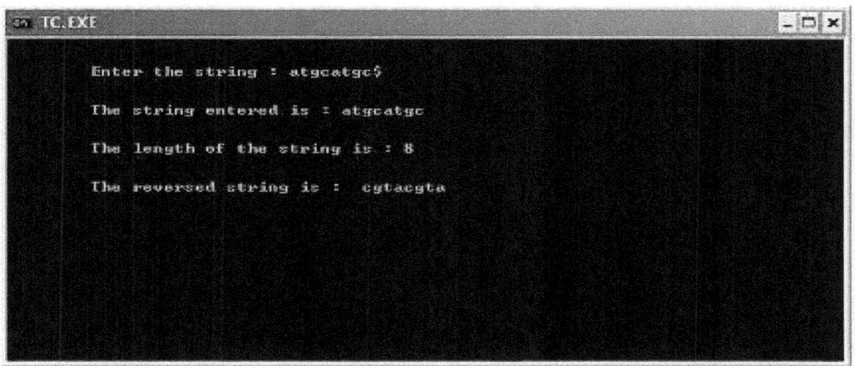

Figure 7.15 Sample run—reversing a string

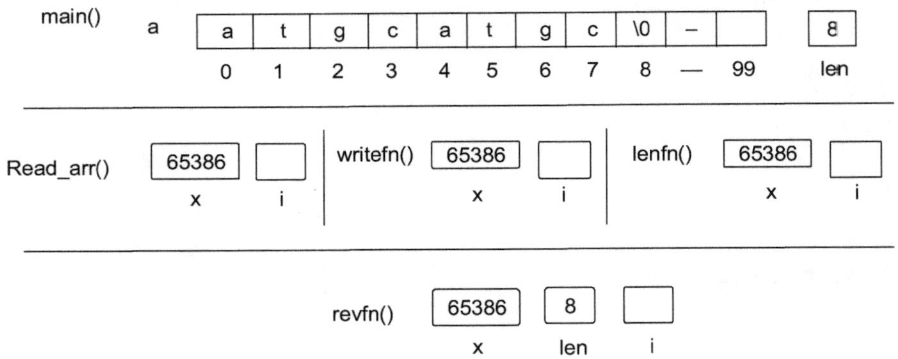

Figure 7.16 Memory picture—reversing a string

Let us discuss one more program using functions, in which we would find the number of times a particular pattern exists in the given string. The following is the program coding for the same.

```
/* Check for the occurence of a pattern in the given string/text */
#include<stdio.h>
main()
{
char a[100],b[100];
int count;
void readfn(char[]);
void writefn(char[]);
int check(char[],char[]);
clrscr();
printf("\n\n\tEnter the text : ");
readfn(a);
fflush(stdin);
printf("\n\tEnter the pattern to check : ");
readfn(b);
fflush(stdin);
printf("\n\n\tThe string entered is : ");
writefn(a);
count=check(a,b);
printf("\n\nThe given pattern \"%s\" occurs in the given text \"%s\" \"%d\"times",b,a,count);
```

```
getch();
}
void readfn(char x[])
{
int i;
for(i=0;(x[i]=getchar())!='$';i++);
x[i]='\0';
}
void writefn(char x[])
{
puts(x);
}

int check(char x[],char y[])
{
int i,j,k,l=0;
for(i=0;x[i];i++)
{
for(j=0,k=i;y[j]==x[k]&&y[j];k++,j++);
if(y[j]=='\0')
l++;
}
return l;
}
```

In the program above, we are using a new function fflush(stdin). This function is essentially needed when one manipulates with multiple character arrays. In this program we use two character arrays 'a' and 'b'. Normally when we give the input strings through the keyboard, we give an enter or a space or a tab character between the two inputs (default delimiters that we had discussed in Chapter 2). Every device attached with the computer has its own memory and this memory is called as a 'buffer' or a temporary memory. When we operate with more than one character array, any one of the abovesaid delimiter character will appear in between the two input strings and hence when we check the existence of a pattern with the given text, only the 'enter' character (which is in the array 'b') gets checked with the given text in the array 'a' and hence you will get the output much against your anticipation. I would suggest, you please check with this by removing the function **fflush(stdin)** given between the two inputs and after the second input. You will not get the right answer. Figure 7.17 shows the output when the above-quoted function is ignored from the program (just remark it to ignore it from the program and do not delete it). Observe the last line of the Figure 7.17. You can see that it is spread over two lines. The reason is very simple. When we ignore the fflush(stdin) function, the 'enter' character that is available in the keyboard buffer (the 'enter' character that is given after the first text) will be appended automatically with the first input text. Hence we are getting the last line spread over two lines. The output is also wrong, since it is obvious from the input text **atgcatcg** that the test string **at** occurs two times, whereas the output indicates that it occurs '0' times. The reason is that when we ignore the function fflush(stdin) from the program the 'enter' character that is given after the first text (atgcatcg) is checked for its occurrence and not the string 'at'. Now include the function fflush(stdin) and repeat the execution with the same inputs. You will get the correct result in this execution. This is because, the usage of the function fflush(stdin) will clear off the keyboard buffer and when the input text (atgcatcg) is checked for the occurrence of the pattern 'at', it gives a right result. Figure 7.18 shows the execution of the program with the function fflush(stdin)included in the program.

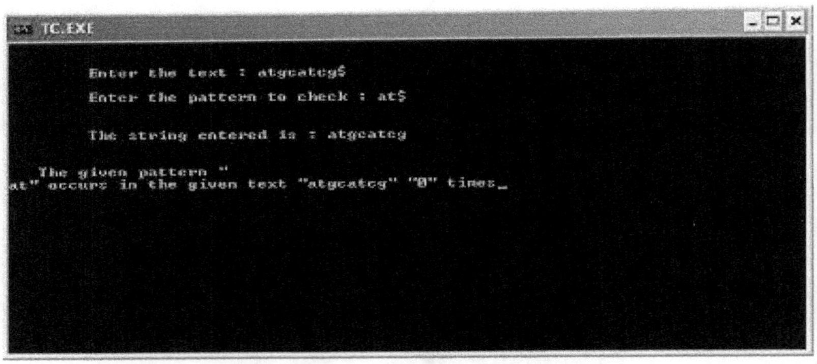

Figure 7.17 Pattern matching—O/p when fflush(stdin) not included

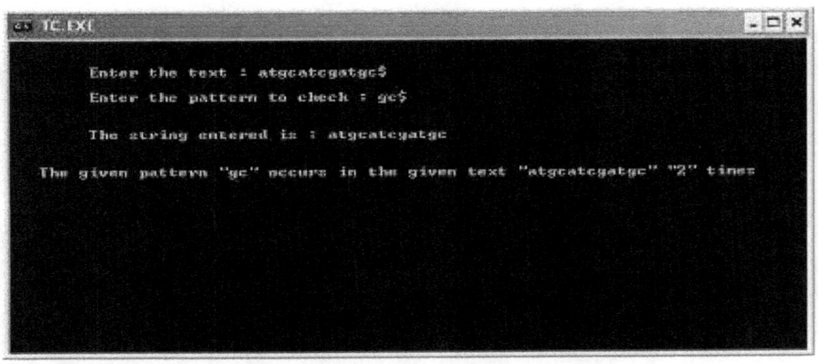

Figure 7.18 Pattern matching—O/p when fflush(stdin) included

So, we can conclude that whenever we operate with multiple character arrays, we must use the function fflush(stdin) to clear off the keyboard buffer and get the anticipated result.

When you run the program for different sets of inputs, you might find a discrepancy. For example, test the program with the inputs "hello world" and "hell". As such the word "hell" does not occur in the input text, but it is an integral part of the word "hello" and hence when you execute the above program with the inputs "hello world" and "hell", you will get the result that the word is occurring once in the given text. So, our logic not only checks the exact occurrence of a pattern in the given input text, but also the integral part of the input text. The output shown in Figure 7.19 justifies this. Since in Bioinformatics applications, this discrepancy is not so significant, we go with this logic. But I leave it as an exercise to the learners of the book to identify the most suitable logic that would work with any input text and test pattern.

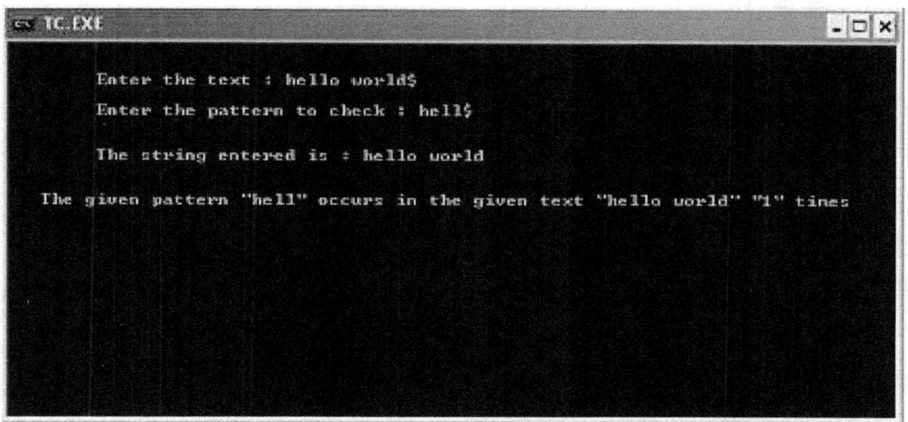

Figure 7.19 Part of the input text is also matched with given pattern

I would appreciate if the learners figure out the memory picture for the above program, which will surely increase the gravity of their understanding.

With this we conclude the discussion of how the character arrays work with the functions. I am sure the programs that I have discussed would be of immense use for the learners. Try to solve the exercises given at the end of this chapter so that you can work more confidently with the functions. What you have understood till now in functions would be of great significance when you learn the concepts in the forthcoming chapters.

7.7 FUNCTIONS AND POINTERS

We have been discussing functions earlier in this chapter and in our discussion we did not include pointers while we wrote simple functions. We also know that any function at a given time can return only one value or no value at all. No function can return more than one value after processing. This simply means that using the *return* statement, one can make a function only one value after the process. But think of those situations when a programmer wants to get more than one value from the function. Is this achievable? Does C have a provision to do this? Yes. Of course, it has. When we write functions with pointers we would possibly get more than one value returned from the function and that too without explicitly using the *return* statement. This method of calling functions and getting the result is called as call by value. This is the most effective method especially when the functions are used with arrays. We shall see the concept of using the arrays with functions and pointers later in this chapter. Let us see how the functions work with the pointers. Consider the program where we would like to swap two values of type integer. We would be writing a simple function swap() to achieve this task. The following is the program code for the same.

```
/* Swapping two values of type integer */
#include<stdio.h>
main()
{
int a,b;
void swap(int*,int*);
clrscr();
printf("\n\tEnter the values to swap : ");
scanf("%d%d",&a,&b);
printf("\n\tThe values of a and b before swapping : %d\t%d",a,b);
swap(&a,&b);
printf("\n\n\tThe values of a and b after swapping : %d\t%d",a,b);
getch();
}
void swap(int *x,int *y)
{
int temp;
temp=*x;
*x=*y;
*y=temp;
}
```

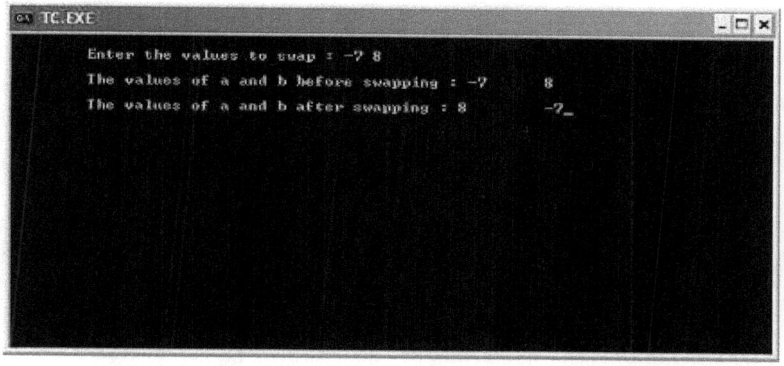

Figure 7.20 Swapping two integers using functions and pointers

We see from the program that we have a function **swap()** that swaps the values of two integers. We see that no values are returned from the function. But still the values would be swapped. How is this possible? If you observe the function prototype **void swap(int*,int*)** and the function call **swap(&,a,&b)**, we pass the address of the variables a and b to the function swap() and hence whatever changes are made to the variables in the function, they are reflected in the function main(). Earlier when we used call by value method of accessing functions, we just copied the values of the variables that we passed from the function main() to the called function. In other words, we created dummy variables for copying the values that are passed. When we made changes in the copies, they are not reflected in the original variables. In the above program we have used **call by reference** method of accessing functions.

The output of the above program is shown in the Figure 7.20. The memory picture shown in the Figure 7.21 explains how the call by reference method of function is effectively used.

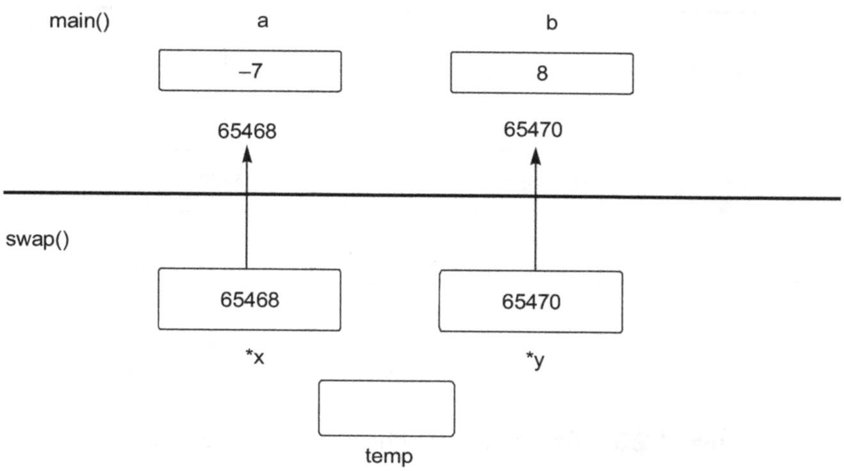

Figure 7.21 Memory picture for swapping program

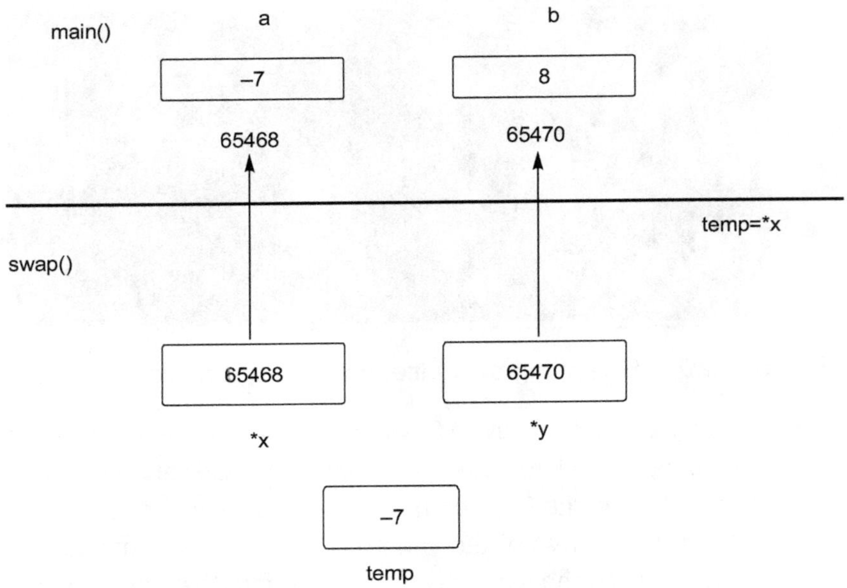

Figure 7.22 Execution of the code temp=*x in the function swap()

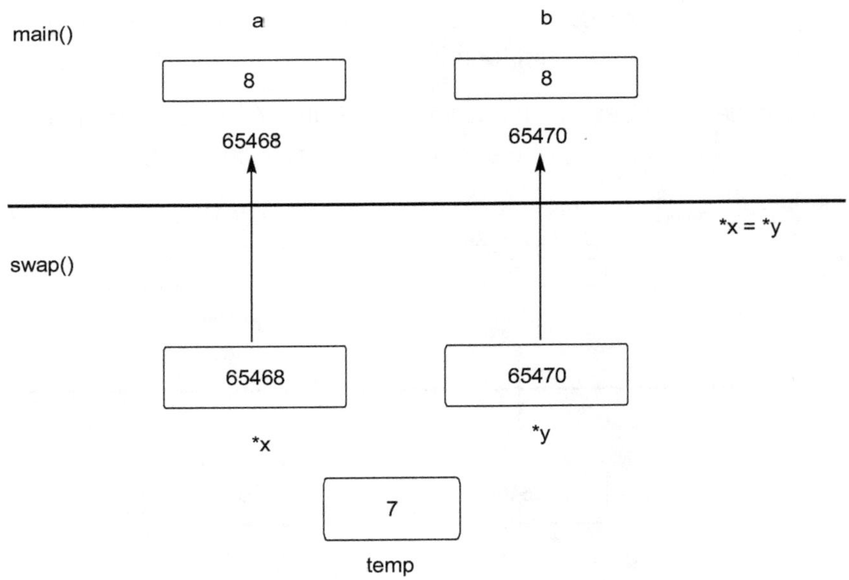

Figure 7.23 Execution of the code *x=*y in the function swap()

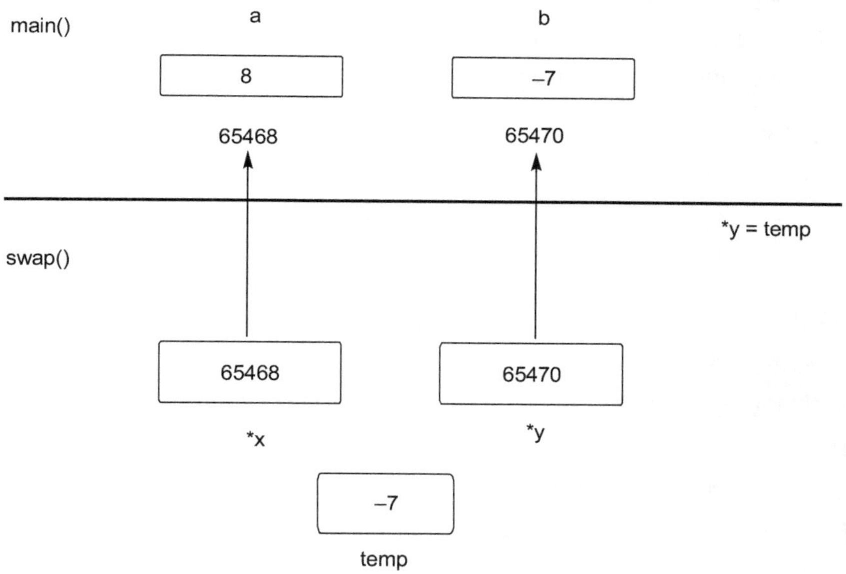

Figure 7.24 Execution of the code *y=temp in the function swap()

Analyse the function swap() closely. Since the pointer variables x and y in the function swap() have the addresses of the variables a and b in the function main(), when we process the pointer variables in the function swap(), they make changes in the variables in the function main(). The execution of the three lines of coding in the functions swap() is shown in Figure 7.22, 7.23 and 7.24.

From the memory pictures shown in Figure 7.22, 7.23 and 7.24, we can easily observe that the changes are made in the called function swap() and these changes get reflected in the calling function, the main(). Earlier, this was not the case, since there is no relationship between the variables in the calling function and the called function, except that the variables in the called function just act as a copy (dummy) of those in the calling function. But in the above program, the addresses of the variables in the calling function are available with the variables in the called function and hence they become the pointers for them and so, we can say that a relationship is established. When the code temp=*x is encountered by the compiler, it is converted internally to temp=*65468 which is equivalent to temp=-7. Thus we are able to get the value of a variable in the function main() to the function swap() by call by reference method. The other two lines of coding work in a similar way. The line of coding *x=*y is converted internally as *65468=*65470. Due to this conversion, we get the value of 'b' (8) in the address *65468, which is the address of 'a'. The line of coding *y=temp is internally taken as *65470=-7 and hence the value of 'a' occupies 'b'. Thus at the end of these three lines, we get the values of a and b in the function main() swapped.

Let us discuss another program for a better understanding of the concept. Consider the program in which we are finding the factorial of the given unsigned integer value using functions and pointers. The following is the program coding for the same.

```
/* Factorial of a given unsigned integer */
#include<stdio.h>
main()
{
unsigned n;
long unsigned f=1;
void fact(unsigned,long unsigned*);
clrscr();
printf("\n\tEnter the number for which factorial has to be found out : ");
scanf("%u",&n);
if(n==0)
{
printf("\n\tThe factorial of %u is %lu",n,f);
getch();
return;
}
fact(n,&f);
printf("\n\tThe factorial of %u is %lu",n,f);
getch();
}
void fact(unsigned n, long unsigned *pf)
{
int i;
for(i=n;i>1;i--)
*pf*=i;
}
```

We see that the value of the number for which the factorial value has to be found out is obtained as an unsigned number (since factorial for negative values is meaningless. Also, when the value for 'n' is given as '0', the function fact need not be called at all, since the factorial value for '0' is defined as '1'. Hence we check this in the beginning itself and terminate the program in the function main() itself, if the input is '1'. The variable that is supposed to hold the result of the factorial is 'f' and it is initialized as '1' and when we call the function fact(), we pass the value of 'n' and the address of 'f', since the variable 'f' alone is getting manipulated in the whole program. Once we pass the address of 'f' it is collected in the pointer variable 'pf' and as we manipulate 'pf' in the function swap(), the change is done in the original variable 'f' and hence when we print the result in the function main(), the correct value of factorial for the given number is printed. Observe also that the variable 'f' is declared as **long unsigned**, since the result of the factorial will be a bigger number in the case of input exceeding '7' and an ordinary integer cannot hold that big value, as it will exceed the range of it. I would suggest you to go through the list of data types and their permissible range of values for more clarity on this point. Figure 7.25 shows a sample run of the program (for the inputs '0' and another valid value for 'n').

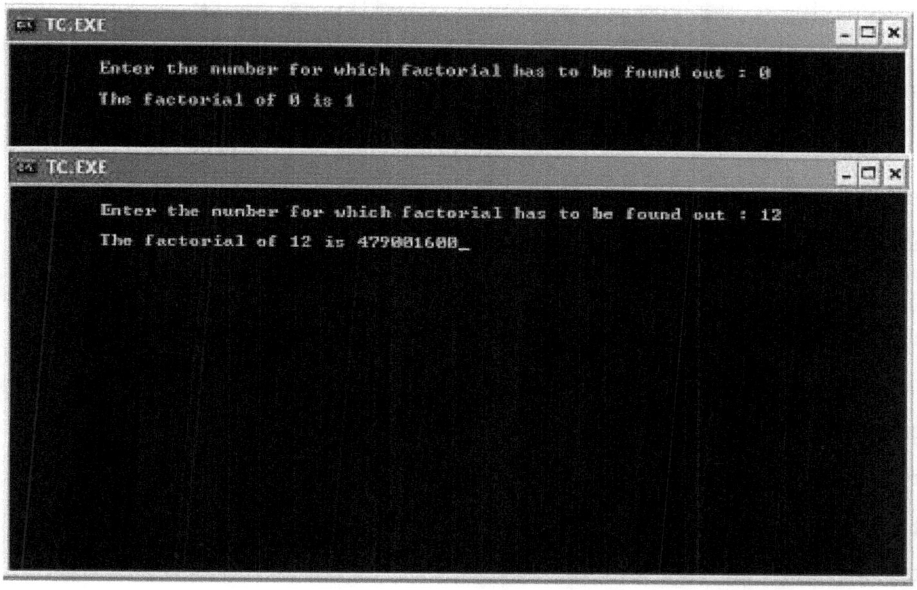

Figure 7.25 Finding the factorial of the given unsigned integer

The memory picture for the above program is shown in Figure 7.26. From the memory picture, we see that since the variable 'pf' in the function swap() holds the address of the variable 'f' in the function main(), the changes that we make with the variable 'pf' reflects in the variable 'f'. Initially the variable 'f' had the value '1' and when it is getting manipulated through 'pf', its value changes according to the input value for 'n'. Thus when we give 12 as the input for 'n', the result of the factorial, which is 47,90,01,600, is correctly shown as in the displayed figure above. The program works fine till the input value of 32, after which the range of long unsigned integer is exceeded and hence we will start getting different values, which are not anticipated.

I would suggest you to practise all the programs that are covered in the initial chapters with pointer and function notation and those which are given at the end of this chapter, to gain efficiency in working with functions and pointers.

7.8 FUNCTIONS, POINTERS AND ONE-DIMENSIONAL NUMERICAL ARRAYS

After having gained a thorough knowledge of how pointers are used in association with the functions, we are ready to advance ourselves to the next idea of using pointers and arrays with functions. To begin with we will discuss how pointers work with one-dimensional numerical arrays when they are used in functions. We know how pointers work with arrays from the previous chapter and how functions work with numerical arrays in this chapter in the section 7.5.1. We will combine those ideas in this section and will try to understand the way all the three work together. For that let us consider a program where we find the biggest and smallest of 'n' integers and swap their positions in the array and display the new array. The following is the program coding for the same.

```
/* Finding the biggest and smallest of 'n' integers and swap their positions using functions */
#include<stdio.h>
main()
{
int a[100],small,big,n,*posbig,*possmall;
void readfn(int*,int);
void writefn(int*,int);
void bigfn(int*,int*,int**,int);
void smallfn(int*,int*,int**,int);
void swapfn(int**,int**);
clrscr();
printf("\n\tEnter the number of elements in the array : ");
scanf("%d",&n);
printf("\n\n\tEnter the array elements : ");
readfn(a,n);
printf("\n\n\tThe array elements are : ");
writefn(a,n);
bigfn(a,&big,&posbig,n);
```

```
printf("\n\n\tThe biggest of the %d integers is %d",n,big);
smallfn(a,&small,&possmall,n);
printf("\n\n\tThe smallest of the %d integers is %d",n,small);
swapfn(&posbig,&possmall);
printf("\n\n\tThe array after swapping the biggest and smallest integers is:\n\n\t");
writefn(a,n);
getch();
}
/* Reading the elements of the array */
void readfn(int *pa,int n)
{
int *pt=pa;
for(;pa<(pt+n);pa++)
scanf("%d",pa);
}
/* Printing the elements of the array */
void writefn(int *pa,int n)
{
int *pt=pa;
for(;pa<(pt+n);pa++)
printf("%d\t",*pa);
}
/* Finding the biggest of the 'n' elements and its position */
void bigfn(int *pa,int *pbig,int **posbig,int n)
{
int *pt=pa;
*pbig=*pa;
*posbig=pa;
for(;pa<(pt+n);pa++)
if(*pa>*pbig)
{
```

```
*pbig=*pa;
*posbig=pa;
}
}
/* Finding the smallest of the 'n' elements and its position */
void smallfn(int *pa,int *psmall,int **possmall,int n)
{
int *pt=pa;
*psmall=*pa;
*possmall=pa;
for(;pa<(pt+n);pa++)
if(*pa<*psmall)
{
*psmall=*pa;
*possmall=pa;
}
}
/* Swapping the biggest and smallest using their positions */
void swapfn(int **posbig, int **possmall)
{
int temp;
temp=**posbig;
**posbig=**possmall;
**possmall=temp;
}
```

We see from the program that we have used five user-defined functions, to read the array, to display it, to find the biggest of its elements with its address and the smallest of its elements with its position and finally swap their positions using the addresses obtained. Even though the program seems to be complicated, the logic used is quite easy to understand. Here we use not just pointers, we use pointers to pointers since we need the addresses of the biggest and smallest numbers once they are found out from the respective functions.

Note the prototype of the functions readfn() and writefn(). We pass the base address of the array (by mentioning the name of the array) and the value of the number of elements (which is obtained and stored in 'n'). Thus we have a mixed type of function call, i.e., we use both the call by value and call by reference in the same function.

Note that the prototype of the functions bigfn() and smallfn() are similar and we are passing the base address of the array from the calling function, addresses of the variables that hold the biggest and smallest values (big and small), the addresses of the pointers posbig and possmall (in order to collect the addresses of biggest and smallest of 'n' integers in the array) and the value of 'n'.

Finally the prototype of the function swapfn() is simple, where we pass the addresses of the pointers from the function main() and once we manipulate the values in those addresses, the biggest and smallest values in the array get interchanged and the array when reprinted, shows the biggest number in the smallest number's position and vice versa.

The output of the above program code is shown in Figure 7.26 and the memory pictures of the same are shown in Figure 7.27 through Figure 7.31.

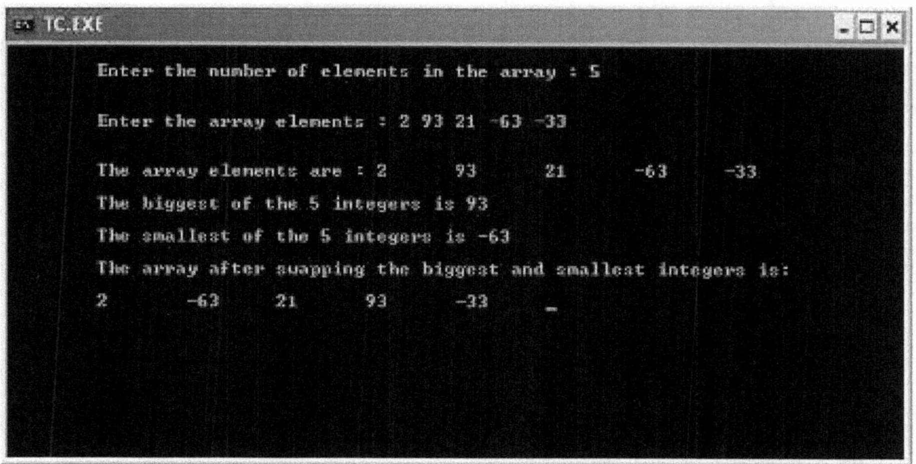

Figure 7.26 Finding the biggest and smallest and swapping them

From Figure 7.26, we can observe that the biggest number in the array is 93 and the smallest number in the array is –63 and their positions (not addresses) in the array are 2 and 4 respectively. Once we find out the addresses of these two elements, we pass them to the function swapfn(), which swaps them.

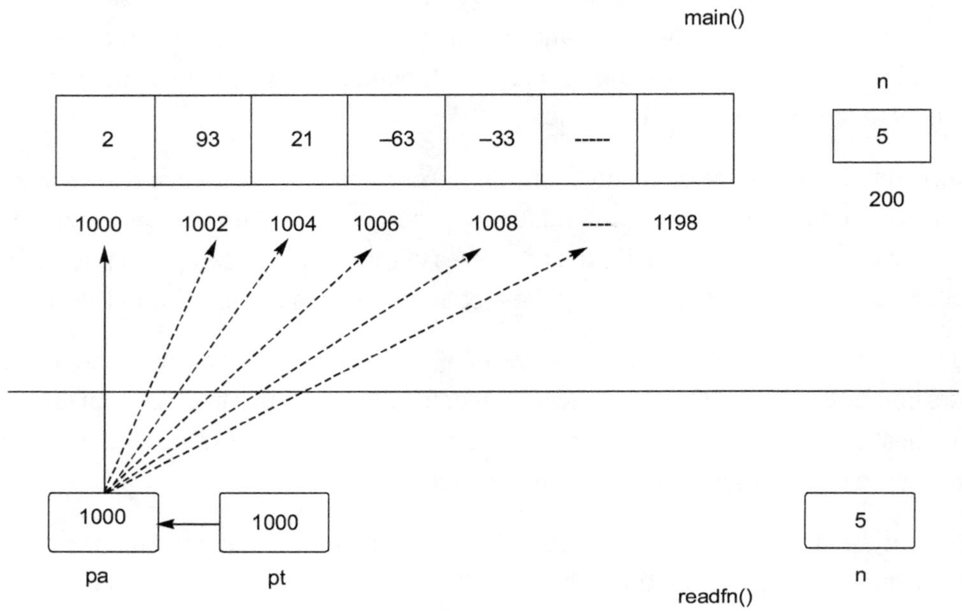

Figure 7.27 Reading the array elements using readfn()

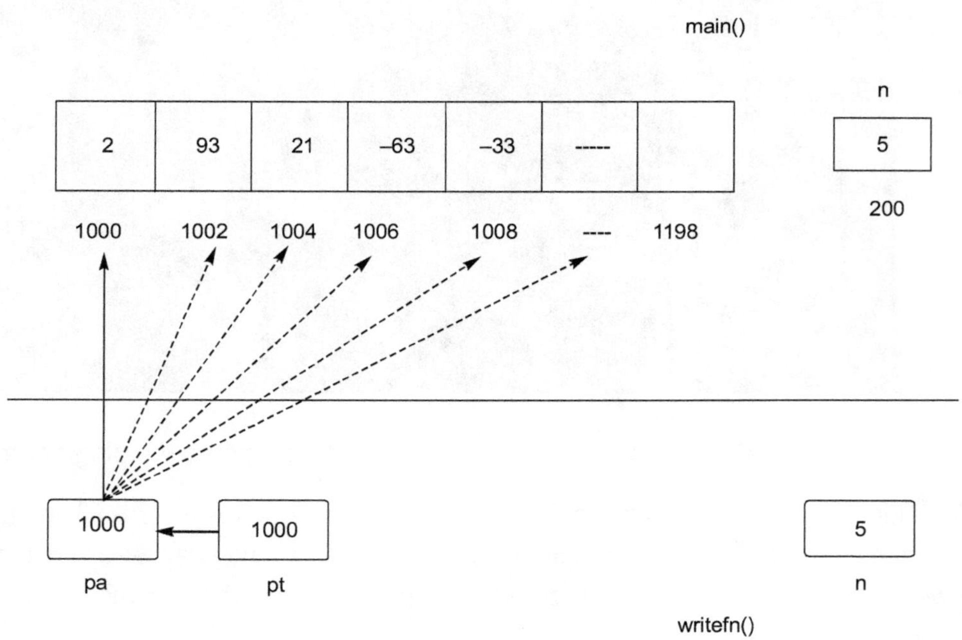

Figure 7.28 Displaying the array elements using writefn()

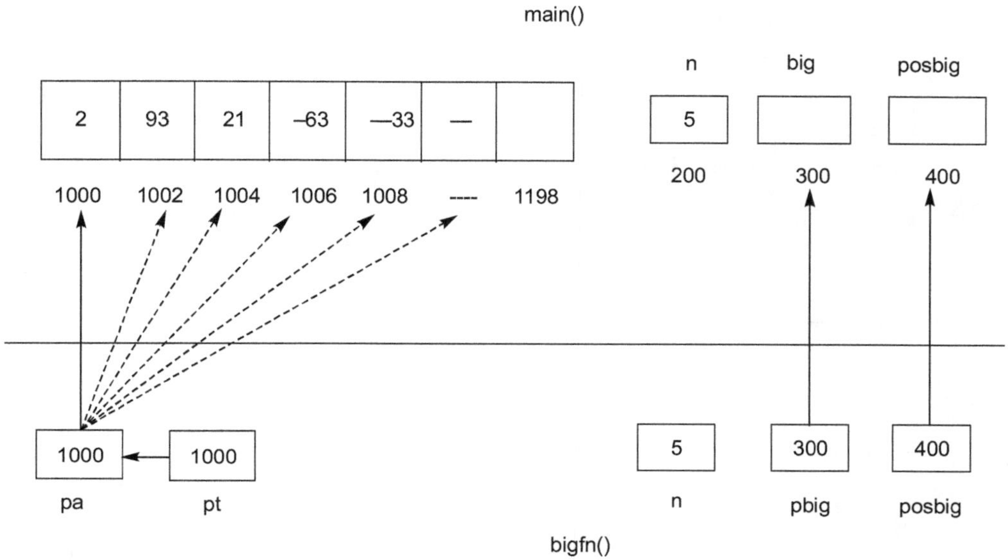

Figure 7.29 Finding the biggest number and its address

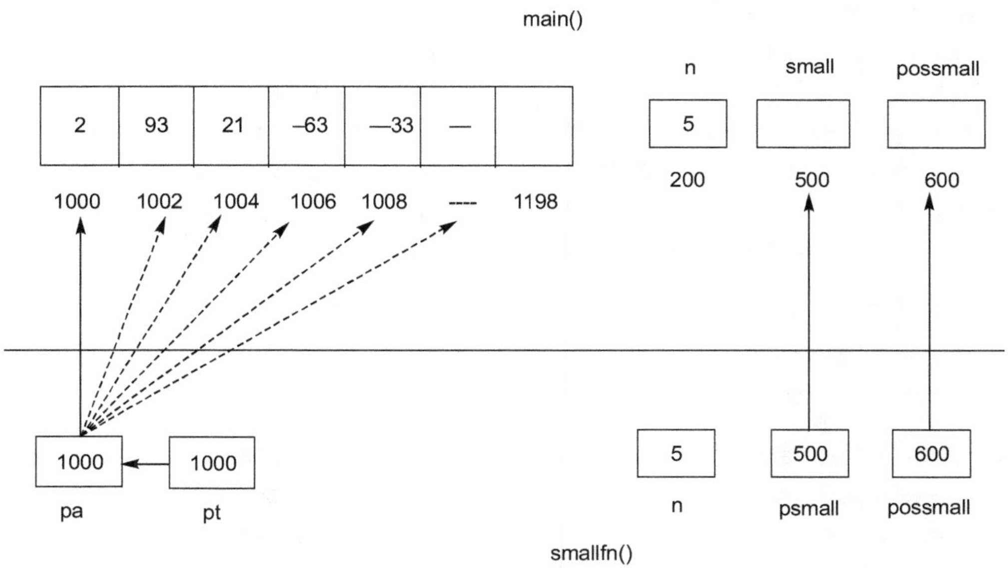

Figure 7.30 Finding the smallest number and its address

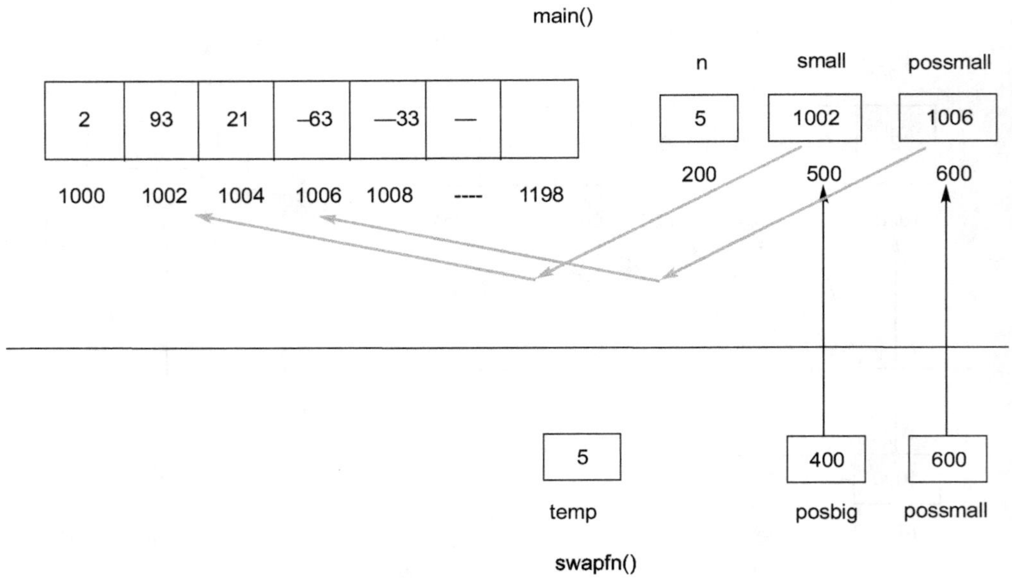

Figure 7.31 Swapping the biggest and smallest numbers

From Figure 7.31, we see that the addresses of the variables posbig and possmall are passed from the function main() and they are collected in the pointer variables posbig and possmall in the function swapfn(). We also note that the variables posbig and possmall in the function main() are pointers and they contain the addresses of the biggest and smallest numbers updated through the functions bigfn() and smallfn(). Hence the variables posbig and possmall in the function swapfn() are not just pointers, they are pointers to pointers. This is indicated with grey arrows in the figure. So, once the addresses of posbig and possmall are received in the respective pointers of the function swapfn(), the biggest and smallest numbers are swapped through the variable temp in the function swapfn(). The following are the equivalent lines of coding for those in the function swapfn().

```
temp=**posbig;              temp=*(*(400))=*(1002)=93
**posbig=**possmall;        *(*400))=*(*(600))=*(1006)=-63
**possmall=temp;            *(*(600))=temp=93
```

Thus at the end execution of these three lines, the value at the address of biggest number would be swapped with that at the address of the smallest number. In other words, the address of the biggest number now would have become that of the smallest number and vice versa. The pointer variable pt is used in all the functions (except the function swapfn()) so as not to lose the base address of the array 'a' received in 'pa'. Since 'pa' in all the functions is manipulated in the loops, we would lose the base address of the array 'a'.

I am sure the above program, explanation, screenshots and the memory pictures would have been a great deal of understanding for the learners of the book. In the discussion, we have not just covered the call by reference method of calling the function. We have covered how the base address of the array is passed on to the functions and how the concept of 'pointer to pointer' is getting implemented. I would advice the learners, to go through the entire discussion again to get the concepts in a more lucid manner.

7.9 FUNCTIONS, POINTERS AND ONE-DIMENSIONAL CHARACTER ARRAYS

We have already discussed how pointers work with character arrays in the chapter 6 (Section 6.4.2) and we have also discussed how pointers work with functions in this chapter (Section 7.7). This section would be just a combination of both and above all, we had just completed our discussion on how pointers work with functions and arrays. So, I do believe that understanding and working the programs in the discussion that follows would be very much easier. We would straight away start with a program where we find whether a given string is a palindrome or not, to explain the concept. The following would be the program for that.

```
/* Checking whether the given string is a palindrome or not */
#include<stdio.h>
main()
{
char a[100];
int flag=1,len=0;
void readfn(char*);
void writefn(char*);
void lenfn(char*,int*);
void palinfn(char*,int*,int);
clrscr();
printf("\n\tEnter the string to check : ");
readfn(a);
printf("\n\n\tThe string you have entered is : ");
writefn(a);
lenfn(a,&len);
```

```
printf("\n\n\tThe length of the string is : %d",len);
if(len%2==0||len==1)
{
printf("\n\n\tThe string \"%s\" is not compatible for the test",a);
getch();
return;
}
palinfn(a,&flag,len);
if(flag)
printf("\n\n\tThe string %s is a palindrome",a);
else
printf("\n\n\tThe string \"%s\" is not a palindrome",a);
getch();
}
/* Reading the string */
void readfn(char *pa)
{
for(;(*pa=getchar())!='$';pa++);
*pa='\0';
}
/* Displaying the string */
void writefn(char *pa)
{
for(;*pa;pa++)
putchar(*pa);
}
/* Finding the length of the string */
void lenfn(char *pa,int *plen)
{
```

```
for(;*pa;pa++)
(*plen)++;
}
/* Checking for the palindromic nature */
void palinfn(char *pa, int *pflag,int len)
{
char *pt,*pj,*pk;
pt=pa;
pa=pa+len/2;
pj=pa-1;
pk=pa+1;
for(;pj>=pt&&*pk;pk++,pj--)
if(*pj!=*pk)
{
*pflag=0;
break;
}
}
```

We have four functions in the program, readfn(), writefn(), lenfn() and palinfn(). The string to be checked is read and displayed using the functions readfn() and writefn(), the length is found out using the function lenfn() and it is checked for its palindromic character using the function palinfn(). We see that for the functions readfn() and writefn(), the base address of the character array 'a' is passed from the calling function main(). For the function lenfn(), we pass the address of the variable 'len' (whose initial value is zero) along with the base address of the array 'a'. When we do so, the length gets updated in the calling function and hence when we print the same in the calling function, it is getting printed correctly. The length of the string is essential for checking the palindromic nature of the input string, as obvious from the program coding. We use a variable 'flag' (treated as a Boolean variable even though declared as an integer), whose initial value is '1'. When we call the function palinfn(), the base address of the array 'a', address of the variable 'flag' and the value of the length of the string, which is available in the variable 'len' are passed to it.

We can also see that when the function palinfn() is coded, we use a character pointer 'pt' in which we store the base address of the array that is received in the pointer 'pa', since we would not lose it when we code the condition part of the *for* loop. We see that in the function main(), once the length of the string is found out, immediately we check whether the length is odd or even and if it is even, we skip the call to the function palinfn(), since the palindromic strings are always of odd length. Thus we make the program work only for strings which contain odd number of characters, including the spaces. Again while the string has odd number of characters or the length of the string is an odd number, we skip the value '1', which again is not valid for the palindrome test, as no palindromes would be having the length as '1'.

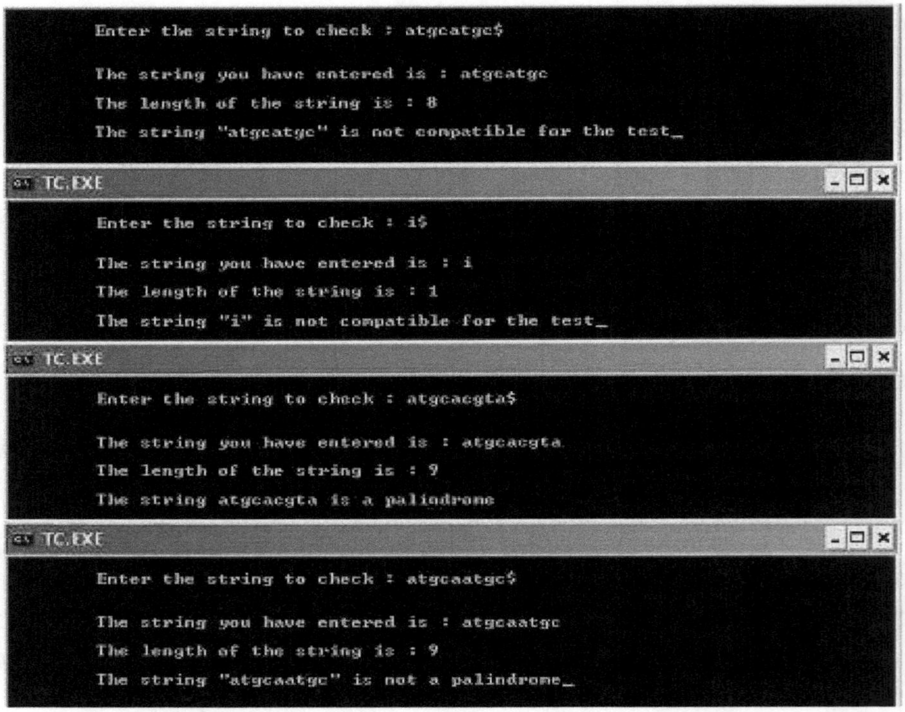

Figure 7.32 Checking the palindromic nature of a string

Figure 7.32 shows the sample run of the program for four different kinds of inputs (For a string whose length is odd, for a string whose length is '1', for a string whose length is odd and which is a palindrome, and for a string whose length is odd and which is not a palindrome).

I hope the program and output screenshot is clear to you. For more clarity, do go through the memory pictures given for each function call, in Figures 7.33–7.36 and the explanation given at the end.

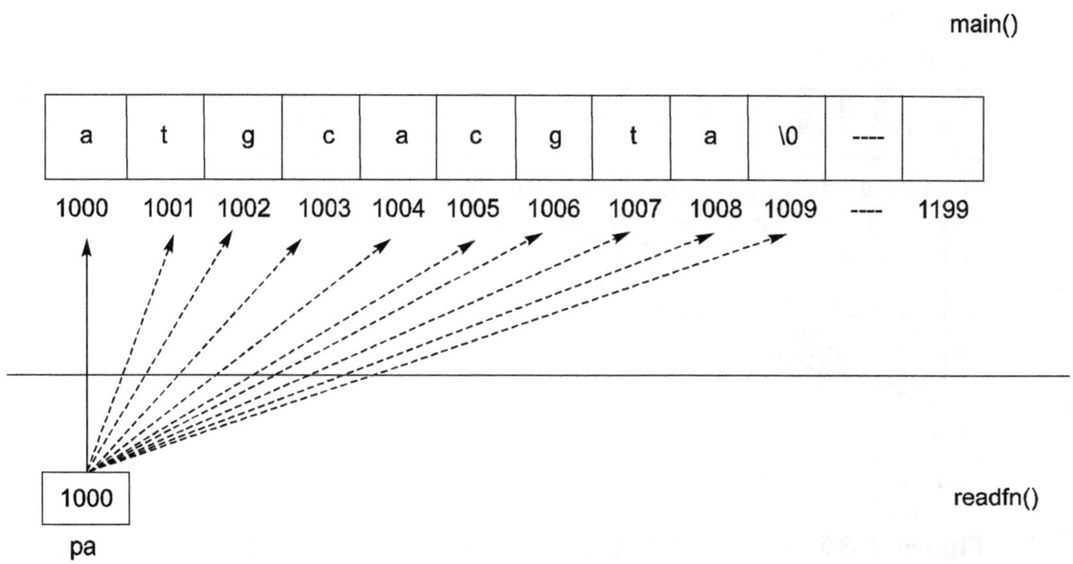

Figure 7.33 Reading the string through the function readfn()

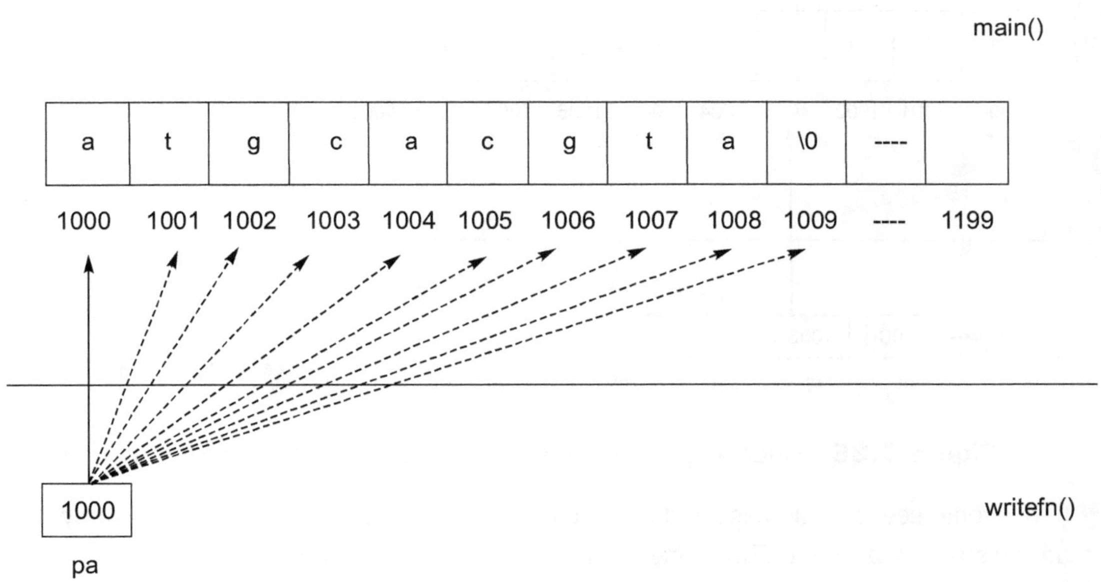

Figure 7.34 Displaying the string through the function writefn()

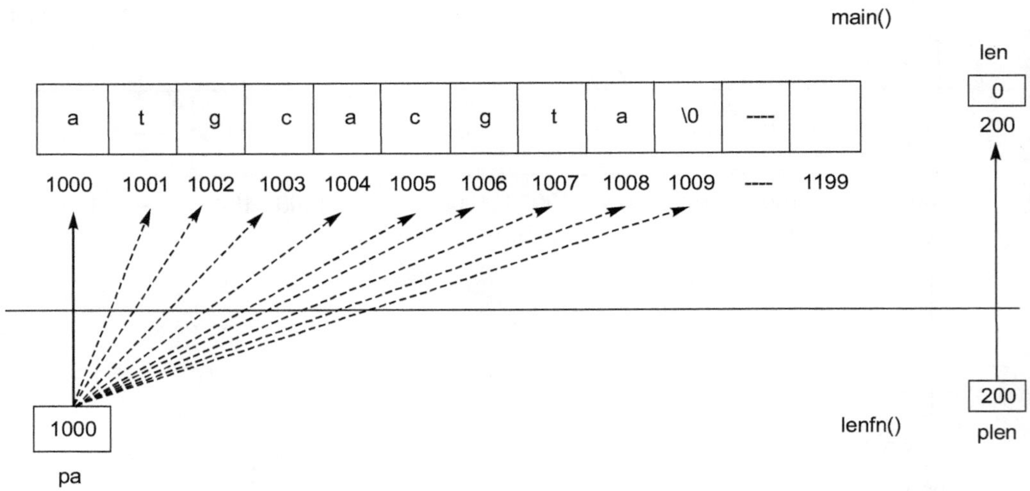

Figure 7.35 Finding the length of the string using the function lenfn()

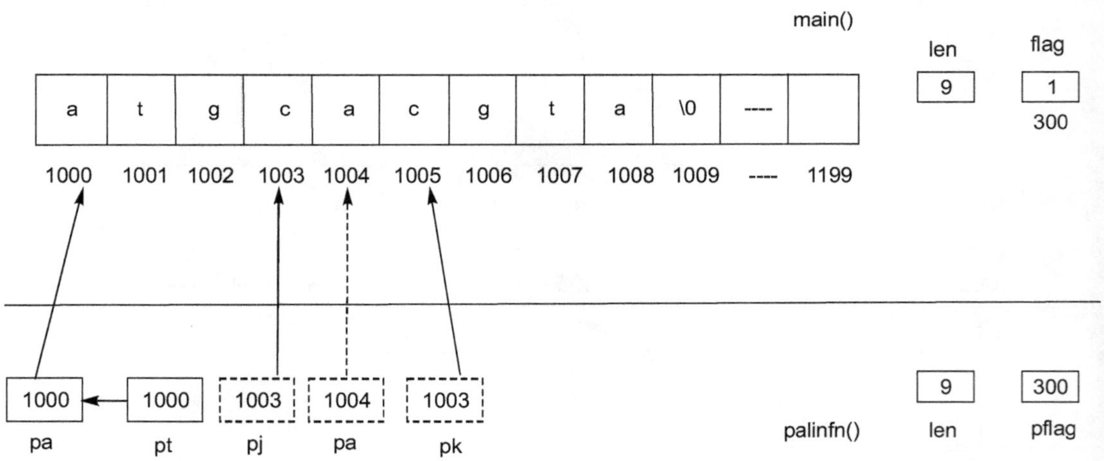

Figure 7.36 Checking the nature of the string using function palinfn()

We could see that the base address is passed while we read and display the array. Once we read the string, we store NULL character at the end of the array. So, when we read the array back and display the string, we check for the occurrence of NULL character. For the function lenfn(), we pass the base address of the array along with the address of the variable 'len', whose value is '0' from the function main(). As we read the array in the function lenfn(), we update the variable 'len' in the function main() through the pointer 'plen' in the function lenfn(). When we check the nature of the string (whether it is a palindrome or not), we pass the base address of the array along with the address of the variable 'flag' (Boolean variable) whose initial value is '1' and the value of length obtained using the earlier function. The base address of the array is collected in the pointer 'pa' and the address of the variable 'flag' is collected in the pointer 'pflag'. We have three more pointers declared in the function palinfn(), namely 'pt' for storing the base address so that we won't lose it, 'pj' for positioning one character towards left of the middle character of the given string and 'pk' for positioning one character towards right of the middle character of the given string. As we go with the coding, 'pj' is decremented and 'pk' is incremented and as we do so, we check for the equality of the characters pointed by them and if so, we proceed. Even if one mismatch occurs, then we stop by setting the value of the variable 'flag' to '0' (this is done through the pointer to flag 'pflag') and break the loop. In the function main(), we print the nature of the string (whether palindrome or not) based on the value of the variable 'flag'. If the value is '1', the string given is a palindrome and if it is '0', it is not.

Thus we are able to associate functions and pointers with arrays and build an efficient relationship among these three. You may try out how the multi-dimensional arrays work with functions and pointers. The discussion of it, however, would be beyond the scope of this book.

7.10 RECURSIVE FUNCTIONS

Recursive functions are a significant feature of C language. Any function that calls itself several times is called as a recursive function. Consider the following simple code.

```
main()
{
printf("Good Morning");
main();
}
```

What would be the result of executing the above program code? Yes, you are right. We will get the message Good Morning infinite number of times and the reason for this is the function main() is being called from within its coding. So there would be no termination for the execution of the program, unless the stack overflows and you get an error message as shown in the Figure 7.37.

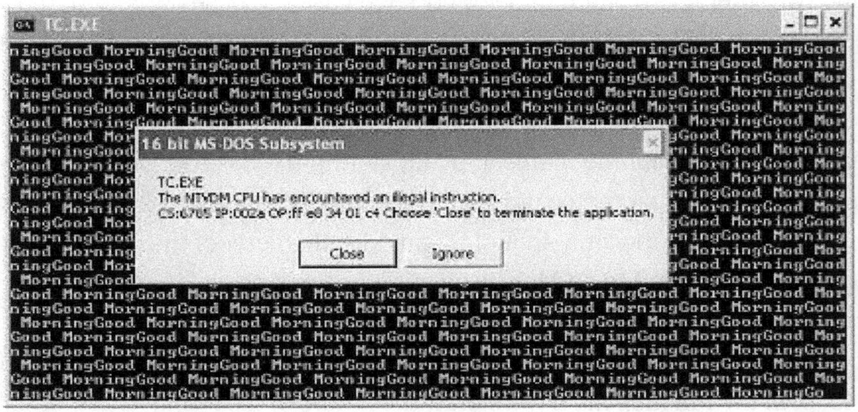

Figure 7.37 Error message displayed for the recursive program

Recursive functions are used in many applications, when we want to pass a string through a cycle and iterate it and print it as many times we would like to without really using the idea of the loops. Unlike the above program, any recursive function shall have criteria for terminating it. Otherwise, it will never terminate and you will get error as shown in the figure. The key advantage of using recursive functions is that, as said earlier, they simply replace loops as they recursive functions themselves are iterative in nature. The disadvantage of using recursive functions is that computationally they are very expensive since more number of interrupts will occur during the program execution.

But, in most of the applications involving calculations and string processing, recursion is very useful. The most important thing one should keep in mind before using recursive logic is that, the function has to terminate at some point of time once a proper condition is given in the function coding. Be aware of the fact that what can be done with recursion, could be done away even without it. So use recursive logic in your programs only when their need is essential. Let us consider a program where we find out the factorial of a given unsigned integer using recursive function. The following is the program code for that.

```
/* Factorial of a given number using recursion */
#include<stdio.h>
#include<conio.h>
void main()
    {
    unsigned long f;
```

```
        int n;
unsigned long fact(int);
        clrscr();
        printf("\n\tEnter the number : ");
        scanf("%d",&n);
        f=fact(n);
        printf("\n\tThe factorial of the number %d is %lu",n,f);
        getch();
        }
unsigned long fact(int x)
        {
        unsigned static long f1;
        if(x<=1)
        return 1;
        else
        f1=x*fact(x-1);
        return f1;
    }
```

We see that in the function fact(), we have the condition for termination as if(x==1). When the condition stated above is true, the control comes back to the calling function and prints the value of the factorial. If we give an input '5' for the variable 'n' in the function main(), it is processed in the function fact() recursively and finally when the control returns to the function main(), it would have the factorial of '5', which is '120' and it would be printed then. We also note from the function fact() that, the variable 'f1' is declared as unsigned static long. When we declare something as static, the initial value of that variable would be '0' and it will retain the values between the function calls. Figure 7.38 shows the factorial value for the input '12'. The flow of the control inside the function fact() while it is executed is shown in the Figure 7.39. For the sake of convenience, we have taken the value of 'n' as '5'.

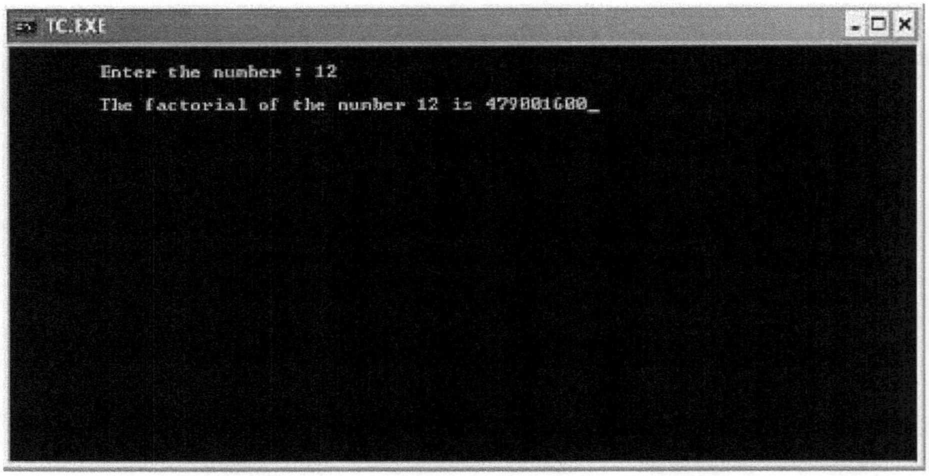

Figure 7.38 Factorial of '12' using recursion

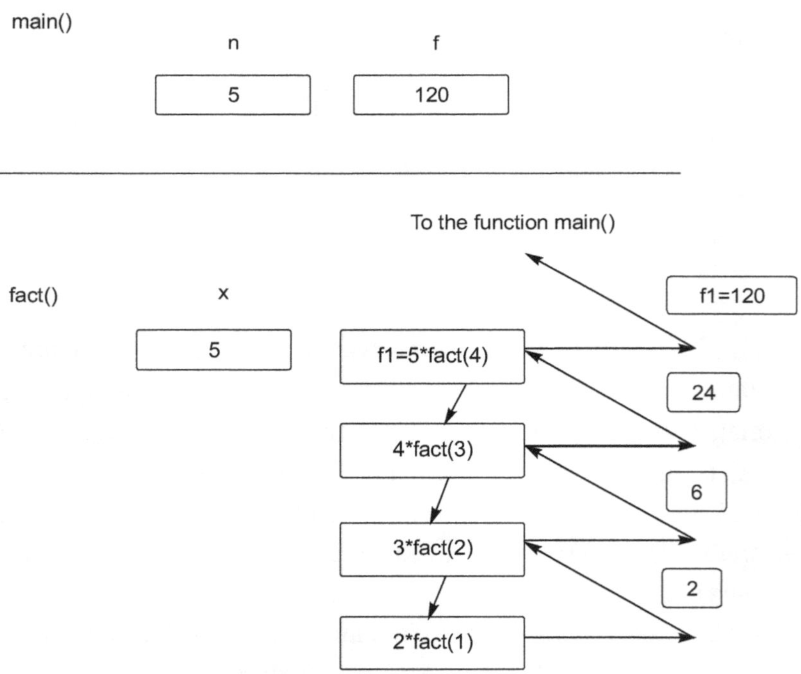

Figure 7.39 Factorial of '5' using recursion—flow of control

We note that when the value of 'x' reaches '1', a value of '1' is returned to the calling function. Thus the expression 2*fact(1) will evaluate to 2, which is returned to its calling function, which is recursively the function fact() and it is multiplied with the value '3' and the result '6' is returned to its calling function which is again fact() and this goes on and on and finally the control comes back to the calling function main() with the final result '120' and this gets printed in the output.

Let us work out another program with string, where we would print the reversed form of the given input string. The following is the program code for that.

```
/* Reversing a string using recursion */
#include<stdio.h>
#include<conio.h>
int i;
char a[100];
void main()
{
void rev(int);
clrscr();
printf("Enter the string : ");
for(i=0;(a[i]=getchar())!='$';i++);
a[i]='\0';
i--;
printf("\nReversed string : ");
rev(i);
getch();
}
void rev(int j)
{
if(j<0)
return;
else
{
printf("%c",a[j]);
j--;
rev(j);
}
}
```

The output of the above program is shown in the Figure 7.40.

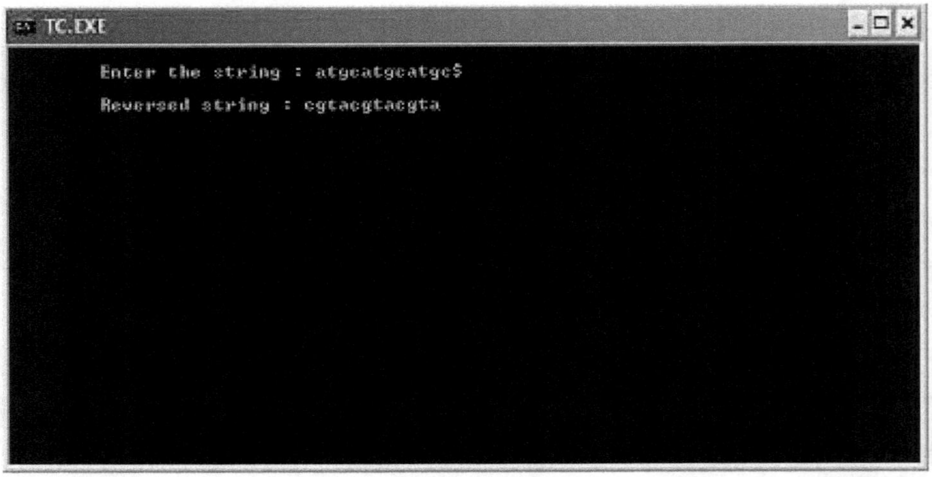

Figure 7.40 Reversing a string using recursion

I would suggest you to make out the flow of control inside the function rev() and hence draw the memory picture. With this program we would wind up the discussion of recursive functions and their usage.

For recursion, additional work has to be done in the form of saving return addresses, so that the correct statement is executed next when the *return* is executed. Compared to iteration, recursion is more efficient.

SUMMARY

This entire chapter has dealt with the concept of functions with a special attention to the user-defined functions. We started with explaining the reasons for having user-defined functions and we discussed the categories of the functions. We also discussed the methods of accessing the functions namely call by value and call by reference. Following that we discussed what function prototyping is and how significant is prototyping in C language. Different types of function headers were discussed in brief and we wrote some simple functions and validated them with C programs.

Based on our understanding of the basic usage of functions, we started discussing the way the functions work with one-dimensional and two-dimensional numerical arrays. We had an elaborate discussion on how the numerical arrays are passed on to the functions and how they are accessed and manipulated in the functions. We continued the discussion with the one-dimensional character arrays and we did a few programs to understand how the character arrays work with the functions.

We then followed the discussion of functions and pointers and what is the importance of using pointers with functions. We discussed a few programs that demonstrated the importance of passing addresses to the functions and thereby return more than one value from the functions. This section has provided the basis for the subsequent sections, where we discussed how the pointers and functions work with the numerical and character arrays. We discussed elaborately the way the pointers and functions work with the arrays. We had screenshots and memory pictures in the sections for a better understanding of the concepts and the technical details involved in a program.

We had a discussion on recursive functions and their applications in the concluding part of this chapter. We illustrated two programs for getting the concept in a much better way, with necessary screenshots and memory pictures.

I am sure the concepts that are provided in this chapter would have been a great amount of significance to the learners and obviously the concepts would be a good foundation for those Bioinformaticians who aspire to develop the best algorithms.

CRITICAL THINKING QUESTIONS

What will be the output of the following programs?

1.
```
small(int,int);
main()
{
Int a,b;
scanf("%d%d",&a,&b);
small(&a,&b);
}
small(int p,int q)
{
 if(p==r)
```

```
printf("The numbers are equal");
else
if(p<r)
printf("The smaller number is : %d",p);
else
printf("The smaller number is : %d",q);
}
```

2.

```
main()
{
int a,b,diff;
printf("Enter the integers :  ");
scanf("%d%d",&a,&b);
diff=sub(a,b);
printf("The difference is %d",diff);
int sub(int p, int q)
return p-q;
}
```

3.

```
main()
{
printf("The following message is displayed");
message();
}
message();
{
printf("Good Morning");
}
```

4.

```
main()
{
Int n,sqr;
```

```
Printf("Enter the number :  ");
scanf("%d",&n);
sqr=sqrvalue(&n);
printf("The square of %d is %d",n,sqr);
}
int sqrvalue(int a)
{
return a*a;
}
```

5.

```
main()
{
int x,y,z,sum  ;
printf("Enter the numbers :  ");
scanf("%d%d%d",&x,&y,&z);
sum=add(x,y,z);
printf("The total is %d",sum);
}
int add(int p, int q, int r)
{
return(p,q,r);
}
```

6.

```
main()
{
int num=10;
fn(num);
printf("%d",num);
}
fn(int num)
{
```

```
return ++num;
}
```

7.
```
main()
{
printf("\nUVCE");
main();
}
```

8.
```
void main(int x, int y)
{
printf("%d",x) ;
}
```

9.
```
main()
{
char x=67, y='c';
fn(x,y);
}
fn(x,y)
{
printf("x=%d, y%d", x, y);
}
```

10.
```
main()
{
int m=50, n=60, x;
x=fn(m,n) ;
printf("x=%d",x);
}
fn(int m, int n)
{
int y, z;
```

```
y=m+n;
z=m*n;
return (z,y,z+y);
}
```

SELF-LEARNING EXERCISES

1. Rewrite all the programs that you have been given within self-learning exercises till the last chapter using functions.
2. Write a function to find out the GCD of two numbers.
3. Write a recursive function to generate the first 'n' terms of Fibonacci series.
4. Write a recursive function to find the minimum and maximum of numbers.
5. Write a function to swap two numbers without using a third variable.
6. Write a program to concatenate two strings using functions and pointers.
7. Write a function to search for a given string and if it is available, replace it with another string, wherever it appears in the input string.
8. Write a function to find the length of the given string using pointers.
9. Write a function that finds out the number of palindromes in the given text.
10. Write a function that checks whether the given string is a substring of another using pointers.
11. Write a function to convert the lowercase array to uppercase array using pointers.
12. Write a function to find out the number of words, lines and characters in the given text using pointers.
13. Write a function to find out the number of uppercase and lowercase alphabets, digits and special characters in the given text using pointers.
14. Write a function to count the number of prime numbers between the given limits using functions.
15. Write a function to compare two given strings using functions and pointers.

FILES AND COMMAND LINE ARGUMENTS

8.1 INTRODUCTION TO FILES

Files are places for reading data from or writing data to. This includes disk files and it devices such as the printer or the monitor of a computer. C treats all information which enters or leaves a program as though it were a stream of bytes: a file. The most commonly used file streams are stdin (the keyboard) and stdout (the screen), but more sophisticated programs need to be able to read or write to files which are found on a disk or to the printer, etc.

An operating system allows a program to see files in the outside world by providing a number of channels or **portals** (inlets and outlets) to work through. In order to examine the contents of a file or to write information to a file, a program has to open one of these portals. The reason for this slightly indirect method of working is that channels/portals hide operating-system-dependent details of filing from the programmer. Think of it as a protocol. A program which writes information does no more than pass that information to one of these portals and the operating system's filing subsystem does the rest. A program which reads data simply reads values from its file portal and does not have to worry about how they got there. This is extremely simple to work in practice.

The standard library in C has many file I/O functions. The file-handling functions, like the other functions, are easy to use and above all they are powerful and complete. This chapter elaborates the various file operations and file manipulations that are important for any person who uses C to develop an application. The following is the sequence of I/O operations while using files.

- Open the file
- Read from or write to the file
- Close the file

A program opens a file by calling a standard library function and is returned a file pointer by the operating system, which allows a program to address that particular file and to distinguish it from all others.

C provides two levels of file handling; these can be called **high level** and **low level**. High level files are all treated as text files. In fact, the data which go into the files are exactly what would be seen on the screen, character by character, except that they are stored in a file instead. This is true whether a file is meant to store characters, integers or floating point types. Any file, which is written to by high level file handling functions, ends up as a text file, which could be edited by any text editor.

High level text files are also read back as character files, in the same way that input is acquired from the keyboard. This all means that high level file functions are identical in concept to keyboard/ screen input/output.

The alternative to these high level functions is obviously low level functions. These are more efficient, in principle, at filing data as they can store data in large lumps, in raw memory format, without converting to text files first. Low level input/output functions have the disadvantage that they are less **programmer-friendly** than the high level ones, but they are likely to work faster.

8.2 FILE POSITIONS

When data are read from a file, the operating system keeps track of the current position of a program within that file so that it only needs to make a standard library call to 'read the next part of the file' and the operating system obliges by reading some more and advancing its position within the file, until it reaches the end. Each single character which is read causes the position in a file to be advanced by one.

Although the operating system does a great deal of task regarding file positions, a program can control the way in which that position changes, if need be. In most cases it is not necessary and it should be avoided, since complex movements within a file can cause complex movements of a disk drive mechanism which in turn can lead to wear on disks and the occurrence of errors.

8.3 HIGH LEVEL FILE HANDLING FUNCTIONS

Most of the high level input/output functions which deal with files are easily recognizable in that they start with the letter f. Some of these functions will appear strikingly familiar. For instance:

```
fprintf()
fscanf()
fgets()
fputs()
```

These are all generalized file handling versions of the standard input/output library. They work with generalized files, as opposed to the specific files stdin and stdout which printf() and scanf() use. The file versions differ only in that they need an extra piece of information: the file pointer to a particular portal. This is passed as an extra parameter to the functions. They process data in an identical way to their standard I/O counterparts. Other filing functions will not look so familiar. For example:

```
fopen()
fclose()
getc()
ungetc();
putc()
fgetc()
fputc()
feof()
```

Before any work can be done with high level files, these functions need to be explained in some detail. This is exactly what is done in the following sections.

8.4 OPENING A FILE

Consider the following program coding:

```
/* Demonstrating the file operations in C */
#include <stdio.h>
main( )
{
FILE *fp;
```

```
char name[100];
int index;
clrscr();
fp = fopen("sample.txt","w"); /* Open the file for writing */
/* Some lines of coding go here */
strcpy(name,"ATGCATGCATGCAATGCTATAGCATGC");
for (index = 1;index <= 10;index++)
fprintf(fp,"%s Line number %d\n",name,index);
fclose(fp); /* Close the file before ending the program */
printf("\nThe line is written specified number of times in the file you have specified!!!");
getch();
}
```

We begin as before with the "#include" statement for "stdio.h", then define some variables for use in the example including a rather strange-looking new type. The type **FILE** is used for a file variable and is defined in the "stdio.h" file. It is used to define a file pointer for use in file operations. The definition of C contains the requirement for a pointer to a **FILE**, and as usual, the name can be any valid variable name (follow the variable-naming conventions here).

Before we can write to a file, we must open it. What this really means is that we must tell the system that we want to write to a file and what the filename is. We do this with the function **fopen()** as illustrated in the program. The file pointer, **fp** in our case, points to the file and **two arguments** are required in the parentheses, the filename first, followed by the mode of accessing the file.

The **first argument** is the filename, which is any valid filename, and can be expressed in upper or lower case letters, or even mixed if you so desire. It is enclosed in double quotes. For this example we have chosen the name **sample.txt**. This file should not exist on your disk at this time. If you have a file with this name, you should change its name or move it because when we execute this program, its contents will be erased. If you don't have a file by this name, that is good because we will create one and put some data into it.

The **second argument** is the file attribute and can be any of three letters, "r", "w", or "a", and must be lower case. When an "r" is used, the file is opened for reading, a "w" is used to indicate a file to be used for writing, and an "a" indicates that you desire to append additional data to the data already in an existing file.

Opening a file for **reading** requires that the file already exist. If it does not exist, the file pointer will be set to NULL and the same can be checked by the program. When a file is opened for **writing**, it will be created if it does not already exist and it will be reset if it does resulting in deletion of any data already there. When a file is opened for **appending**, it will be created if it does not already exist and it will be initially empty. If it does exist, the data input point will be the end of the present data so that any new data will be added to any data that already exists in the file.

8.5 OUTPUTTING TO A FILE

The job of actually outputting to the file is nearly identical to the outputting we have already done to the standard output device. The only real differences are the new function names and the addition of the file pointer as one of the function arguments. In the programs that access files, the function **fprintf()** replaces our familiar function printf() and the file pointer defined earlier is the first argument within the parentheses. The remainder of the statement looks like, and in fact is identical to, the function printf().

8.6 CLOSING A FILE

To close a file, you simply use the function **fclose()** with the file pointer in the parentheses. Actually, in this simple program, it is not necessary to close the file because the system will close all open files before returning to DOS. It would be good programming practice for you to get in the habit of closing all files in spite of the fact that they will be closed automatically, because that would act as a reminder to you of what files are open at the end of each program. You can open a file for writing, close it, and reopen it for reading, then close it, and open it again for appending, etc. Each time you open it, you could use the same file pointer, or you could use a different one. The file pointer is simply a tool that you use to point to a file and you decide what file it will point to.

8.7 WRITING SIMPLE PROGRAMS IN FILES

Compile and run the above program. You will only get the message that is shown in the Figure 8.1. To get the contents of the actual file, look at your directory for a file named **sample.txt** and use the DOS command **type** to view its contents. That is where your output will be. Compare the output with that specified in the program. It should agree. If you desire to choose the DOS prompt from within C editor, you may select the option **File** from the editor (by pressing Alt+f) and then select the option **OS shell** (by pressing the letter O from the keyboard) as shown in Figure 8.2. When you view the contents it shall be the one as shown in the Figure 8.3. Do not erase the file named sample.txt as we will use it later in this Chapter.

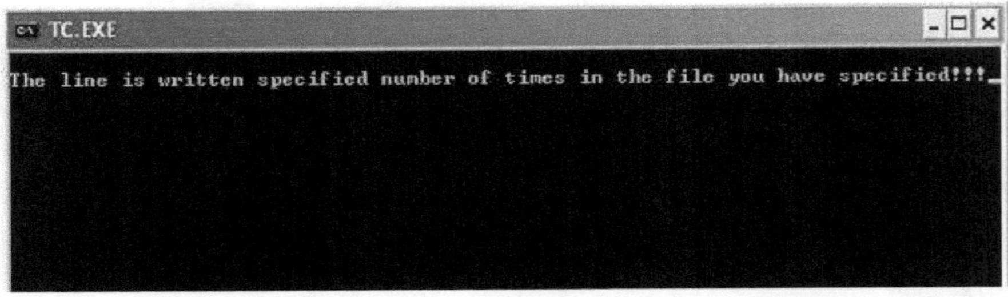

Figure 8.1 Output for the program that writes data to the file

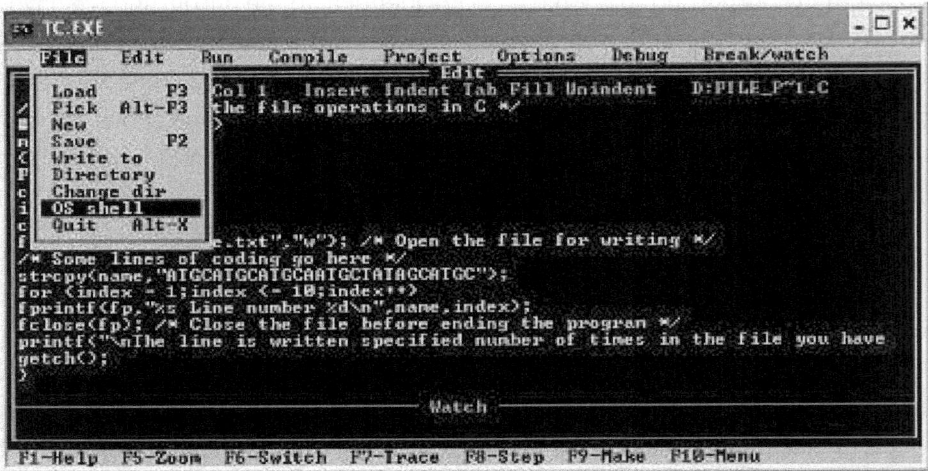

Figure 8.2 Getting into the DOS prompt from within the C editor

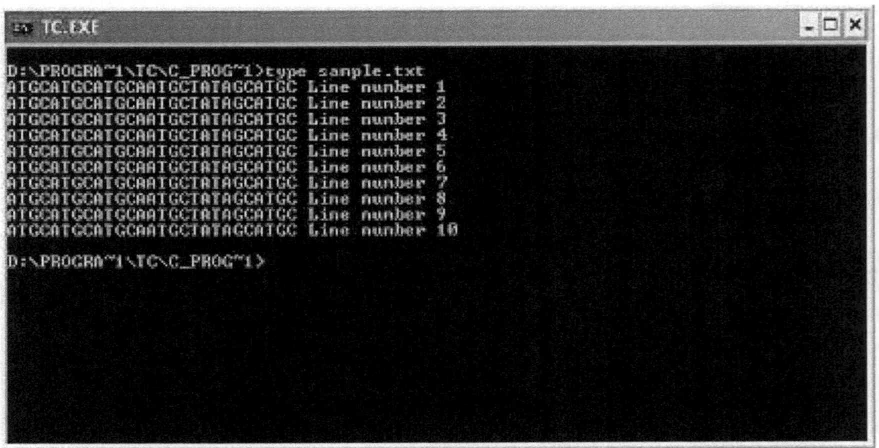

Figure 8.3 Contents of the file sample.txt in the directory

Consider the following program coding:

```
/* Appending contents to an existing file */
#include <stdio.h>
main()
{
FILE *fp;
char name[100];
int index,count; clrscr();
strcpy(name,"AAAATTTGGGCCCCCCTTTTAAAGGGG");
fp = fopen("SAMPLE.TXT","a"); /* Open for appending */
for (count = 1;count <= 10;count++)
{
for (index = 0;name[index];index++)
putc(name[index],fp); /* Output a single character to the file */
putc('\n',fp); /* Output a linefeed to the file */
}
fclose(fp);
printf("\nThe given sequence is appended successfully");
getch(); }
```

Figure 8.4 shows the output for the above program. We view the contents of the file **sample.txt** in which the appending is carried out.

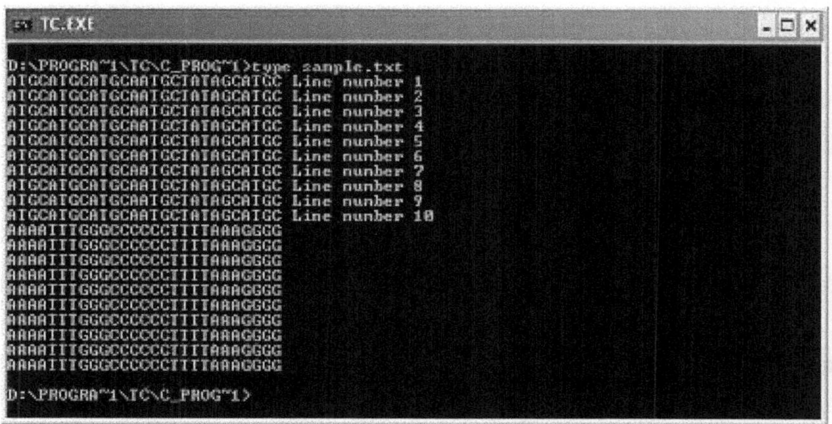

```
D:\PROGRA~1\TC\C_PROG~1>type sample.txt
ATGCATGCATGCAATGCTATAGCATGC Line number 1
ATGCATGCATGCAATGCTATAGCATGC Line number 2
ATGCATGCATGCAATGCTATAGCATGC Line number 3
ATGCATGCATGCAATGCTATAGCATGC Line number 4
ATGCATGCATGCAATGCTATAGCATGC Line number 5
ATGCATGCATGCAATGCTATAGCATGC Line number 6
ATGCATGCATGCAATGCTATAGCATGC Line number 7
ATGCATGCATGCAATGCTATAGCATGC Line number 8
ATGCATGCATGCAATGCTATAGCATGC Line number 9
ATGCATGCATGCAATGCTATAGCATGC Line number 10
AAAATTTGGGCCCCCCTTTTAAAGGGG
AAAATTTGGGCCCCCCTTTTAAAGGGG
AAAATTTGGGCCCCCCTTTTAAAGGGG
AAAATTTGGGCCCCCCTTTTAAAGGGG
AAAATTTGGGCCCCCCTTTTAAAGGGG
AAAATTTGGGCCCCCCTTTTAAAGGGG
AAAATTTGGGCCCCCCTTTTAAAGGGG
AAAATTTGGGCCCCCCTTTTAAAGGGG
AAAATTTGGGCCCCCCTTTTAAAGGGG
AAAATTTGGGCCCCCCTTTTAAAGGGG

D:\PROGRA~1\TC\C_PROG~1>
```

Figure 8.4 Appended contents of the file sample.txt

The program begins with the "include" statement, then defines some variables including a file pointer. We have named the file pointer as fp. We then define a string of characters to use in the output function using the function strcpy(). We are ready to open the file for appending and we do so in the function fopen(), except this time we use the upper cases for the filename and its extension. This is done simply to illustrate that DOS does not care about the case of the filename. Notice that the file will be opened for appending so we will add to the lines inserted during the last program. The program is actually coded with two nested *for* loops. The outer loop is simply a count to ten so that we will go through the inner loop ten times. The inner loop calls the function putc() repeatedly until a character in the array **name** is detected to be a zero.

The part of the program we are interested in is the function putc(). It outputs one character at a time, the character being the first argument in the parentheses and the file pointer being the second and last argument. Why the designer of C made the pointer first in the function fprintf(), and last in the function putc() is a good question for which there may be no answer. It seems like this would have been a good place to have used some consistency. When the array **name** is exhausted, a new line is needed because a new line was not included in the definition above. A single **putc** is then executed which outputs the "\n" character to return the carriage and do a linefeed. When the outer loop has been executed ten times, the program closes the file and terminates. Compile and run this program but once again there will be no output to the monitor. Following execution of the program, view the contents of the file using the DOS command **type** and you will see that the 10 new lines were added to the end of the 10 that already existed. If you run it again, yet another 10 lines would be added.

Now that we have been able to write contents to the file and append the same, we are ready to read the contents that we have written and then appended. For that we may have a coding like the one shown below.

```
/* Program to read the contents of a file – One character at a time */
#include <stdio.h>
main( )
{
FILE *fp;
int c;
clrscr();
fp= fopen("sample.txt","r");
if (fp == NULL)
{
```

```
printf("File doesn't exist\n");
getch();
return;
}
do
{
c = getc(fp); /* Get one character from the file */
putchar(c); /* Display it on the monitor */
} while (c != EOF); /* Repeat until EOF (end of file) */
fclose(fp);
getch();
}
```

This program begins with the familiar "include", some data definitions, and the file opening statement which should require no explanation except for the fact that an "r" is used here because we want to read it. In this program, we check to see that the file exists, and if it does, we execute the main body of the program. If it doesn't, we print a message and quit. If the file does not exist, the system will set the pointer equal to NULL which we can test.

The main body of the program is the *do-while* loop in which a single character is read from the file and output to the monitor until an EOF (end of file) is detected from the input file. The file is then closed and the program is terminated. Figure 8.5 shows the output for the above program coding (see that we have read the file sample.txt and the contents of it are displayed on the screen). We could use any loop other than a *do-while* and I would appreciate the learners if they use the other two loops to achieve the same result.

At this point, we have the potential for one of the most common and most perplexing problems of programming in C. The variable returned from the function getc() is a character (based on the prototype), so we could use a "char" variable for this purpose. There is a problem with that however, because on some, if not most, implementations of C, the EOF returns a minus one which a "char" type variable is not capable of containing. A "char" type variable can only have the values of zero to 255, so it will return a 255 for a minus one on those compilers that use a minus one for EOF. This is a very frustrating problem to try to find because no diagnostic is given. The program simply can never find the EOF and will therefore never terminate the loop. This is easy to prevent, always use an "int" type variable for use in returning an EOF. You can tell what your compiler uses for EOF by looking at the "stdio.h" file where EOF is defined. That is the standard place to define such values.

After you compile and run this program and are satisfied with the results, it would be a good exercise to change the name of "sample.txt" and run the program again to see that the NULL test actually works as stated. The output when the file name is changed is shown in the Figure 8.6.

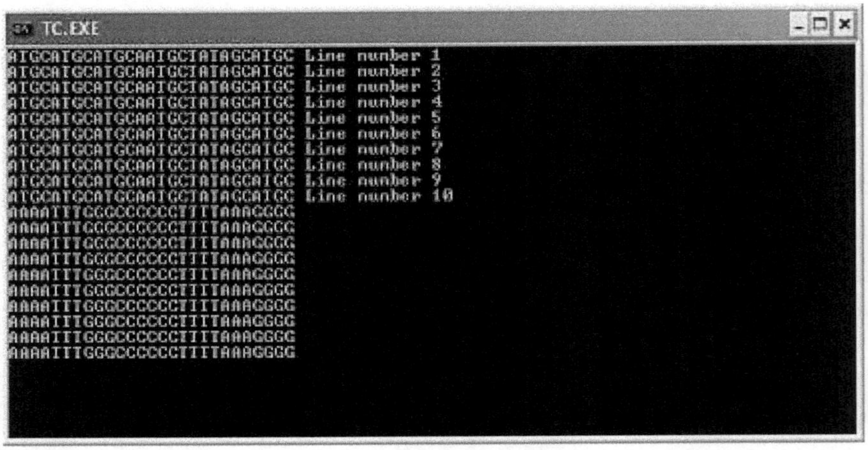

Figure 8.5 Reading the contents of sample.txt character by character

In the above program, we have read one character at a time from the file and displayed the same on the monitor. But if we want to read one word at a time, we could do it and the following would be the modified coding for the same.

```
/* Program to read the contents of a file - One word at a time */
#include <stdio.h>
main( )
{
FILE *fp;
int c;
char word[100];
clrscr();
fp= fopen("sample.txt","r");
if (fp == NULL)
{
printf("File does not exist\n");
getch();
return;
}
```

```
do
{
c = fscanf(fp,"%s",word); /* Get one word from the file */
printf("%s\n",word); /* Display it on the monitor */
} while (c != EOF); /* Repeat until EOF (end of file) */
fclose(fp);
getch();
}
```

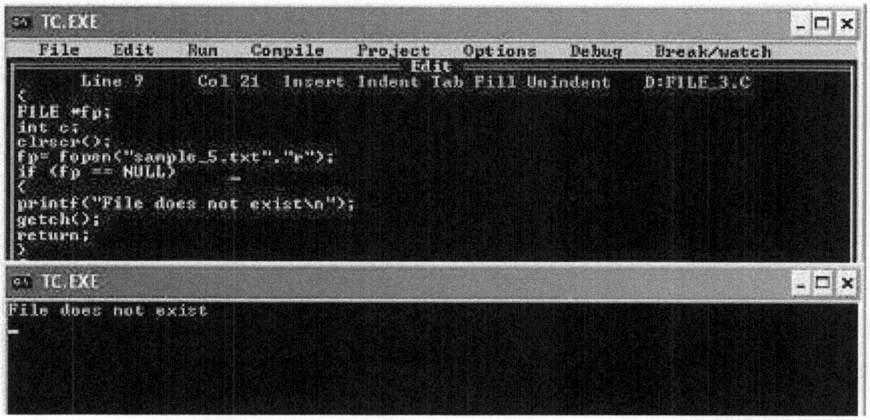

Figure 8.6 Reading the contents of the file sample_5.txt

("File does not exist error" is indicated)

The output of the above program is shown in the Figure 8.7 below.

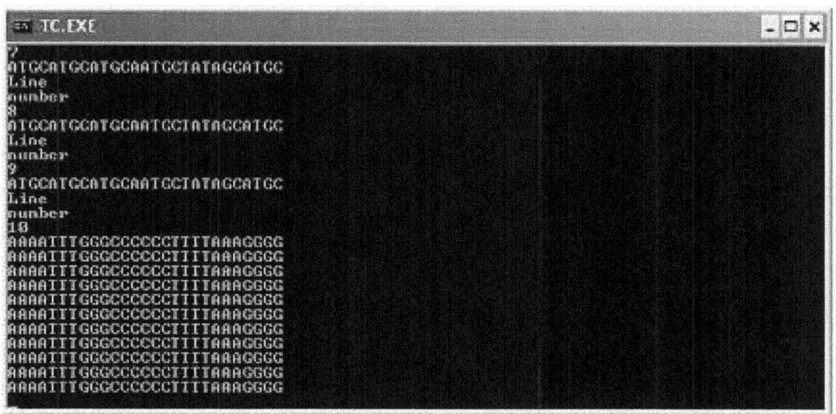

Figure 8.7 Reading the contents of sample.txt word by word

Inspection of the program output (from Figure 8.7) will reveal that when we read data in and detect the EOF, we print out something before we check for the EOF resulting in an extra line of printout. What we usually print out is the same thing printed on the prior pass through the loop because it is still in the buffer of the array word. We therefore must check for EOF before we execute the function printf(). Compile and execute the above program and observe the output. There will be additional lines displayed because of the function printf() before checking for EOF. Thus in the following program coding we use NULL instead of checking EOF.

In the earlier two examples, we read the file character by character and then word by word. Now we will try reading the contents line by line. The following would be coding required for the same.

```
/* Program to read the contents of a file - One line at a time */
#include <stdio.h>
main( )
{
FILE *fp;
char *c;
char word[100];
clrscr();
fp= fopen("sample.txt","r");
if (fp == NULL)
{
printf("File does not exist\n");
getch();
return;
}
do
{
c = fgets(word,100,fp); /* Get one line from the file */
if(c!=NULL)
printf("%s",word); /* Display it on the monitor */
} while (c != NULL); /* Repeat until NULL */
fclose(fp);
getch();
}
```

Figure 8.8 is the output of the above coding and we see that there are no additional lines as in Figure 8.7. Due to the presence of additional lines, we get a partial output in Figure 8.7 and this is rectified in the above program and hence we get the complete contents of the file.

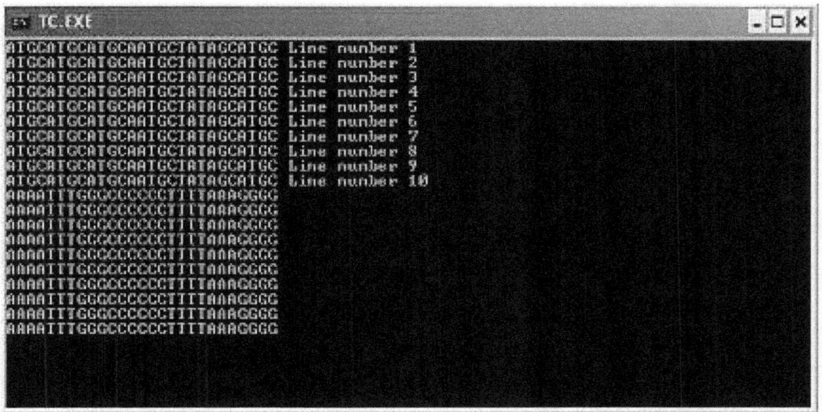

Figure 8.8 Reading the contents of sample.txt line by line

8.7.1 How to use a Variable File Name?

So far we have been opening the file by specifying the name with extension directly as one of the arguments to the function fopen(). What if we want to open the file by giving the name with extension through the keyboard? This is possible by just doing a minor modification in the above program code. I have presented the new coding below and it should be clear to you in all the aspects, since we have dealt with them already.

```
/* Program to read the contents of a file - One line at a time – Getting File Name from the keyboard using the function scanf() */
#include <stdio.h>
main()
{
FILE *fp;
char *c;
char word[100], filename[100];
clrscr();
printf("\nEnter the file name -> " );
scanf("%s",filename); /* Read the name of the desired file to be opened */
fp= fopen(filename,"r");
```

```
if (fp == NULL)
{
printf("File does not exist\n");
getch();
return;
}
do
{
c = fgets(word,100,fp); /* Get one line from the file */
if(c!=NULL)
printf("%s",word); /* Display it on the monitor */
} while (c != NULL); /* Repeat until NULL */
fclose(fp); getch();
}
```

If you compile and execute this program, it will request the file name from you. Enter the desired file to be opened. The file you desire could be even a .c file and not necessarily .txt file. The output for the above program is shown in Figure 8.9.

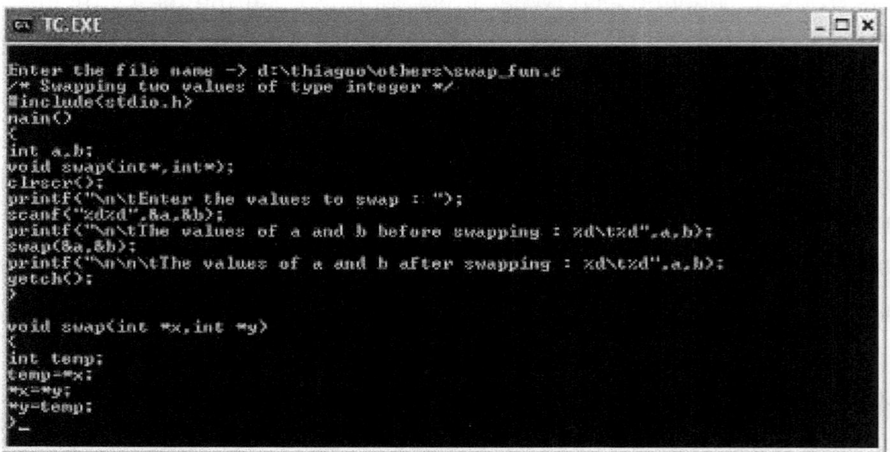

Figure 8.9 Reading the file whose name is given dynamically

There is another point the learners must observe and understand here. When we enter the name of the file, we are supposed to give the complete path of the file if we are not displaying the contents of that file which is available in the working directory of C. In the output shown above, we have chosen to display a .c file which is available in the directory specified from the prompt and it is not the one in the working directory of C.

8.8 COMMAND LINE ARGUMENTS—WHAT AND WHY?

When we use files, if we incorporate the idea of command lines, the program would be more efficient. What are these command line arguments and what is a command line after all?

If you have observed the example programs that we have used in all the Chapters, we have used the function main() invariably in all the programs and as mentioned in the Chapter 2, it is the heart throb of any C program. We know from the concept of functions that a function could be called by passing values or addresses as arguments to it (from the Chapter 7 on functions). Have we ever passed any arguments to the function main()? The answer is very clear. We have never used such idea in any of our programs. Is it possible for us to pass arguments to the function main() as we do with the other functions (both built-in and user-defined)? The answer is yes. If you recall the function prototyping discussion, we had used two terms namely **caller** and **callee**. The **caller** function is that which calls some other function from within its code and the **callee** function is that which is called from another function. Invariably all the functions do have a **caller** for them. So, who is the caller for the function main()? The operating system is the caller for the function main() and one can pass arguments from the operating system to the function main() and make the program work. In our discussion, we have used DOS as the platform and hence we would say DOS is the **caller** for the function main(). Hence if we want to pass arguments to the function main(), we could do so from the DOS prompt, which is called as the **Command Line**. I am sure that by now you would have understood what command line arguments are. In simple words, the arguments that are passed from the command line are called the command line arguments and they are passed to the function main(). They are significant in programming in the sense that the user can execute such programs from the command line and he/she need not get into the coding of the program while validating the program. The following is a simple example of a program where we have used the concept of command line arguments. Here we read the contents of the file and display the same on the monitor.

```
/* Program to read the contents of a file - Command Line Arguments */
#include <stdio.h>
main(int argc, char *argv[])
{
FILE *fp;
char *c;
char word[100], filename[100];
```

```
/*clrscr();*/
if(argc<2)
{
printf("\nUsage Error");
getch();
return; }
fp= fopen(argv[1],"r");
if(fp==NULL)
{
printf("File does not exist\n");
getch();
return;
}
/* Printing the addresses of the arguments */
printf("\n\n\nAddress of the first argument : %u\nAddress of the second argument :
%u\n\n\n",argv[0],argv[1]);
do
{
c = fgets(word,100,fp); /* Get one line from the file */
if(c!=NULL)
printf("%s",word); /* Display it on the monitor */
}while (c!=NULL); /* Repeat until NULL */
fclose(fp);
getch();
}
```

We see that we pass the arguments to the function main() from the DOS prompt and hence get the display of the contents of the file that we want to view. This program shall be compiled first using the key F9 as usual, but the execution shall not be done by pressing the key combination ctrl+F9 as in the earlier cases. Choose the option **File** from the menu and choose the option OS shell. Then type the .exe file name of the presently executed file (invariably .exe file name shall be the same as the original .c source file) as the first argument and the second argument shall be the file name or the path of the filename whose contents are going to be displayed. The command that is to be given from the DOS prompt or the OS shell is the one as given below for the above program.

```
d:\progra~1\tc\bin>file_7 d:\swap_fun.c
```

Assume that the program is saved as file_7.c and the file whose contents are to be viewed is swap_fun.c and it is stored in the D drive of the system. When the .c file is compiled to success, we would get the corresponding .exe file and hence we are ready to execute the program from the command line. Once we press the enter key after the commands are given, we would get the display of the contents of the file swap_fun.c on our screen. Figure 8.10 shows the output along with the command line command. We can display any type of file compatible for the display using these command line parameters. The programmers and those who execute the above program must keep in mind the fact that the command line execution is possible with the .exe file only from the path where it is located and not necessarily the path of the present working directory. Let me explain this idea. We might have installed the C compiler in D drive inside the sub-folder tc. But for our convenience, we might have created two sub-folders namely source_programs and bin inside the folder tc for storing all the C programs and their corresponding .exe files respectively. So, the present working directory in our case is D:\tc. All the source files would be stored in D:\tc\source_programs and all the corresponding .exe files would be stored in D:\tc\bin. Thus when we choose the option OS Shell from the File menu, we would be in the path D:\tc\source_programs. We cannot execute the program from here, since the .exe file is in the path D:\tc\bin. Hence, before we execute the program, we must get into the path D:\tc\bin. This is illustrated in the Figure 8.11.

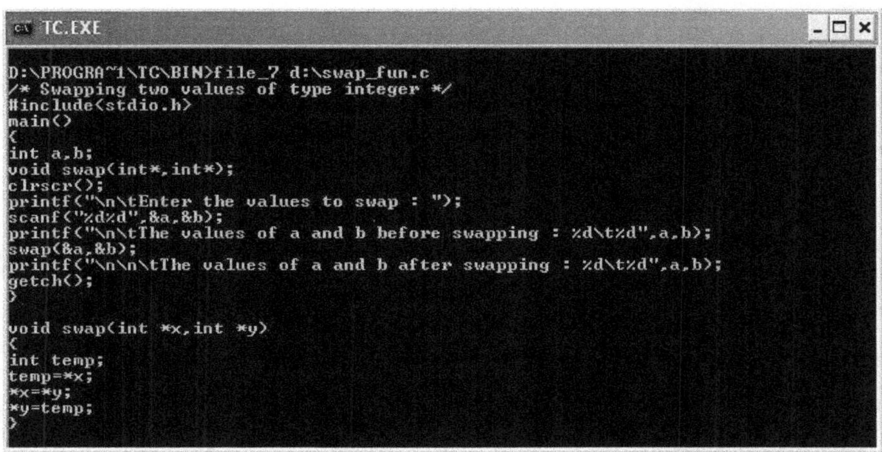

Figure 8.10 Displaying the contents of a file through command line

If we do not follow the above procedure, then we would get the error, which is also illustrated in Figure 8.11.

Thus we can conclude that command line parameters could be executed provided we have the corresponding .exe files in the path from where we pass the parameters. I am sure the learners are now ready to work with more of such programs. We will do some programs in the next section to understand the way the command line arguments work. But before we proceed, one must have a clear understanding of the two arguments argc and argv and hence the way the command line arguments work.

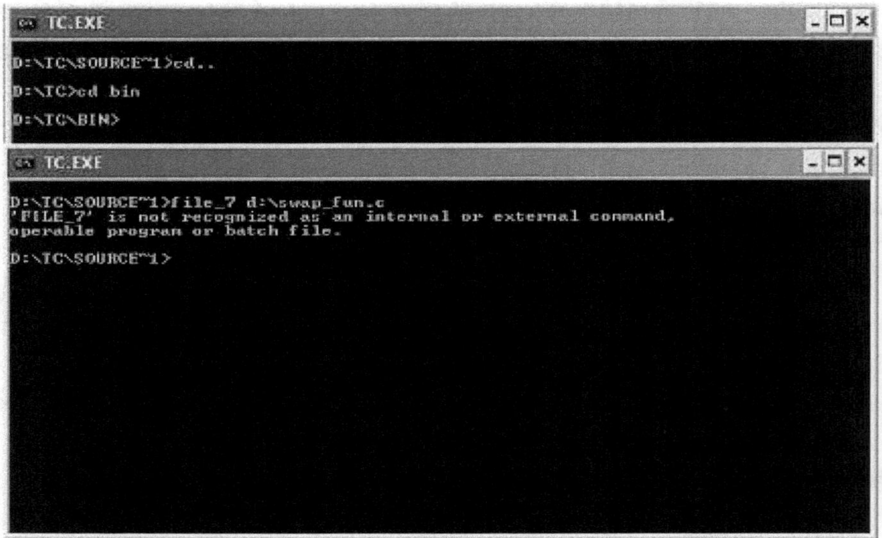

Figure 8.11 Executing the program through the command line

The argument argc is the argument counter and argv is the argument vector. The argument counter keeps a check on minimum required number of arguments that the user must pass through the command line. If you observe the program closely, we have checked this minimum requirement through *if* condition and if the minimum number of arguments required is not supplied from the command line, we immediately flash the error and terminate the program. Since the number of arguments is always counted as a whole number, its type is an integer.

The argument argv as mentioned is the argument vector that keeps the base addresses of the arguments that is passed through the command line. That is the reason why it is declared as an array of pointers. You must be wondering why the base addresses of the arguments are stored and what must be the significance of doing so. If you again recall the method of passing the arguments from the command line, those arguments are nothing but the names of different files and all these files are located within the computer's hard disk and hence they must possess base address. The file names are nothing but strings and when they are accessed through their names, it is enough if we knew the base address of these strings (which will invariably be the base address of the corresponding files). When we collect these addresses and store them in a separate array, the array becomes the array of pointers. In our case, the array is that of **character pointers** and hence the declaration **char *argv[]**. The size of the array is not specified, which means that it would be dynamic. Thus argv[0] would contain the base address of the first argument, argv[1] the address of the second argument and so on. In the program coding, we do not need the argv[0] at all since it is this argument that drives the program and gets the output for us. One can print the addresses of the arguments inside the function main() to ensure the above explanation. Figure 8.12 shows the output that contains the **addresses of the arguments** that we pass from the command line.

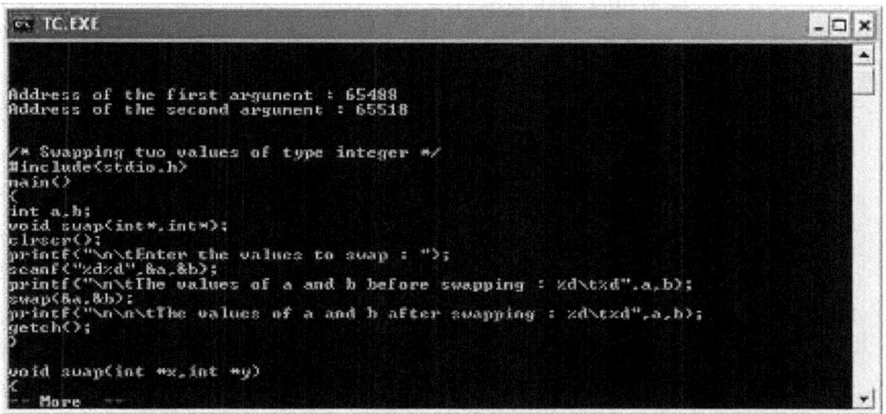

Figure 8.12 Addresses of the command line arguments printed

Another point the user must understand here is that the names of the arguments **argc** and **argv** could be changed while programming and there is absolutely no issue in doing so. I would appreciate if the programmers adhere to the same naming convention, as it would not defeat the spirit of the language.

8.9 SIMPLE PROGRAMS USING COMMAND LINE PARAMETERS

Let us write simple programs using the ideas that we have learnt in the earlier sections with respect to the command line arguments.

8.9.1 File Copying

To start with let us write a simple program for copying the content of one file to another. This aspect is called **file copy** and this is achieved through the copy command in DOS. The following is the coding for the same.

```
/* Program to copy data from one file to another */
#include<stdio.h>
main(int argc,char *argv[])
{
        char a;
        FILE *fp1,*fp2;
        if(argc<3)
            {
            printf("Usage Error");
            return;
            }
        fp1=fopen(argv[1],"r");
        fp2=fopen(argv[2],"w");
        if(fp1==NULL||fp2==NULL)
            {
            printf("File Opening Error");
            return;
            }
        a=getc(fp1);
        while(a!=EOF)
            {
            putc(a,fp2);
            a=getc(fp1);
            }
        fclose(fp1);
        fclose(fp2);
    }
```

Let us execute the above program and copy the nucleotide sequence (DNA sequence) of *Drosophila melanogaster* which is stored as a text file in FASTA format (To know more about FASTA format, you may refer any standard beginner's book on Bioinformatics. It is a four-letter format of any nucleotide of a protein or DNA. The four letters indicate the four nucleotide namely adenine, thymine, guanine, cytosine).

The source file **drosophila.txt** is available in the directory **SOURCE** in the drive D: and the target file **seq_1.txt** would be created in the directory **TARGET** in the same drive. While executing the program, the programmers are advised to be cautious about the file-naming convention. The DOS (Disk Operating System) follows an eight-letter naming convention for the files and hence when the name of the file in the **SOURCE** directory has more than eight characters, the programmers shall make sure that only seven characters are mentioned through the command line. For example, if the directory **SOURCE** has the file name **drosophila.txt** and if the users specify the full name of the file as it is, then **File Opening Error** will be displayed while executing the program. Instead, if the file name is specified as **drosop~1.txt** from the command line, the anticipated output will be generated. The generation of the error as well as the output is illustrated in Figure 8.13. The name of the above program is specified as **p182.c** and it is the **first of the three arguments** passed through the command line. The name of the source file is **drosophila.txt** and it is the **second argument** and the name of the target file is **seq_1.txt**, which is the **third argument**. In the program coding we check the minimum number of arguments that must be passed through the command line and the minimum required is three. We also see that the target file is read and its contents are displayed using the program to read and display the contents of a file that we had written in the previous section. The name of that program is **file_7.c** which was the **first argument** passed through the command line.

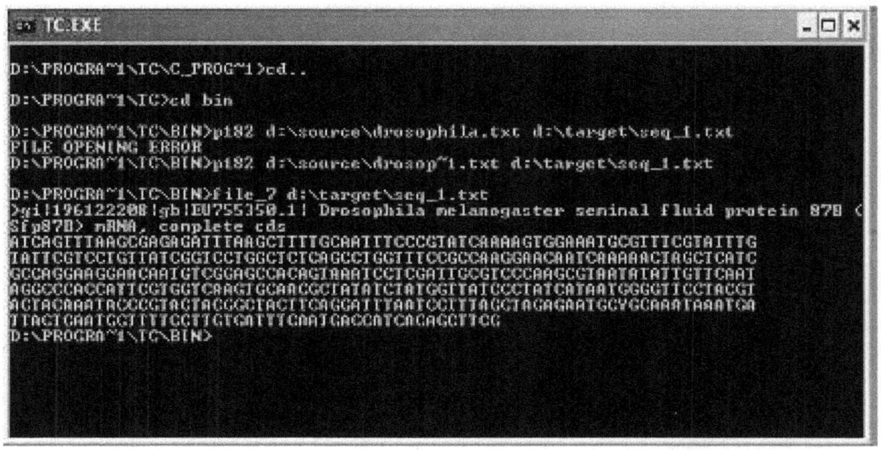

Figure 8.13 File copying through command line

8.9.2 File Merging

Let us write a program to merge the contents of two files together and display the contents of the merged file. For that we will take the DNA sequences of *Drosophila melanogaster* and *Rungwecebus kipunji* and merge their sequences into a single file and call the merged file as seq_2.txt. The source files are in the directory SOURCE and the merged file is in the directory TARGET and both the directories are available in the drive D. The following is the program code for the merging process. The output is shown in Figure 8.14. The contents of the merged file are viewed through the program, file_7.c.

```
/* Program to merge the contents of two files */
#include<stdio.h>
main(int argc,char *argv[])
        {
        char a,b;
        FILE *fp1,*fp2,*fp3;

        if(argc!=4)
            {
            printf("USAGE ERROR");
            return; }
        fp1=fopen(argv[1],"r");
        fp2=fopen(argv[2],"r");
        fp3=fopen(argv[3],"w");
        if(fp1==NULL||fp2==NULL||fp3==NULL)
                {
                printf("FILE OPENING ERROR");
                return;
                }
        for(;(a=getc(fp1))!=EOF;)
        putc(a,fp3);
        for(;(b=getc(fp2))!=EOF;)
        putc(b,fp3);
        fclose(fp1);
        fclose(fp2);
        fclose(fp3);
}
```

Figure 8.14 File merging through command line

8.9.3 Case Conversion

Let us write a program to convert the contents of a file in uppercase to lowercase. The coding for the same is given below.

```
/* Program to copy uppercase content in one file as lowercase */
#include<stdio.h>
main(int argc,char *argv[])
{
        char a;
        FILE *fp1,*fp2;
        if(argc!=3)
            {
            printf("USAGE ERROR");
            return;
            }
        fp1=fopen(argv[1],"r");
        fp2=fopen(argv[2],"w");
        if(fp1==NULL||fp2==NULL)
```

```
                {
                printf("FILE OPENING ERROR");
                return;
                }
        for(;(a=getc(fp1))!=EOF;)
                {
                if(a>='A'&&a<='Z')
                a+=32;
                putc(a,fp2);
                }
        fclose(fp1);
        fclose(fp2);
}
```

The above program is executed with the file **musculus.txt** in the directory source in the D: drive and the contents of the file that are in uppercase (The sequence containing the four letters A, T, G and C) are copied to another file seq_3.txt in the directory target in the D: drive. The output of the program is shown in the Figure 8.15. This kind of program could be useful when we want to convert the uppercase nucleotide sequence to lowercase which could be used for some other purpose. We can have the otherwise logic too, wherein we would check for the existence of lowercase characters and convert the same to uppercase. We just have to make minor changes in the above program to achieve this.

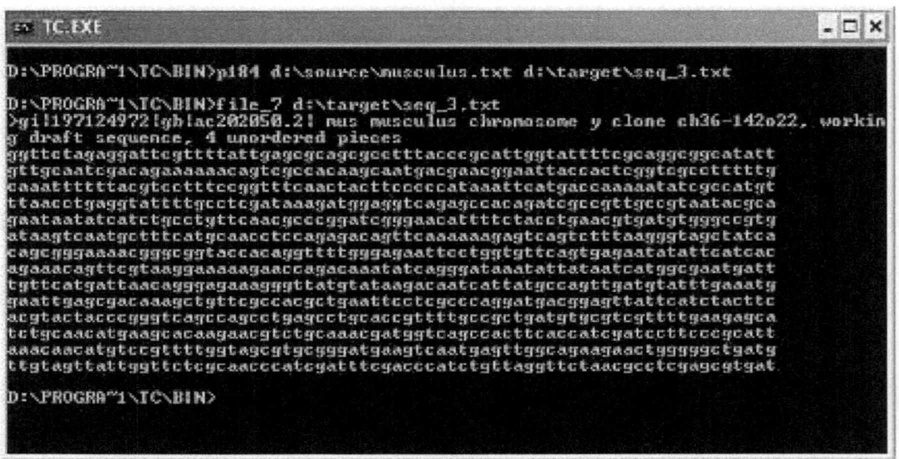

Figure 8.15 Case conversion

8.9.4 Sequence Count

```
/* Program to get the word through command line and check the number of times it is used in the particular
file */
#include<stdio.h>
main(int argc,char *argv[])
        {
        char a[1000];
        int i,j,k,w=0;
        FILE *fp;
        if(argc!=3)
            {
            printf("USAGE ERROR");
            return;
            }
        fp=fopen(argv[1],"r");
        if(fp==NULL)
            {
            printf("FILE OPENING ERROR");
            return;
            }
        for(i=0;(a[i]=getc(fp))!=EOF;i++)
                {
                if(a[i]=='\n')
                    {
                    a[i]='\0';
                    for(i=0;a[i];i++)
                        {
        for(k=i,j=0;argv[2][j]==a[k]&&argv[2][j]!='\0';k++,j++);
                        if(argv[2][j]=='\0')
                        w++;
                        }
                    i=-1;
```

```
                        }
                }
printf("The sequence you have entered occurs %d time(s)",w);
        fclose(fp);
        }
```

The above program is used for checking the occurrence of the sequence ATGGA in the file zea_mays.txt which has the DNA sequence of *Zea mays*. The output is shown in the Figure 8.16. We can check the existence of any big sequence in the file. In the next program, we go one step ahead, where we display the lines in which the given sequence occurs, along with the line number. The program coding is shown below followed by its output in the Figure 8.17.

8.9.5 Displaying the Lines in which the Sequence Occurs

```
/* Program to get the word through the command line argument and search for that word in the file and display the lines in which the word occurs along with the line number */
#include<stdio.h>
main(int argc,char *argv[])
        {
        char a[1000];           int i,j,k,l=0;          FILE *fp;
        if(argc!=3)
                {
                printf("USAGE ERROR");
                return;
                }
        fp=fopen(argv[1],"r");
        if(fp==NULL)
                {
                printf("FILE OPENING ERROR");
                return;
                }
        for(i=0;(a[i]=getc(fp))!=EOF;i++)
                        {
                        if(a[i]=='\n')
```

```
                        {
                        l++;
                        a[i]='\0';
                        for(i=0;a[i];i++)
                        {
        for(k=i,j=0;argv[2][j]==a[k]&&argv[2][j];k++,j++);
                                {
                                if(argv[2][j]=='\0')
                                        {
                                        printf("%d\t",l);
                                        puts(a);
                                        }
                                } }
                        i=-1;
                        }
}
fclose(fp);
}
```

Figure 8.16 Sequence count

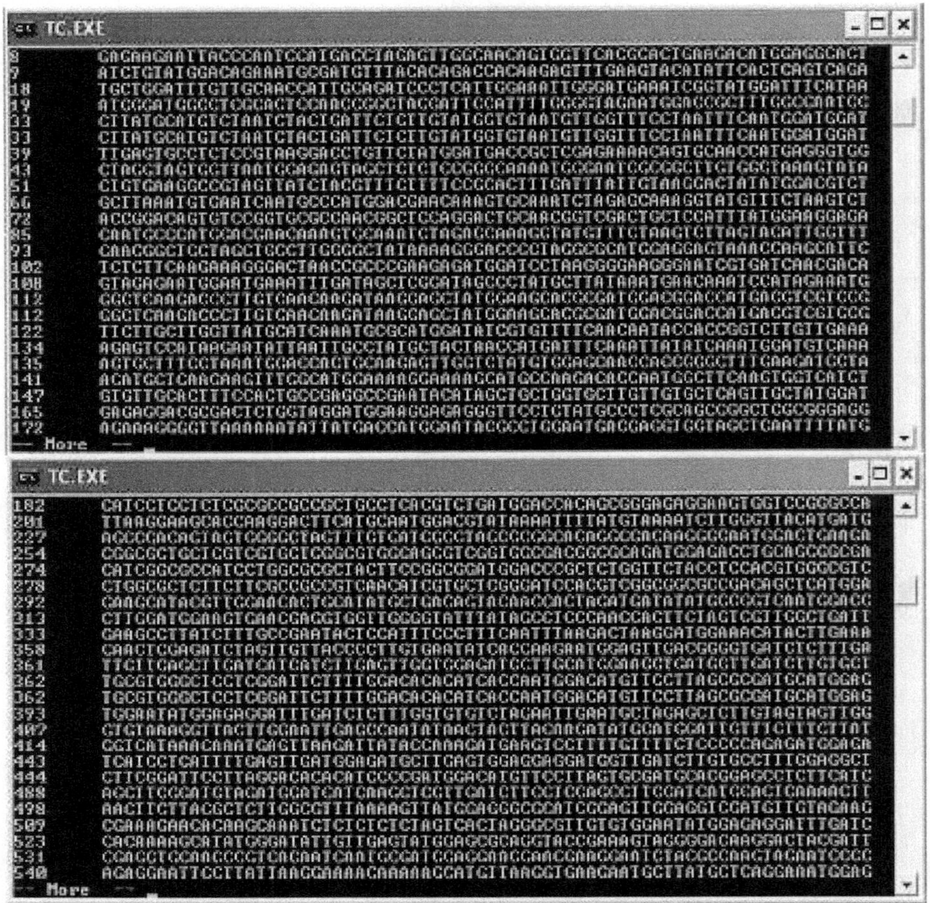

Figure 8.17 Lines in which the sequence occurs (Partial Output)

The output in the Figure 8.17 shows only a partial output since the file size (zea_mays.txt) is bigger and the given sequence (ATGGA) occurs 165 times in it as is obvious from the Figure 8.16.

8.9.6 Displaying the Lines in which the Sequence doesn't Occur

The following program coding illustrates how to get the display of the lines in which the given sequence does not occur. We have used the DNA sequence of *Drosophila melanogaster* for the program and it is available in the directory source of the drive D. The test sequence given was AAGG and we could notice from the output shown in the Figure 8.18 that the lines 1, 2, 5 and 6 does not have the given sequence and hence are displayed. The line number 1 is just the description of the sequence, but still it is considered by the program for the availability of the pattern.

If the programmer wants to skip the first line of the program from being considered, he/she may suitably modify the logic.

```
/* To print the lines in which the pattern does not exist */
#include<stdio.h>
main(int argc,char *argv[])
    {
    char a[1000];
    int i,j,k,l,f;
    FILE *fp;
    if(argc!=3)
        {
        printf("USAGE ERROR");
        return;
        }
    fp=fopen(argv[1],"r");
    if(fp==NULL)
        {
        printf("FILE OPENING ERROR");
        return;
        }
    l=0;
    for(i=0;(a[i]=getc(fp))!=EOF;i++)
        {
        if(a[i]=='\n')
            {
            f=0;
            l++;
            a[i]='\0';
            for(i=0;a[i];i++)
```

```
{
for(j=0,k=i;argv[2][j]==a[k]&&argv[2][j]!='\0';k++,j++);
    {
    if(argv[2][j]=='\0')
        {
        f=1;
        continue;
        }}    }
i=-1;
if(f==0)  {
  printf("%d\t",l);
  puts(a);
  }   }     }           fclose(fp);           }
```

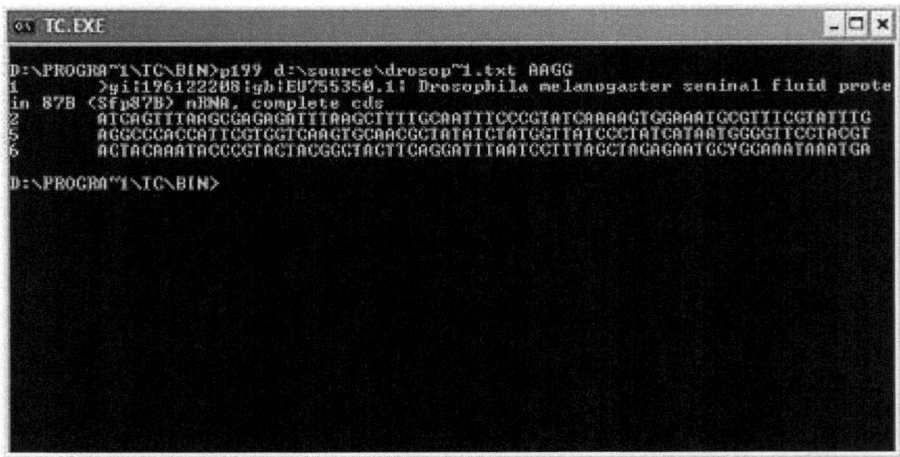

Figure 8.18 Lines in which the sequence does not occur

I am sure the programs that are discussed above using the command line parameters would be an eye opener for the learners of this book and I would also suggest them to do more such exercise to get full confidence in using the command line arguments efficiently.

SUMMARY

We began this chapter with an introduction to files and the various operations that are possible with them. We had a brief discussion on the file positions and high level file handling functions. With sufficient illustrations, we understood the concepts of opening and closing a file, writing into a file and reading from a file. We also discussed how to append contents in the existing file (without overwriting the existing contents). We then discussed how to get the file name as a variable name from the keyboard and open the same for the display on the monitor.

We then came to an important aspect of the files, namely the command line arguments, which are considered to be the key idea in efficiently handling the files. We had a brief introduction to command line arguments and how significant it is in accessing the files using it. We followed the discussion with some simple programs, where we had used DNA sequence of some of the species like *Drosophila melanogaster*, *Mus musculus*, *Zea mays*, etc.

CRITICAL THINKING QUESTIONS

Identify the outputs of the following programs

1.
```
main()
{
FILE *fp;
char c;
clrscr();
fp=fopen("myfile.txt","r");l
fclose(myfile);
}
```

2.
```
main()
{
FILE fptr;
clrscr();
```

```
fptr=fopen("file_1.txt","r");
if(fptr=NULL)
{
puts("file opening error");
exit();
}
fclose(fptr);
}
```

3.

```
main()
{
FILE *fp;
char str[30], str2[8];
fp=fopen("myfile.txt","r");
printf("Enter a string :  ");
gets(str);
fputs(str,fp);
fputs("\n",fp);
fclose(fp);
}
```

4.

```
main()
{
FILE *fp;
char str[30], str2[8];
fp=fopen("myfile.txt","r");
while(fgets(str,fp)!=NULL)
puts(str); fclose(fp);
}
```

5.
```
main()
{
FILE *fp;
char var[20];
fp=fopen("sample.c","r");
while(fgets(var,10,fp))!=EOF)
printf("%c",var); }
```

SELF-LEARNING EXERCISES

(Do all the programs using Command Line Parameters)

1. Write a program to convert the given DNA sequence to RNA sequence by replacing T (Thymine) by U (Uracil). Use a source file and a target file for the process.
2. Write a program to skip a particular sequence from the given DNA sequence of a species.
3. Write a program to check for the number of occurrences of the pattern TATA (technically called as TATA Box) in the given DNA sequence.
4. Write a program to encrypt the contents of a given file and display the encrypted file.
5. Write a program to decrypt the contents of a given file and display the same (use the output file of the above program as input for this program).
6. Write a program to count the number of uppercase, lowercase alphabets, special characters and digits in a given file.
7. Write a program to display the contents of more than one file.
8. Write a program to check for the number of occurrences of a given pattern in more than one file at a given time.
9. Write a program to count the number of words, lines and characters in a given file and display the statistics.
10. Write a program to display the contents of a lengthy file, page by page (Consider a lengthy DNA sequence as input to this program).

ROLE OF PROGRAMMING LANGUAGES IN BIOINFORMATICS

9.1 PROGRAMMING LANGUAGES USED

There are six commonly used programming languages in Bioinformatics—C, C++, C#, Java, PERL and Python. These languages are used under two different operating systems and all these languages have their own impact on Bioinformatics applications. The overall analysis of these languages shall surely enable the user to select a particular language for the task he/she is allotted. The choice depends on the overall efficiency of the language and the library available with the language.

Bioinformatics analysis involve a range of tasks and processes. Diverse programs have been written for various Bioinformatics applications using every available language. Because of the size of Bioinformatics datasets, computation time is not trivial, and efficiencies in computational speed are desirable. However, it is possible that the same program, written in different languages, or running under different operating systems, may exhibit significant differences in speed and efficiency. There is, at present, little direct data on the underlying speed and efficiency of equivalent algorithms written in different languages. While languages themselves have been benchmarked, such comparisons have not been done using algorithms that are relevant to Bioinformatics.

A typical Bioinformatics program reads FASTA files, holds the DNA sequences in memory, performs different computing tasks on the sequences, and finally writes the results to a file. Another common task in Bioinformatics is text mining or text parsing. Large amounts of data can be generated in different formats. Because file formats can be different, linking programs in a pipeline is difficult, hence scripts are written to act as interfaces between programs performing the sequential parts of an analysis. Scripts are also used to extract information from large data files, thus enhancing the presentation of results. These quick scripts are usually implemented in Perl or Python. Consequently, any Bioinformatics procedure has a number of areas where programming might be improved, these being: the space required to temporarily store data, the speed of computation, linkage between programs, and presentation of analyses.

Python and Perl are often called script languages and when executed, are compiled in an intermediate representation without creating an intermediate file (syntax tree in Perl and byte code in Python) and then interpreted. Both languages use automatic memory management and have large free libraries. C and C++ are fully compiled languages, suitable for system-intensive tasks. Java and C# are semi-compiled languages using automatic memory management. A Java program is compiled in an intermediate-level code or byte code, then it is run by either an interpreter or compiler at runtime, in this case, the Java Virtual Machine (JVM). In C# the intermediate-level code is called Microsoft Intermediate Language and is run on the .NET Common Language Runtime engine.

9.2 HIGHLIGHTING FEATURES OF THE LANGUAGES USED

Language philosophies often explain differences in the relative expressiveness and readability of languages. For example, the philosophy of Python is to take the clearest, simplest and most straightforward approach to writing a program, and to accept the resulting performance penalty. Perl gives more freedom to the programmer, resulting in some cases, in programs that are unreadable for non-Perl programmers.

Factors such as performance and memory usage are important, but need not be the sole determinant when choosing a language. Since time management is also an important factor, a language can be chosen for its library, future scalability, active community and interface to other languages.

While it is hard to define a learning curve for each language, advantages and disadvantages of each language can be found. Memory management such as memory allocation makes tasks easier for Java, C#, Perl, Python and Perl programmers, even though memory usage should always be under scrutiny. The use of pointers in C and C++ can be overwhelming for a learner, and it may take some time to master their use. Python is designed to be a highly readable language, frequently using English keywords, whereas the five other languages use punctuation.

Platform independence can be a factor for choosing a language and can also facilitate its learning. Java uses a virtual machine to run on different operating systems and since Perl and Python are stored in a text file, and not in a binary file, these scripts can be executed on any computer having the appropriate interpreter.

The standard library diversity and size of Java, Python and C# are a major advantage compared to the other languages, including sets of classes to create graphical interfaces, data structures (vectors, hash tables, stacks, queues), regular expression, database access and networking.

Although the Perl standard library appears smaller compared to Java, Python and C#'s, it benefits from a large community of programmers creating modules gathered by the CPAN (Comprehensive Perl Archive Network).

C has a very limited standard library supporting input and output streams, memory allocation, mathematics, character strings, and time values. The C++ library contains the C library as well as a class to manipulate string of characters and the Standard Template Library containing containers or data structures such as algorithms, an improved String library and input/output stream libraries.

Some Bioinformatics tasks, such as loading a FASTA file or parsing a BLAST file, are so frequently used that it makes sense to reuse bits of code by creating libraries or modules. Many programmers produce modules for the Bioinformatics community and make them available through their websites. The most widely used libraries are BioPerl, BioPython and BioJava. These Open Source projects belong to the Open Bioinformatics Foundation and they provide toolkits with multiple functionality that makes it easier to create customized pipelines or analyses.

Unfortunately libraries with this level of organization and diversity are not available for C, C++ and C# even though some small libraries are available independently.

In most cases of Bioinformatics applications, C happens to be the proved performer in terms of both speed and memory usage. However, the best choice of language for a task would be according to the original philosophy, keeping in mind that Java is a portable web-oriented language, Perl is a powerful script language, Python is an easily coded language and C and C++ are efficient languages used in operating systems and drivers.

SUMMARY

This chapter has just analysed the basic principles involving the choice of a suitable programming language for Bioinformatics. We have discussed in brief the different programming languages used (we have chosen six frequently used languages) in the Bioinformatics applications and their highlighting features, which are unique to themselves. We have concluded with the fact that C has the greatest impact on the Bioinformatics applications, when compared with the other languages, even though it has a limited library. I really mean it and it must be obvious to all the learners of this book after having gone through the length and breadth of the programming language and the concepts involved.

SELF-LEARNING EXERCISES

1. Take a case study from Bioinformatics, such as, an algorithm for parsing the BLAST (Basic Local Alignment Search Tool) file outputs and apply all the six programming languages and hence compare their efficiency in handling them.
2. Make a study on the open source projects available with the Open Bioinformatics Foundation and explain how much contribution these projects have given to the growth of Bioinformatics.
3. "Bioinformatics projects, tools and applications shall be kept open and the users must have the necessary freedom to use them. This freedom of usage will only be the greatest contributing factor for the growth of Bioinformatics". Suggest your opinions on this statement.
4. Do you think Bioinformatics, which is a decade-old field, is still in its infancy?
5. "Bioinformatics, since its emergence as a field, has helped the research community to explore a lot". Comment.

10 ROLE OF C++ AND PERL IN BIOINFORMATICS

10.1 ESSENTIALS OF C++

10.1.1 Why Learning C++ is Important?

In all the earlier chapters, we have been discussing the essential features of C language and we have learned to write good applications at almost all the levels. But then there are a few shortcomings with respect to C language. There are absolutely no issues with the efficiency of language. No programming language can arguably have the power of C language. Still the language has a slip, namely lack of object orientation. Since most of the applications and tools that are in use today in the scientific community possess object-oriented features, the programmers and the users expect the same from C. The designers of the language, at the time of designing C, did not show a major concern over the incorporation of object-oriented features and the language went on to be used by the programming community. When there was much desire among the programmers, Bjarne Stroustrup, one of the colleagues of Dennis Ritchie, started to build the object-oriented features in the language and with the help of structures and unions in C, he could add more features and came out with the concept of class. He inherited all the procedure-oriented features of C to C++. He derived the object-oriented features from the language Simula 67. Initially he called the new language as C with Classes. But later on, he renamed it as C++ as it was an advanced version of C (The name being derived from the increment operator ++ in C). The name C++ (read as C Plus Plus) was coined by Rick Mascitti.

C++, as mentioned in the previous paragraph, is essential for the programmers due to its efficient object-oriented features. What actually is object-oriented programming and what are its significant features? We will explore some of the important object-oriented programming features and also the real meaning of it in the forthcoming section.

10.1.2 Object-oriented Programming—A Brief Introduction

The concept of object orientation dates back to the 1960s. The word **object-oriented** was coined by Alan Kay in the year 1967. The main reason behind developing object-oriented languages is to improve the quality of the software. The programming language **Simula** was the first to introduce the concepts underlying object-oriented programming such as objects, classes, subclasses, virtual methods, coroutines, garbage collection and discrete event simulation and it was considered as a superset of Algol. The fundamental application of Simula was in physical modelling and it was initially used to monitor the movement of ships and their content through cargo ports. In object-oriented programming languages, each object is capable of receiving messages, processing data and sending messages to other objects. The jargon involved in object-oriented programming language are briefed in the following paragraphs.

Class It defines the abstract features of an object, including its characteristics such as its attributes, fields or properties and its behaviours such as what it can do, the methods that can be applied on it and the operations that it can perform. One might say that a class is a template that describes the nature of an object. For example, consider the class "person", which may have attributes such as name, age, sex, height, weight, complexion, address, etc. When we consider a special case of the person, such as an employee or a student, one might require additional attributes apart from the abovesaid attributes. In the case of an employee, we might require additional details such as employee id, department, designation, basic salary, etc. In the case of a student, we require register number, marks in various subjects, total, average, grade and result. In simple words, a class is an **abstract data type**. The properties and methods defined by a class are called its members.

Object It is a pattern of a class. It is otherwise termed as an instance of a class. If we want to implement the features of a class, we must declare objects for it. The process of declaring an object for a class is called as **instantiation** (declaring an instance of a class).

Method It is the ability of an object. Methods are nothing but the actions performed by the objects which are termed as **verbs** in object-oriented terminology. For example, in the case of a class of students, calculation of his/her grade could be one of the methods.

Message passing It is the process by which an object sends data to another object or asks the other object to invoke a method. This aspect is also known to some programming languages as interfacing. For example, if we want to cross the river from one bank to another, we might use various methods such as a boat, a bridge or we may swim. All these methods are nothing but interfaces.

Abstraction It is the act of representing essential features without including the background details or explanations. In simpler words, abstraction is the concept of simplifying a real world concept into its essential elements.

For example, let us consider that you are driving a car. You know only the essential features to drive a car such as handling the gear, steering, clutch, accelerator and brakes. But while driving, do you ever get into the internal details of the car such as wiring, working of the motors, etc.? You just apply the gears and brakes and what is happening internally is hidden from you. This is called as abstraction, where you know the essential things and the background details are hidden from you.

Encapsulation It is the most important concept of any object-oriented programming language. It is the way of combining both data and the functions that operate on that data under a single unit. The wrapping up of data and functions into a single unit is known as data encapsulation. The only way to access data is provided by an interface and that interface in C++ is the member functions inside a particular class. The data cannot be accessed directly and is hidden and so is safe from any accidental manipulation or malicious usage. Consider for example, a company in which the departments access and work with their data on their own. One department cannot access data of another department directly. Rather a request would be made from one department for the required data and then the required data would be handed over by the members of the other department to those who requested them. Thus, it can be said that the data pertaining to one department and the employees of that department are encapsulated or wrapped up into a single entity called as department.

Modularity It is the act of dividing a task into individual modules or components. This will reduce the program complexity to a greater extent and would create a number of well-defined, documented boundaries within the program. Modules work independently and contribute a lot to achieve the final result of the program. For example, consider a music system. Let us assume that our program represents the entire music system and it would comprise of speakers, disk player, recorder, tuner, etc. Each of these contributes to the entire system and enables the user to hear the music.

Inheritance It is the capability of one class acquiring the properties of another. This works just like genetics in biology. In genetics, we work with different species, produce offsprings and analyse the properties of the parent and the children. For example, we all are human beings and biologically termed as *Homo sapiens* and we inherit the properties from the class *Homo sapiens* and those properties are again inherited from yet another class mammal and the properties of a mammal are inherited from the class animal. Thus the parent class is the animal and the child class for the class animal is the mammal and the child class of the class mammal is *Homo sapiens*. Thus the class mammal is the base class and the class *Homo sapiens* is the derived class. Based on the gravity of its usage, one can classify inheritance as single, multiple, multi-level, hierarchical and hybrid.

Polymorphism It is the ability to process the data in more than one form. Poly means "many" and morphing means "forms". In simplest terms, polymorphism means **single interface multiple methods**. For example, consider a thermos flask, which can be used for storing, either hot or cold. Thus the interface is only one, whereas the number of ways of utilizing the interface is more than one. In thermos flask, one can keep things hotter or colder, without having to look for another interface when the nature of storage changes.

Once you are familiar with the above concepts of OOP languages, it would be easy to use them and develop applications using their syntax and semantics.

10.1.3 OOP in Scripting

Just as procedural programming led to refinement of techniques such as structured programming, modern object-oriented software design methods include refinements such as the use of design patterns, design by contract, and modelling languages.

In recent years, object-oriented programming has become especially popular in scripting programming languages. Python and Ruby are scripting languages built on object-oriented principles, while Perl and PHP have been adding object-oriented features. BioPerl and Biopython are some of the object-oriented languages that are very much useful in developing applications in Bioinformatics.

10.2 ESSENTIALS OF PERL

Currently, the most popular use of computer programming languages in the biological sciences is in the field of Bioinformatics. A frequent task needed in Bioinformatics is to find certain patterns in databases. The problem is that these databases often contain information in various formats. A solution is to use regular expressions to specify these patterns so that the information can be mined from the databases. Scripting languages are often used in Bioinformatics due to their ability to work with regular expressions.

BioPerl is a collection of more than 500 Perl modules for Bioinformatics that have been written and maintained by an international group of volunteers. BioPerl is free and you can find it at http://bioperl.org. BioPerl is one of the most successful projects launched by the Open Bioinformatics Foundation.

One of the most difficult things about BioPerl is getting started using it, since it won't be just like using any other programming language. One reason for this may be due to a scarcity of good documentation as well as the sheer size of the BioPerl module library. The modules in BioPerl are written in the object-oriented style and hence it would be justified and fitting if we use this scripting language for developing applications in Bioinformatics.

Perl became more and more popular for biology applications as its popularity started growing. The support for object-oriented programming in the version 5 of Perl really accelerated the development of reusable modules for biology at many research centres around the world. Initially the users of BioPerl were skeptical about the efficiency of the language (as is the case with any new programming language), but as time went on, its usage prevailed and almost all those including researchers offered using BioPerl to develop stand-alone applications. Today in the stream of programming languages that are used in Bioinformatics, Perl stands tall next only to C. One important difference between C and Perl is that one can create .exe file in C, but not in Perl. Due to this, the scripts that are developed in Perl are easily sharable and one cannot easily commercialize the applications developed using the Perl scripts. This is why most of the applications developed using the Perl scripts are open to the users.

10.3 OPEN SOURCE IN BIOSCIENCES

The idea of using open source software in Bioinformatics is still in infancy and necessary awareness has to be created to make the idea reach more people, especially researchers. The very purpose of giving stress on using open source applications here is to let the discipline grow and reach the heights by sharing the ideas freely. Researches would equally contribute a lot to the growth of a field.

For example, initially when biotechnology came into the academy as one of the disciplines, people had reservations about it but in less than half a decade it became very popular and is now the backbone of biosciences. Similarly, if we want to make Bioinformatics to be another backbone, we would share the ideas that are gathered at the same time, not letting down the privileges of a programmer or an application.

Open source software and applications would surely contribute to the growth of Bioinformatics. The unknown advantages of the open source in Bioinformatics and other biosciences are threefold, namely, security, strategic ownership, and extensibility.

Until recently, open source has often appeared to the life scientists as some sort of novelty, or, in the worst case, a serious threat to Intellectual Property protection. In the last few years, though, solid achievements in clustering, genomic data management and web publication have established open source as a serious technical alternative.

Even big pharmaceutical companies and research institutes have started believing and realizing the significance of open source applications as to how they can systematically cut costs, improve security and quality. I can confidently say that this is really the apt time for establishing the open source ideas in Bioinformatics and the initiative must come from the level of the students and shall be advanced to the level of researchers.

SUMMARY

We started the discussion of this chapter with the reason for a programmer in Bioinformatics to learn C++. The reasons were quite obvious since it is an object-oriented programming language and has a lot of credits to it in being so. We followed with a brief section on exploring the various features of object-oriented languages with necessary examples as and when needed. We also discussed in brief on how object-oriented languages help in writing good scripts. This discussion was followed with an example of a scripting language namely Perl, which is most frequently used in writing the scripts while developing simple applications in Bioinformatics. We had a very crispy discussion on the use of BioPerl in developing scripts in Bioinformatics applications. We concluded the chapter with the discussion on the advent of open source in Bioinformatics and other biosciences. We could appreciate the threefold advantage of open source in Bioinformatics and other biosciences. It is also recommended towards the end of the section that more of open source shall come into being in Bioinformatics so that the field will expand both vertically and horizontally among all the academic community.

SELF-LEARNING EXERCISES

1. Make a spade work on the various open source applications available in the various disciplines of biosciences and give a substantiating view on the opinion that "Bioinformatics would flourish in an extraordinary manner" with these applications.
2. What is the need for having open source applications? Give your views and suggestions on this.
3. Discuss some of the scripting languages used in Bioinformatics and compare them with BioPerl.
4. As a Bioinformatician, which language would you suggest as the best general-purpose language for developing applications—C or C++? Give valid reasons for your answer.
5. "Bioinformatics community needs the mindset of sharing of knowledge than giving much importance to Intellectual Property Rights."—Would you accept this argument? Give valid reasons for accepting or not accepting this.

PART II

'OMICS IN BIOLOGY

1

INTRODUCTION TO MOLECULAR BIOLOGY

The cell is the structural and functional unit of life. It is the place where all the activities take place. Some organisms, such as bacteria, are unicellular, consisting of a single cell. Other organisms, such as humans, are multicellular. Each cell is an amazing world unto itself: it can take in nutrients, convert these nutrients into energy, carry out specialized functions, and reproduce as necessary. Even more amazing is that each cell stores its own set of instructions for carrying out each of these activities.

CELL

We all know that the cell is the basic unit of all living organisms. Certain living organisms exist only as a single cell. This is the case for bacteria, as well as certain animals and plants. You will be surprised to know that an average-sized human contains from 60 to 100 trillion cells. We all know that cells do their work for themselves like nourishing, produce energy, exchange information with their neighbours, multiply, and they even know their last day and eventually die when their time has come.

1.1 TYPES OF CELLS

One should know the make-up of a cell, which is categorized based on its inherent make-up. In this manner, it might be placed into one of two groups: the prokaryotes and the eukaryotes. Cells can also be classified into three domains: Eukaryota, Eubacteria, and Archaea where Eubacteria and Archaea is a split of the prokaryotes based on genetic differences. All multicellular animals (and some unicellular organisms) are Eukaryota, including humans.

1.1.1 Prokaryotic Cell

Prokaryotic cells (Figure 1.1) are generally smaller and vary in size in different members. Prokaryotes differ from eukaryotes by the lack of a nuclear membrane and a cell nucleus. Prokaryotes also lack most of the intracellular organelles and structures that are seen in eukaryotic cells. The functions of most organelles such as mitochondria, chloroplasts, and the Golgi apparatus, are taken over by the prokaryotic cell's plasma membrane itself. Prokaryotic cells have three architectural regions: appendages called flagella and pili—proteins attached to the cell surface; a cell envelope consisting of a capsule, a cell wall, and a plasma membrane; and a cytoplasmic region that contains the cell genome called as DNA and ribosomes and various sorts of inclusions. Other differences include:

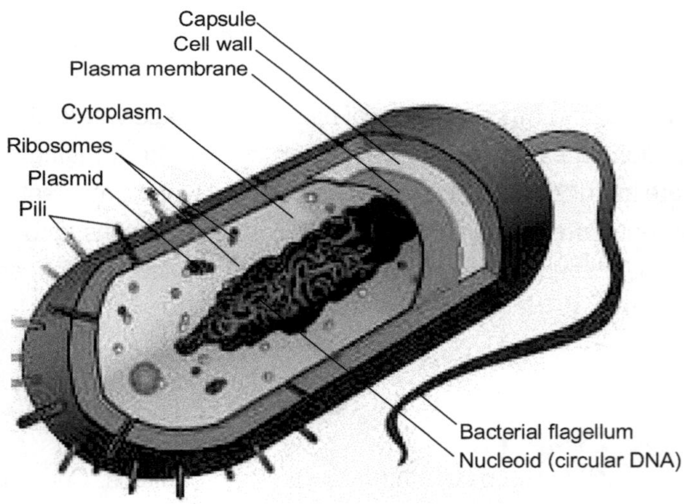

Figure 1.1 A prokaryotic cell

The plasma membrane separates the interior of the cell from its environment and serves as a filter and communications beacon. Most prokaryotes have a cell wall. Some exceptions are Mycoplasma (bacteria) and Thermoplasma (Archaea). This wall consists of peptidoglycan in bacteria, and acts as an additional barrier against exterior forces. It also prevents the cell from exploding from osmotic pressure against a hypotonic environment. A cell wall is also present in some eukaryotes like plants in the form of cellulose and fungi, but has a different chemical composition. A prokaryotic chromosome is usually a circular molecule an exception is that of the bacterium *Borrelia burgdorferi*, which causes Lyme disease. Even without a real nucleus, the DNA is condensed in a nucleoid. Prokaryotes can carry extrachromosomal DNA elements called plasmids, which are usually circular. One important feature of a plasmid is that it can carry additional functions, such as antibiotic resistance.

1.1.2 Eukaryotic Cells

The eukaryotic cells are about 10 times the size of a typical prokaryote and can be as much as 1000 times greater in volume. The major difference between prokaryotes and eukaryotes is that eukaryotic cells contain membrane-bound compartments in which specific metabolic activities take place. Most important among these is the presence of a cell nucleus, a membrane-delineated compartment that houses the eukaryotic cell's DNA. It is this nucleus that gives the eukaryote its name, which means true nucleus.

The plasma membrane resembles that of prokaryotes in function, with minor differences in the set-up. Cell walls may or may not be present. The eukaryotic DNA is organized in one or more linear molecules, called chromosomes, which are associated with histone proteins. All chromosomal DNA is stored in the cell nucleus, separated from the cytoplasm by a membrane. Some eukaryotic organelles also contain some DNA. Eukaryotes can move using cilia or flagella. The flagella are more complex than those of prokaryotes.

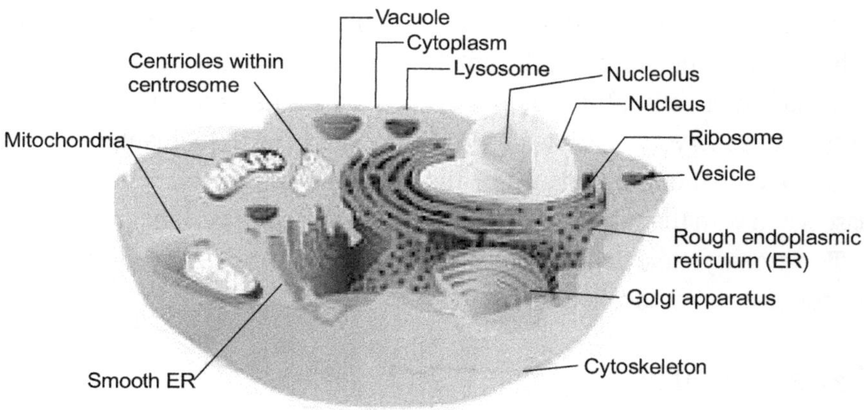

Figure 1.2 Eukaryotic cell

1.2 COMPONENTS OF CELL

From Figure 1.2, we can understand that the prokaryotic or eukaryotic cells have a membrane that protects/envelops the cell, separates its interior from its environment, regulates what moves in and out (selectively permeable), and maintains the electric potential of the cell. Inside the cell membrane, a salty cytoplasm takes up most of the cell volume. All cells possess DNA, the hereditary material of genes, and RNA, containing the information necessary to build various proteins such as enzymes, the cell's primary machinery. There are also other kinds of biomolecules in cells.

1.2.1 Cell Membrane: A Cell's Defining Boundary

The cytoplasm of a cell is surrounded by a plasma membrane. The plasma membrane in plants and prokaryotes is usually covered by a cell wall. This membrane serves to separate and protect a cell from its surrounding environment and is made mostly from a double layer of lipids of hydrophobic fat-like molecules and hydrophilic phosphorus molecules. Hence, the layer is called as phospholipid bilayer. This phospholipid bilayer may also be called a fluid mosaic membrane. Embedded within this membrane is a variety of protein molecules that act as channels and pumps that move different molecules into and out of the cell. The membrane is said to be semi-permeable, in that it can either let a substance (molecule or ion) pass through freely, pass through to a limited extent or not pass through at all. Cell surface membranes also contain receptor proteins that allow cells to detect external signalling molecules such as hormones.

1.2.2 Cytoskeleton: A Cell's Scaffold

As we have a definite shape for our skeleton, the same as the shape of the cell is maintained by the cytoskeleton, acts to organize and maintain the cell's shape; which anchors cell organelles in its apt place; helps during endocytosis, the uptake of external materials by a cell, and cytokinesis, the separation of daughter cells after cell division; and moves parts of the cell in processes of growth and mobility. The eukaryotic cytoskeleton is composed of microfilaments, intermediate filaments and microtubules for the maintenance of the structure of the cell. There is a great number of proteins associated with them, each controlling a cell's structure by directing, bundling, and aligning filaments. The prokaryotic cytoskeleton is less well-studied but is involved in the maintenance of cell shape, polarity and cytokinesis.

1.2.3 Genetic Material

There are two different kinds of genetic material—deoxyribonucleic acid (DNA) and ribonucleic acid (RNA). Most of the organisms use DNA for their long-term information storage, but some viruses like retroviruses have RNA as their genetic material. The biological information, which is to be transferred, is encoded in the form of DNA or RNA sequence. RNA is also used for information transport like mRNA and enzymatic functions like ribosomal RNA in organisms that use DNA for the genetic code itself.

Now let us learn about the prokaryotic genetic material which has organized in a simple circular DNA molecule the one present in bacterial chromosome, in the nucleoid region of the cytoplasm. Then comes the next division of cells, i.e., eukaryotic genetic material which is divided into different, linear molecules called chromosomes inside a discrete nucleus, usually with additional genetic material in some organelles like mitochondria and chloroplasts.

To understand our genetic material, i.e., in humans, the cell has the genetic material in the nucleus which is called as the nuclear genome and in the mitochondria the genetic material is present in the mitochondrial genome. In humans the nuclear genome is divided into 46 linear DNA molecules called as chromosomes.

The mitochondrial genome is a circular DNA molecule separate from the nuclear DNA. Although the mitochondrial genome is very small, it codes for some important proteins. Foreign genetic material (most commonly DNA) can also be artificially introduced into the cell by a process called transfection. This can be transient, if the DNA is not inserted into the cell's genome, or stable, if it is.

1.2.4 Organelles

We know that the human body contains many different organs, such as the heart, lung, and kidney, with each organ performing a different (and its own) function. Cells also have a similar set of little organs, called organelles, that are adapted or specialized for carrying out one or more vital functions. Membrane-bound organelles are found only in eukaryotes.

1.2.5 Nucleus (A Cell Information Centre)

Nucleus is the most important organelle found in a cell. It is the residence for the cell's chromosomes, and is the place where almost all DNA replication and RNA synthesis occur. The nucleus is spherical in shape and separated from the cytoplasm by a double membrane called the nuclear envelope. The nuclear envelope protects a cell's DNA from various molecules that could accidentally damage its structure or interfere with its processing. During processing, DNA is transcribed, or copied into a special RNA, called mRNA. This mRNA is then transported out of the nucleus, where it is translated into a specific protein molecule. In prokaryotes, DNA processing takes place in the cytoplasm.

1.2.6 Mitochondria and Chloroplasts (The Power Generators)

Mitochondria are self-replicating organelles that occur in more numbers, shapes, and sizes in the cytoplasm of all eukaryotic cells. Mitochondria contain their own genome which is separate and distinct from the nuclear genome of a cell. They play a critical role in generating energy in eukaryotic cells. Mitochondria give energy to the cells by the process of respiration. In plants as like mitochondria chloroplasts are seen. Organelles that are modified forms of chloroplasts are broadly called as plastids, and are often involved in storage. Since they contain their own genome, they are thought to have once been separate organisms which later formed a symbiotic relationship with the cells. Chloroplasts are the counterparts of the mitochondria. Instead of giving off CO_2 and H_2O, plants give off glucose, oxygen, and six molecules of water. This process is called photosynthesis.

1.2.7 Endoplasmic Reticulum and Golgi Apparatus (Macromolecule Managers)

The endoplasmic reticulum (ER) acts as the transport network for molecules targeted for certain modifications and specific destinations, as compared to molecules that will float freely in the cytoplasm. The ER has two forms: the rough ER, which has ribosomes on its surface, and the smooth ER, which lacks them. Also the golgi apparatus ends pinch off and become new vacuoles in the animal cell.

1.2.8 Ribosomes (The Protein Production Centres in the Cell)

The ribosome is a large complex composed of many molecules; in prokaryotes they exist freely floating in the cytosol, whereas in eukaryotes they can be found either free or bound to membranes.

1.2.9 Lysosomes and Peroxisomes (of the Eukaryotic Cell)

The destroyers of our cell are lysosomes and peroxisomes which have the destructive enzymes found in a membrane-bound system. These organelles are often called as "suicide bags" because of their ability to detonate and destroy the cell.

1.2.10 Centrosome (The Cytoskeleton Organizer)

The centrosome produces the microtubules of a cell and is a key component of the cytoskeleton. It directs the transport through the ER and the golgi apparatus. Centrosomes are composed of two centrioles, which separate during cell division and help in the formation of the mitotic spindle. A single centrosome is present in the animal cells. They are also found in some fungi and algal cells.

1.2.11 Vacuoles

Once the food is taken by the system it should be stored for the future and it is the vacuoles which store food, water and waste. They are often described as liquid-filled spaces surrounded by a membrane. Some cells, most notably *Amoeba*, have contractile vacuoles, which are able to pump water out of the cell if there is too much water.

1.2.12 Capsule

Capsules are gelatinous structures with a layer of slime of variable composition that covers the surface of some bacteria with loosely associated capsules. The capsule may be a polysaccharide as in pneumococci and meningococci, as a polypeptide as in *Bacillus anthracis* or as hyaluronic acid as in streptococci. Capsules are not stained by ordinary stain; they are detected by special stain. The capsule is antigenic and antiphagocytic in function, and this determines the virulence of many bacteria. It plays an important role in the attachment of the organism to mucous membranes.

1.2.13 Flagella

Flagella are the organs for motility. They arise from the cytoplasm and extrude through the cell wall. They are long and thick threadlike appendages, protein in nature. They are formed from a protein called flagellin, which acts as an antigenic substance. Based on the arrangement of flagella, the organism may be classified as monotrichate, amphitrichate, lophotrichate or peritrichate.

1.2.14 Fimbriae (Pili)

Pili are short and thin hairlike filaments, formed of protein called pilin (antigenic). Fimbriae are responsible for attachment of bacteria to specific receptors of human cell (adherence). There are special types of pili called sex pili involved in the process of conjunction.

NUCLEIC ACIDS

Nucleic acids are the most important constituents of the nucleus, which play a vital role in the inheritance of characters from one generation to the other. There are two types of nucleic acids—deoxyribonucleic acid (DNA), and ribonucleic acid (RNA).

1.3 DEOXYRIBONUCLEIC ACID (DNA)

DNA is the genetic material in almost all organisms except for certain viruses. In prokaryotes like *Escherichia coli*, the genetic material is a single giant molecule of DNA. It is mainly concentrated in the nucleus; it has also been reported in cytoplasmic organelles like chloroplast, mitochondria, etc.

1.3.1 Chemical Composition of DNA

The DNA extract from many living organisms have revealed it to be a complex macromolecule. It is a long chain polymer composed of monomers (units), called nucleotides. Each nucleotide in turn is composed of three subunits. They are nitrogen base, sugar and phosphate.

1.3.2 Nitrogen Bases

The genetic material is made up of the nitrogen bases which are of two kinds, purines and pyrimidines. Both pyrimidines and purines are heterocyclic compounds; with a basic six-membered ring in which one or the other carbon position is replaced by nitrogen. The pyrimidines have a basic six-membered benzene ring. There are two kinds of pyrimidines—cytosine and thymine. Structurally, cytosine has a single six-membered ring with a hydroxyl group at the 2′ position and amino group at the 6′ position. Thymine also has a single six-membered ring with hydroxyl group at the 2′ and 6′ position, and a methyl group at the 5′ position.

The purines have a double-ringed structure. Attached to the six-membered benzene ring is a five-membered imidazole ring. The imidazole ring is joined to the hexagonal ring at 4′ and 5′ position. The benzene ring has nitrogen in the 1′ and 3′ position. There are two types of purines—adenine and guanine. Adenine has an amino group at the 6′ position, while guanine has an amino group at the 2′ position and a hydroxyl group at the 6′ position.

1.3.3 Sugar

The sugar molecule in DNA is basically a ribose sugar. The ribose sugar is a pentose sugar. It is a five-membered ring. The ring consists of four carbon atoms while the 5′ C atom is outside the ring Positions 1′, 3′ and 5′ in the ring have OH groups, while the 2′ carbon position has only H. There is no O; hence the sugar is said to be deoxyribose.

1.3.4 Phosphate

The phosphate present has one central phosphorus with four oxygen atoms.

1.3.5 Nucleosides

A nitrogen base coupled with sugars forms a nucleoside. DNA has four nucleosides—deoxycitidine deoxythymidine, deoxyadenosine and deoxyguanosine. The sugar molecule is attached to the nitrogen base to the 3′ position in the benzene ring in the case of pyrimidines and to the 9′ position of the imidazole ring in the case of purines.

1.3.6 Nucleotides

A nucleotide is derived from the nucleoside by the addition of phosphoric acid molecule to the latter. The phosphorus group is usually linked to the two sugar molecules above and below. The linkage is with the 3′C atom of one deoxyribose sugar and with the 5′C atom of another. DNA has four nucleotides—deoxycytidilic acid, deoxythymidilic acid, deoxyadenylic acid and Deoxyguanilic acid.

1.3.7 Polynucleotide

DNA is said to be a polynucleotide chain with a number of nucleotides joined together with the help of DNA polymerase.

1.3.8 Molecular Model of DNA Structure

The molecular models of DNA have two strands of the double helix which are anti-parallel, which means that they run in opposite directions. The sugar–phosphate backbone is on the outside of the helix, and the bases are on the inside. The backbone can be thought of as the sides of a ladder, whereas the bases in the middle form the rungs of the ladder.

Each rung is composed of two base pairs. Either an adenine-thymine pair that forms a two-hydrogen bond together or a cytosine-guanine pair that form a three-hydrogen bond. The base-pairing is thus restricted. This restriction is essential when the DNA is being copied: the DNA helix is first unzipped in two long stretches of sugar–phosphate backbone with a line of free bases sticking up from it, like the teeth of a comb. Each half will then be the template for a new, complementary strand. Biological machines inside the cell put the corresponding free bases onto the split molecule and also "proof-read" the result to find and correct any mistakes. After the doubling, this gives rise to two exact copies of the original DNA molecule. The coding regions in the DNA strand, the genes, make up only a fraction of the total amount of DNA. The stretches that flank the coding regions are called introns, and consist of non-coding DNA. Introns were looked upon as junk in the early days. Today, biologists and geneticists believe that this non-coding DNA may be essential in order to expose the coding regions and to regulate how the genes are expressed.

1.3.9 Alternative Double-helical Structures

DNA exists in many possible conformations. However, only A-DNA, B-DNA, and Z-DNA have been observed in organisms. The conformation adopted by DNA depends on the sequence of the DNA, the amount and direction of supercoiling, chemical modifications of the bases and also solution conditions, such as the concentration of metal ions and polyamines. Of these three conformations, the "B" form described above is most common under the conditions found in cells. The two alternative double-helical forms of DNA differ in their geometry and dimensions. The A form is a wider right-handed spiral, with a shallow, wide minor groove and a narrower, deeper major groove. The A form occurs under non-physiological conditions in dehydrated samples of DNA, while in the cell it may be produced in hybrid pairings of DNA and RNA strands, as well as in enzyme–DNA complexes. Segments of DNA where the bases have been chemically modified by methylation may undergo a larger change in conformation and adopt the Z form. The strands turn about the helical axis in a left-handed spiral, the opposite of the more common B form. These unusual structures can be recognized by specific Z-DNA binding proteins and may be involved in the regulation of transcription.

1.4 RIBONUCLEIC ACID (RNA)

RNA is a nucleic acid similar to DNA but containing ribose rather than deoxyribose sugar. RNA is formed upon a DNA template. There are several classes of RNA molecules.

1.4.1 Molecular Structure of RNA

The molecular backbone of RNA is made up of altering sugar and phosphate. The single chain folds to enable the base pairs to meet here and there to form the steps. The pairing is between A and U, and C and G. Occasionally, the RNA may be double-stranded but does not form a helix like DNA. Single-stranded RNA is the genetic material in plant viruses, poliomyelitis viruses etc. Double-stranded RNA occurs as the genetic material in viruses.

1.4.2 Types of RNA

There are three types of RNA—mRNA, tRNA, rRNA. Messenger RNA (mRNA) carries information about a protein sequence to the ribosomes, the protein synthesis factories in the cell. It is coded so that every three nucleotides (a codon) correspond to one amino acid. In eukaryotic cells, once precursor mRNA (pre-mRNA) has been transcribed from DNA, it is processed to mature mRNA. This removes its introns—non-coding sections of the pre-mRNA. The mRNA is then exported from the nucleus to the cytoplasm, where it is bound to ribosomes and translated into its corresponding protein form with the help of tRNA. In prokaryotic cells, which do not have nucleus and cytoplasm compartments, mRNA can bind to ribosomes while it is being transcribed from DNA. After a certain amount of time the message degrades into its component nucleotides with the assistance of ribonucleases.

Transfer RNA (tRNA) is a small RNA chain of about 80 nucleotides that transfers a specific amino acid to a growing polypeptide chain at the ribosomal site of protein synthesis during translation. It has sites for amino acid attachment and an anticodon region for codon recognition that binds to a specific sequence on the messenger RNA chain through hydrogen bonding.Ribosomal RNA (rRNA) is the catalytic component of the ribosomes. Eukaryotic ribosomes contain four different rRNA molecules: 18S, 5.8S, 28S and 5S rRNA. Three of the rRNA molecules are synthesized in the nucleolus, and one is synthesized elsewhere. In the cytoplasm, ribosomal RNA and protein combine to form a nucleoprotein called a ribosome. The ribosome binds mRNA and carries out protein synthesis. Several ribosomes may be attached to a single mRNA at any time. rRNA is extremely abundant and makes up 80% of the 10 mg/ml RNA found in a typical eukaryotic cytoplasm.Transfer-messenger RNA (tmRNA) is found in many bacteria and plastids. It tags the proteins encoded by mRNAs that lack stop codons for degradation and prevents the ribosome from stalling.

1.5 AMINO ACIDS

So far, we have learnt about the genetic material and its process from transcription to translation. Translation results in protein production, the protein being comprised of amino acids. An amino acid is a molecule containing both amine and carboxyl functional groups (Figure 1.3). In Biochemistry, this term refers to alpha-amino acids with the general formula $H_2NCHRCOOH$, where, R is an organic substituent. In the alpha-amino acids, the amino and carboxylate groups are attached to the same carbon, which is called the α-carbon. The various alpha-amino acids differ in which side chain (R group) is attached to their alpha carbon. They can vary in size from just a hydrogen atom in glycine through a methyl group in alanine to a large heterocyclic group in tryptophan.Beyond the amino acids that are found in all forms of life, many non-natural amino acids have vital roles in technology and industry. For example, the chelating agents EDTA and nitrilotriacetic acid are alpha-amino acids that are important in the chemical industry.

(a) Amino acids with hydrophobic side groups

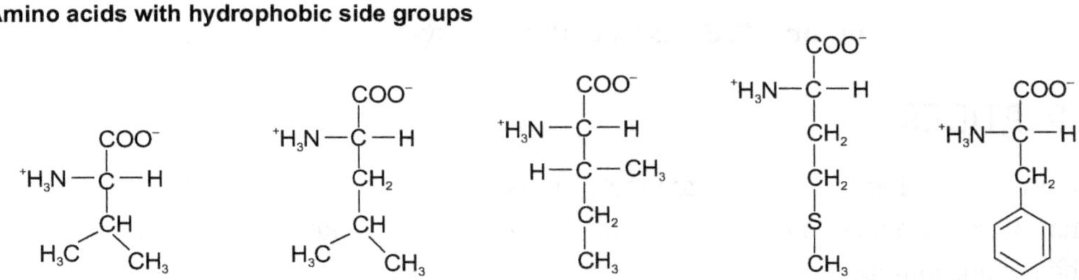

Valine (val) Leucine (leu) Isoleucine (Ile) Methionine (met) Phenylalanine (phe)

Figure 1.3 Structures of twenty amino acids *(Continues)*

(b) Amino acids with hydrophilic side groups

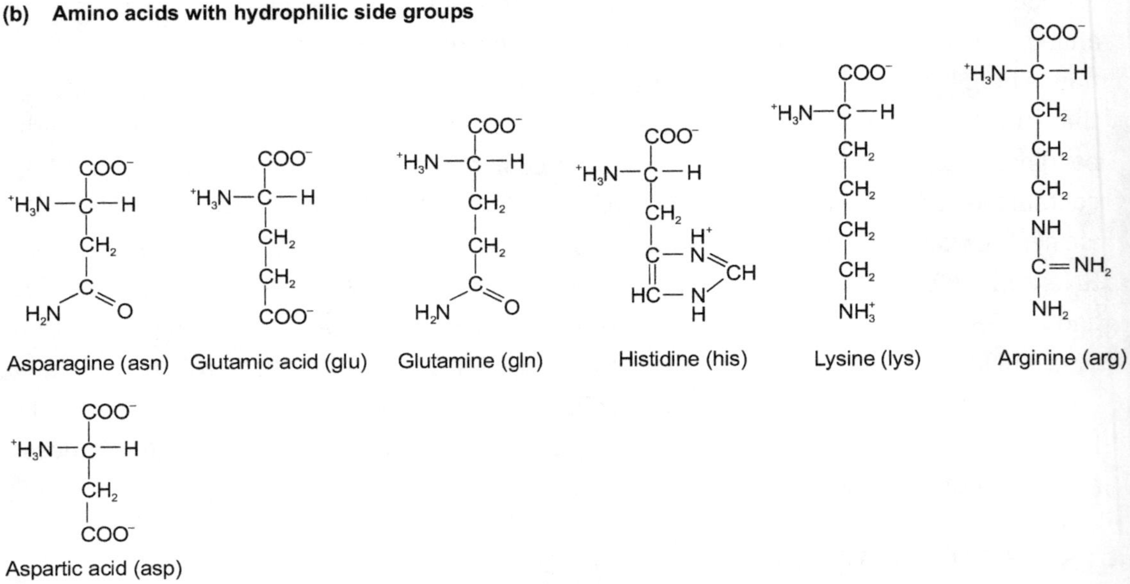

(c) Amino acids that are in between

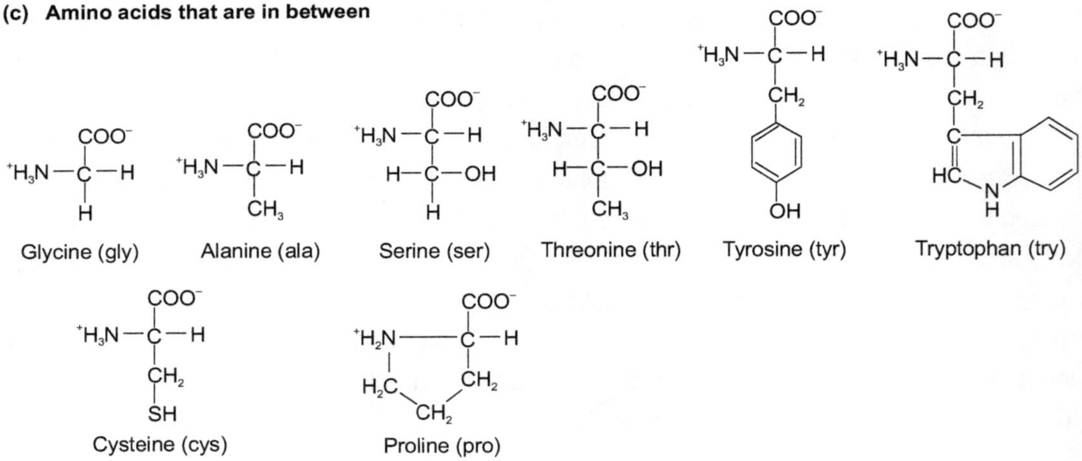

Figure 1.3 Structures of twenty amino acids

1.6 PEPTIDES

A molecule consisting of two or more amino acids is known as a peptide. Peptides are smaller than proteins, which are chains of amino acids. These molecules are small enough to be synthesized from the constituent amino acids, by convention, called peptides rather than proteins. The dividing line is at about 50 amino acids. Depending on the number of amino acids, peptides are called dipeptides, tripeptides, tetrapeptides, and so on.

1.6.1 Ribosomal Peptides

Ribosomal peptides are synthesized by translation of mRNA. They are often subjected to proteolysis to generate the mature form. These functions are seen typically in higher organisms, as hormones and signalling molecules. Some organisms produce peptides as antibiotics, such as microcins.

1.6.2 Non-ribosomal Peptides

Non-ribosomal peptides are assembled by enzymes that are specific to each peptide, rather than by the ribosome. The most common non-ribosomal peptide is glutathione, which is a component of the antioxidant defenses of most aerobic organisms. Other non-ribosomal peptides are most common in unicellular organisms, plants, and fungi and are synthesized by modular enzyme complexes called non-ribosomal peptide synthetases. These complexes are often laid out in a similar fashion, and they can contain many different modules to perform a diverse set of chemical manipulations on the developing product. These peptides are often cyclic and can have highly complex cyclic structures, although linear non-ribosomal peptides are also common. Since the system is closely related to the machinery for building fatty acids and polyketides, hybrid compounds are often found. Oxazoles, thiazoles often indicate that the compound was synthesized in this fashion.

1.6.3 Peptones

Peptones are derived from animal milk or meat digested by proteolytic digestion. In addition to containing small peptides, the resulting spray-dried material includes fats, metals, salts, vitamins and many other biological compounds. Peptone is used in nutrient media for growing bacteria and fungi.

1.6.4 Peptide Fragments

Peptide fragments are proteins which are used to identify or quantify the source protein. Often these are the products of enzymatic degradation performed in the laboratory on a controlled sample, but can also be forensic or palaentological samples which have been degraded by natural effects.

PROTEINS

When the combinations of amino acids occur, there occurs the formation of proteins which are molecules made up of amino acids that are needed for the body to function properly. Proteins are the basis of body structures such as skin and hair and of substances such as enzymes and antibodies.

1.7 STRUCTURE OF PROTEINS

1.7.1 Primary Structure

From the term primary structure you would have understood that it is the amino acid synthesized in the initial stage only with a linear set of sequence. The sequence of the different amino acids is called the primary structure of the peptide or protein. Counting of residues always starts at the N-terminal end (NH_2-group), which is the end where the amino group is not involved in a peptide bond. The primary structure of a protein is determined by the gene corresponding to the protein. A specific sequence of nucleotides in DNA is transcribed into mRNA, which is read by the ribosome in a process called translation. The sequence of a protein is unique to that protein, and defines the structure and function of the protein. The sequence of a protein can be determined by methods such as Edman degradation or tandem mass spectrometry. Often however, it is read directly from the sequence of the gene using the genetic code. Post-transcriptional modifications such as disulphide formation, phosphorylations and glycosylations are usually also considered a part of the primary structure, and cannot be read from the gene.

1.7.2 Secondary Structure

Let us study the next level of the protein structure, i.e., secondary structure. By building models of peptides using known information about bond lengths and angles, the first elements of secondary structure, the alpha helix and the beta sheet, were suggested in 1951 by Linus Pauling and co-workers. Both the alpha helix and the beta sheet represent a way of saturating all the hydrogen bond donors and acceptors in the peptide backbone. These secondary structure elements only depend on properties that all the residues have in common, explaining why they occur frequently in most proteins. Since then other elements of secondary structure have been discovered such as various loops and other forms of helices. The part of the backbone that is not in a regular secondary structure is said to be random coil. Each of these two secondary structure elements have a regular geometry, meaning they are constrained to specific values of the dihedral angles Ψ and ϕ. Thus they can be found in a specific region of the Ramachandran plot.

1.7.3 Tertiary Structure

We all know that the next stage of conformation attained by the secondary structure is the teritary structure. The elements of secondary structure are usually folded into a compact shape using a variety of loops and turns. The formation of tertiary structure is usually driven by the burial of hydrophobic residues, but other interactions such as hydrogen bonding, ionic interactions and disulphide bonds can also stabilize the tertiary structure. The tertiary structure encompasses all the non-covalent interactions that are not considered secondary structure, and is what defines the overall fold of the protein, and is usually indispensable for the function of the protein.

1.7.4 Quaternary Structure

The quaternary structure is the interaction between several chains of peptide bonds. The individual chains are called subunits. The individual subunits are not necessarily covalently connected, but might be connected by a disulphide bond. Not all proteins have quaternary structure, since they might be functional as monomers. The quaternary structure is stabilized by the same range of interactions as the tertiary structure. Complexes of two or more polypeptides (i.e., multiple subunits) are called multimers. Specifically it would be called a dimer if it contains two subunits, a trimer if it contains three subunits, and a tetramer if it contains four subunits. Multimers made up of identical subunits may be referred to with a prefix "homo-" (e.g. a homotetramer) and those made up of different subunits may be referred to with a prefix of "hetero-" (e.g. a heterodimer). Tertiary structures vary greatly from one protein to another. They are held together by glycosidic and covalent bonds.

1.8 CLASSIFICATION OF PROTEINS

As the proteins possess a very complex structure, their classification also is a problem, since many categorizations are possible based on structure, function, etc. The major groups of proteins are recognized based on their solubility and chemical properties.They are classified as simple proteins, conjugated proteins, and derived proteins.

1.8.1 Simple Proteins

Chemically, these have only amino acids. No additional compounds are present in the molecule. Simple proteins are of the following kinds.

- Actin—one of the proteins into which actomyosin can be split; can exist in either a globular or a fibrous form
- Albumin, albumen—a simple water-soluble protein found in many animal tissues and liquids
- Globulin—a family of proteins found in blood and milk and muscle and in plant seed
- Globin, haematohiston—a colourless protein obtained by removing haeme from haemoglobin; the oxygen-carrying compound in red blood cells
- Glutelin—a simple protein found in the seeds of cereals

- Histone—a simple protein containing mainly basic amino acids; present in cell nuclei in association with nucleic acids
- Prolamine—a simple protein found in plants
- Protamine—a simple protein found in fish sperm; rich in arginine; simpler in composition than globulin or albumin; counteracts the anticoagulant effect of heparin
- Scleroprotein, albuminoid—a simple protein found in horny and cartilaginous tissues and in the lens of the eye

1.8.2 Conjugated Proteins

A conjugated protein is a protein that functions in interaction with other chemical groups attached by covalent bonds or by weak interactions.

Many proteins contain only amino acids and no other chemical groups, and they are called simple proteins. However, other kind of proteins yield, on hydrolysis, some other chemical component in addition to amino acids and they are called conjugated proteins. The non-amino part of a conjugated protein is usually called its prosthetic group. Conjugated proteins are classified on the basis of the chemical nature of their prosthetic groups. Some examples of conjugated proteins are lipoproteins, glycoproteins, phosphoproteins, haemoproteins, flavoproteins, metalloproteins, phytochromes, cytochromes and opsins. Glycoproteins are generally the largest and most abundant group of conjugated proteins. They range from glycoproteins in cell surface membranes that constitute the glycocalyx, to important antibodies produced by leucocytes.

1.8.3 Derived Proteins

The breakdown of proteins form derived proteins.These are degradation products of proteins. Primary proteins on hydrolysis with acids or alkalies yield derived proteins—primary derived proteins and secondary derived proteins.

1.8.4 Primary Derived Proteins

There are two categories of primary derived proteins. These are metaproteins and coagulated proteins.

1.8.5 Secondary Derived Proteins

Three types of secondary derived proteins have been identified. These are proteases, peptones, and peptides.

1.9 SUPER-SECONDARY STRUCTURE OF PROTEINS

Some proteins contain an ordered organization of secondary structures that form distinct functional structural motifs. Examples include the helix-turn-helix domain of bacterial proteins that regulate transcription and the leucine zipper, helix-loop-helix and zinc finger domains of eukaryotic transcriptional regulators. These motifs are termed super-secondary structures.

1.9.1 Structural Motif

A structural motif is a three-dimensional structural element or fold within the chain, which appears also in a variety of other molecules. In the context of proteins, the term is sometimes used interchangeably with "structural domain," although a domain need not be a motif nor, if it contains a motif, need not be made up of only one.

1.9.2 Helix-turn-helix Domain

In proteins, the helix-turn-helix (HTH) is a major structural motif capable of binding DNA. It is composed of two α-helices joined by a short strand of amino acids and is found in many proteins that regulate gene expression. It is not to be confused with the basic helix-loop-helix domain.

1.9.3 Leucine Zipper

Like many other transcription factors, the leucine-zipper-containing transcription factors bind DNA as dimers. A leucine zipper is formed by two a helices, one from each monomer. The helices are held together by hydrophobic interactions between leucine residues, which are located on one side of each helix.

1.9.4 Helix-loop-helix

Helix-loop-helix is characterized by two helices connected by a loop. It should not be confused with "Helix-turn-helix". In fact, these two structures are quite different, but the Helix-loop-helix looks like leucine zipper.

1.9.5 Zinc Finger Domains

Motifs in DNA- and RNA-binding proteins whose amino acids are folded into a single structural unit around a zinc atom are called zinc finger domains. In the classic zinc finger, one zinc atom is bound to two cysteines and two histidines. In between the cysteines and histidines are 12 residues, which form a DNA binding fingertip. By variations in the composition of the sequences in the fingertip and the number and spacing of tandem repeats of the motif, zinc fingers can form a large number of different sequence-specific binding sites.

SUMMARY

This chapter introduces its readers to "Omics" and the branches of "Omics" like metabolomics, expressomics, etc. The definition and the concepts of Systems Biology have been given. When learning "Omics" concepts, one should know the basics of biology, like, the type of cells its organelles, functions of the cell. These cell organelles play an important role in analysing the problems in the human system. The heart of the cell, nucleus, is the core in studying the living system. The combination of cells, tissues, organs comprise a system and the analysis of these is systems biology. The expression of the DNA results in the synthesis of proteins. We had also learnt that the proteins have varied structures, such as primary, secondary, tertiary and quaternary.

CRITICAL THINKING QUESTIONS

1. What are the branches in 'Omics?
2. Distinguish between the properties of prokaryotes and eukaryotes.
3. What are macromolecule managers?
4. Which is the cytoskeleton organizer?
5. Define super-secondary structure. Explain the types.

SELF-LEARNING EXERCISES

1. How does systems biology help the researchers in analysing a system?
2. What would happen if the DNA forgets to express its components at a particular time when they want to?
3. Will a slight change in the protein structure cause a drastic effect? Give reasons.

2

INTRODUCTION TO BIOINFORMATICS

In the previous chapter we have seen the biological concepts related to Bioinformatics, now we will discuss what really Bioinformatics is. It is a knowledge derived from computer analysis of biological data (Figure 2.1). It is a rapidly developing branch of Biology and is highly interdisciplinary, using the techniques and concepts from informatics, statistics, mathematics, chemistry, biochemistry, physics and linguistics.

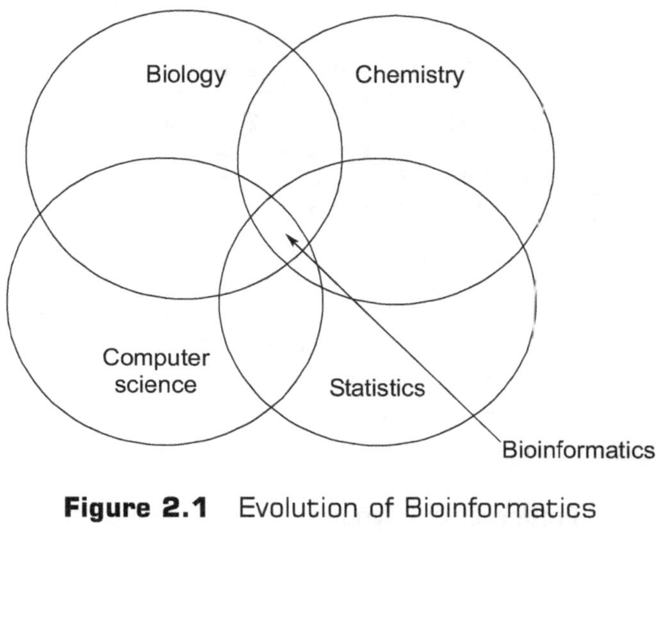

Figure 2.1 Evolution of Bioinformatics

It has many practical applications in different areas of biology and medicine. Bioinformatics is the science of developing computer databases and algorithms for the purpose of speeding up and enhancing biological research. It is the use of mathematical and informational techniques including statistics, to solve biological problems. Bioinformatics is also defined as an intersection of biological, computer and information sciences necessary to manage, process and understand large amounts of data, for instance from the sequencing of the human genome or from large databases containing information about plants and animals, for use in discovering and developing new drugs. One of the main areas of Bioinformatics is the data mining and analysis of the data gathered by the various genome projects. Other areas are sequence alignment, protein structure predictions, systems biology, protein–protein interactions and virtual evolution.

2.1 DYNAMIC PROGRAMMING

I am sure you would have got an idea or programming from Part I of this book. The need of programming is necessary for bioinformatics as a creator. A large class of programming algorithms are based on breaking a large problem down into incremental steps so that, at any given stage, optimal solutions are known sub-problems. Dynamic programming (Figure 2.2) is useful when the length of the query sequences increases and the difference between the lengths of the two sequences increase. More gaps have to be inserted in various places to get the optimal alignment between two sequences.

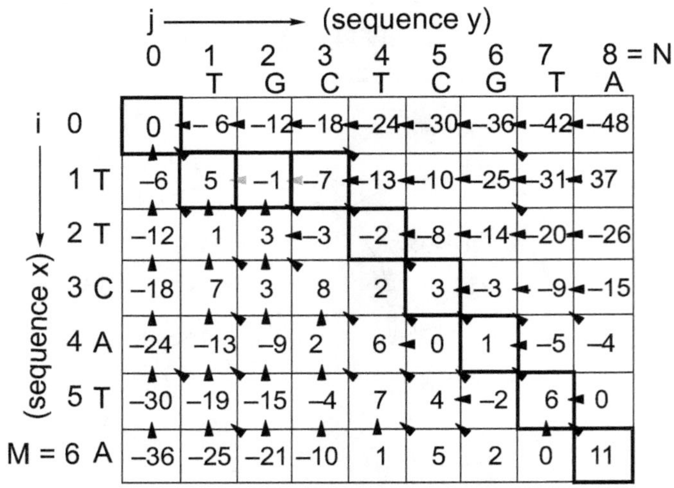

Optimum alignment scores 11:

T – – T C A T A

T G C T C G T A

+5 –6 –6 +5 +5 –2 +5 +5

Figure 2.2 Dynamic programming matrix

2.2 COMPUTER TECHNOLOGY

Computer technology is used/utilized for the analysis of biological information. Perl is commonly used for Bioinformatics calculations as its ability to manipulate character symbols, it is a default CGI language, and it was started out as a scripting language but has become a full-fledged language. In Bioinformatics various databases, different platforms, languages, etc. are being used. GRID—new kid on the block used in many computers to fulfil a single computational task. Bioinformatics is the ideal platform, as it has to deal with a large amount of data in alignment and searches. ORACLE 10g is the world's first GRID database.

DATABASE AND MINING

We know that there are a lot of sequences and sequence databases are available publicly that involve various data mining techniques and are used to pull the data out for the use. As there are a number of literatures and articles available, data mining is currently carried out on them and a lot of research works are also on.

2.3 DATABASES

A database is defined as an organized collection of information, data, or citations stored in electronic format that can be searched for specific information or records by techniques specific to each database.

2.3.1 Biological Database

A collection of large number of biological information under one roof is called as biological database (Table 2.1). A large number of detailed sets of information about the biological sequences are stored in the form of data called as biological database, in which they can be manipulated, retrieved and utilized for further studies.

Table 2.1 Biological databases

Nucleic acid	Protein
EMBL (Europe)	PIR—Protein Information Resource
GenBank (USA)	MIPS
DDBJ (Japan)	SWISS-PROT University of Geneva, now with EBI
	TrEMBL—A supplement to SWISS-PROT
	NRL-3D

As there are too many databases one may not not know which one to search. Some databases are good in some aspects and weak in others.Thus the search can be done in a composite database that has several databases for its base data. These databases are indexed and streamlined so that the same stored sequence is not searched twice in different databases. Some examples for these databases include OWL, SWISS-PROT, PIR, GenBank and NRL-3D.

2.3.2 Major Bioinformatics Resources

Before we go into the core area, we should know some of the major databases like NCBI, EBI and ExPASy. The knowledge of various databases and Bioinformatics tools are available at these resources. The major content of these databases have an important purpose and utility in the field of life sciences for research.

2.3.3 Open Access Bibliographic Resources and Literature Databases

There are certain open access bibliographic resources related to life sciences, based on the significance and need for such resources, the major content of the databases, how to search and use these resources/databases. A few examples are PubMed, MEDLINE, AltaVISTA, PubMed Central, Public Library of Sciences, etc.

PubMed PubMed is an on-line database providing free access to MEDLINE, NLM's database of citations and abstracts in the fields of medicine, nursing, dentistry, veterinary medicine, health care systems, and preclinical sciences. It links to many sites providing full text articles and other related resources. It provides a Clinical Queries search filters page as well as a Special Queries page. It links to related articles for a selected citation.

Figure 2.3 shows the home page of PubMed and Figure 2.4 shows the resultant of a search in PubMed. To get started with PubMed, enter one or more search terms. Search terms may be topics, authors or journals.

MEDLINE MEDLINE is NLM's bibliographic database of citations and abstracts currently from approximately 5,000 biomedical journals and is published in the United States and worldwide. The coverage dates back to 1950.

All citations in MEDLINE are assigned with Medical Subject Headings (MeSH®) from NLM's controlled vocabulary to assist the users in their searches.

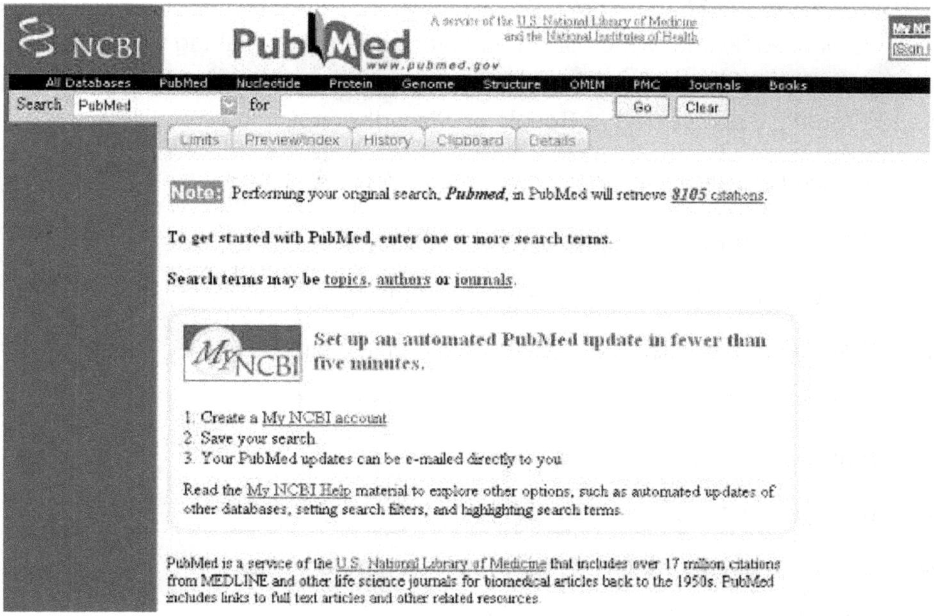

Figure 2.3 Home Page of PubMed

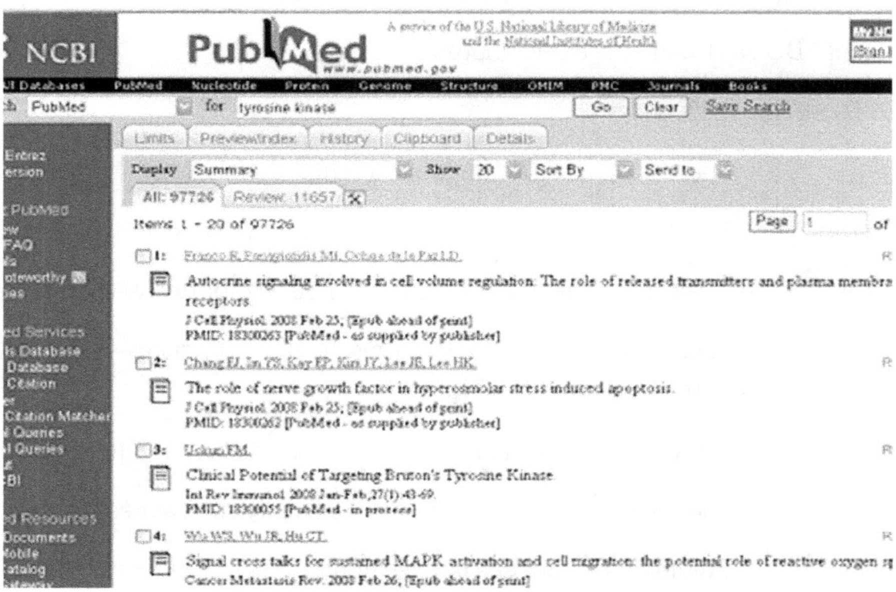

Figure 2.4 Resultant of a search in PubMed

Google advanced scholar search Once you know the basics of Google search, you might want to try Advanced Search, which offers numerous options for making your searches more precise and getting more useful results. Figure 2.5 shows the home page of Google advanced scholar search and Figure 2.6 shows that of Google advanced search.

With advanced search, you can search for sites

- that contain ALL the search terms you type in
- that contain the exact phrase you type in
- that contain at least one of the words you type in
- that DO NOT contain any of the words you type in
- that are written in a certain language
- that are created in a certain file format
- that have been updated within a certain period of time
- that contain numbers within a certain range
- that are within a certain domain, or website
- that are available for anyone to use, share or modify, even commercially
- that DO NOT contain "adult" material

Other advanced search features

Language	Specify in which language you'd like your results.
File Format	Specify the file format you'd like in your results.
Date	Restrict your results to the past three, six, or twelve month periods.
Occurrences	Specify where your search terms occur on the page—anywhere on the page, in the title, or in the URL.
Domains	Search only a specific website, or exclude that site from your search.
Usage Rights	Specify the rights of usage you'd like in your results.
Safe Search	Eliminates adult sites from search results.
Page-specific	Specify pages that are similar or link to your page.
Technology Search	Find information related to Apple Macintosh, BSD Unix, Linux or Microsoft.

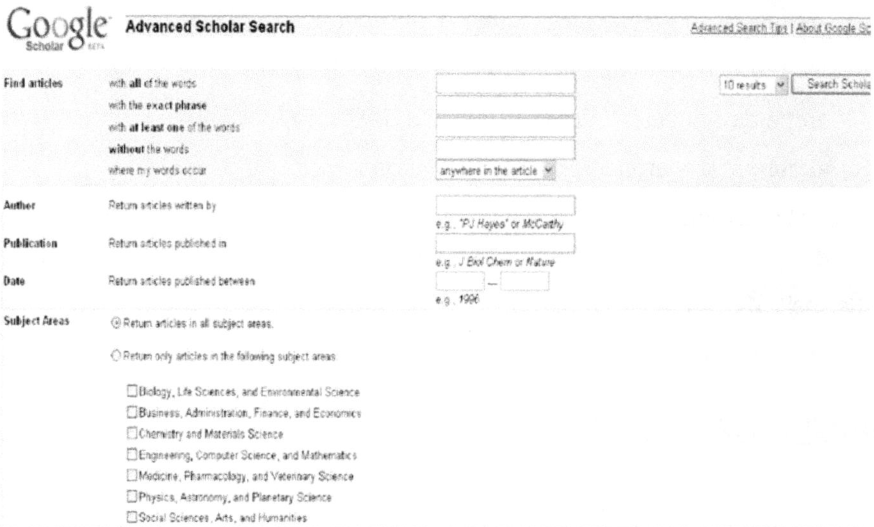

Figure 2.5 Home Page of Google advanced scholar search

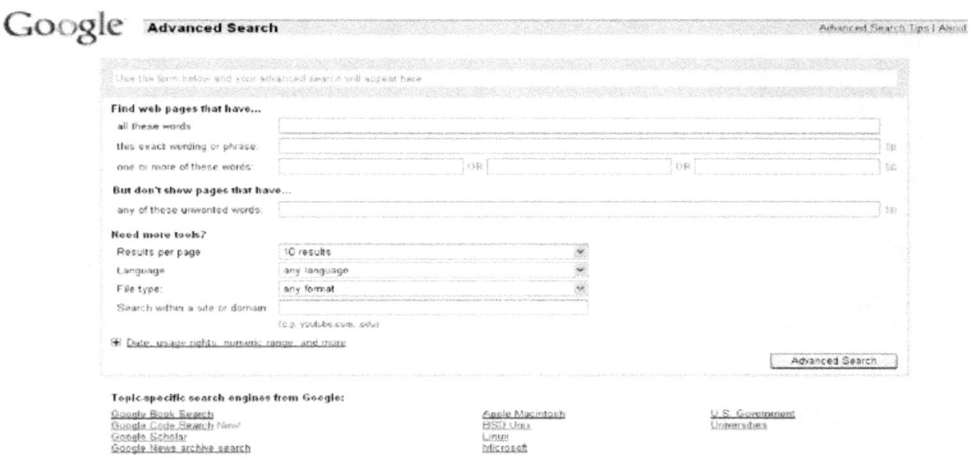

Figure 2.6 Home page of Google advanced search

2.4 Sequence Database

As mentioned earlier in this chapter, there are many number of databases and they have their specificity and restriction. The following is the list of different databases with respect to the organization of data, contents and formats of database entries, retrieval of data using text-based search tools, sources of data (e.g. sequencing projects, individual scientists, patent offices, etc.), method for deposition of data to databases.

- Nucleic acid sequence databases—GenBank, EMBL, DDBJ
- Protein sequence databases—SWISS-PROT, TrEMBL, PIR_PSD
- Genome databases at NCBI, EBI, TIGR, SANGER
- Viral genomes—archaeal and bacterial genomes
- Eukaryotic genomes with special reference to model organisms (Yeast, *Drosophila*, *C. elegans*, rat, mouse), human, plants such as *Arabidopsis thaliana*, rice, etc.
- Repositories for high-throughput genomic sequences—EST, STS GSS, etc.
- Structure databases—PDB, NDB, CCSD

2.5 DERIVED DATABASES

Derived databases are for bio-molecular sequences and structures. The following are databases with respect to basic concept of derived databases, sources of primary data and basic principles of the method for deriving the secondary data, organization of data, contents and formats of database entries, identification of patterns in given sequences and interpretation of the same.

ProSite	PRODOM
Pfam	PRINTS
CATH	SCOP
DSSP	FSSP
DALI	

2.6 SEQUENCE ANALYSIS

Sequence analysis refers to analysing the given sequence for further classification. The determination of the sequence of a nucleic acid, protein, or carbohydrate is based on the fragmentation of the original molecule, the analysis of sequence, and the assembly and interpretation of the sequence information. The various file formats for bio-molecular sequences include GenBank, FASTA, GCG, MSF, NBRF-PIR, etc. The basic concepts underlying the sequence similarity are identity, homology, orthologs and paralogs.

2.6.1 Scoring Matrices

The basic idea behind forming a scoring matrix is often based on the observed substitution rates derived from the substitution frequencies seen in multiple alignments of sequences. The two most commonly used scoring matrices are PAM and BLOSUM. These are used for nucleic acid and protein sequences. The differences between the distance and similarity matrix form the basis for deriving these matrices.

2.6.2 Sequence-based Database Searches

The sequence-based database searches include BLAST and FASTA algorithms and their various versions. The use of these methods for sequence analysis include the online use of the tools and interpretation of results.

2.6.3 Pairwise Sequence Alignment

The concept of sequence alignment is based on Needleman–Wuncsh and Smith–Waterman algorithms defined for pairwise alignments. Pairwise alignments for the analysis of nucleic acid and protein sequences are used only for a pair of sequences and interpretation of the results are used for further analysis.

2.6.4 Multiple Sequence Alignments (MSA)

There was a need for MSA when more than two sequences were to be aligned and compared which was based on a progressive and hierarchical procedure. Based on this, ClustalW and ClustalX were designed and used. The results of the interpretation are in the form of dendograms. These tools are used for evolutionary analysis, and they establish phylogenetic relationship between the species.

2.7 TAXONOMY AND PHYLOGENY

Systematics is the area of biology that deals with the diversity of living organisms (through evolution), their relationships to each other, (phylogeny and molecular evolution) and their classification (taxonomy). The data used in taxonomy and phylogeny are the nucleotides or protein sequences where any relevent data pertaining to their origin is determined.

2.7.1 Sequence Patterns and Profiles

A sequence pattern is defined by the nucleotide sequences of a set of aligned binding sites or by a common protein structure, motifs and profiles. Various types of pattern representations are consensus and regular expressions (prosite-type). The profile-based database searches are carried out using PSI-BLAST, and the results obtained are analysed and interpreted. The properties of protein and nucleic acid sequences are analysed on the tools available at the ExPASy server, GCG utilities and EMBOSS.

2.8 NCBI

NCBI was established in 1988 as a national resource for molecular biology information. NCBI creates public databases, conducts research in computational biology and develops software tools for analysing genome data. Figure 2.7 represents the home page of NCBI and Figure 2.8 lists some of the research options in it.

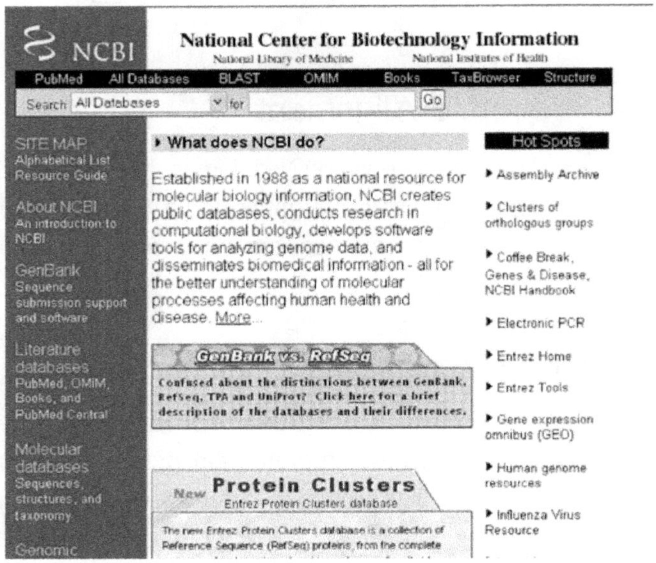

Figure 2.7 Home Page of NCBI

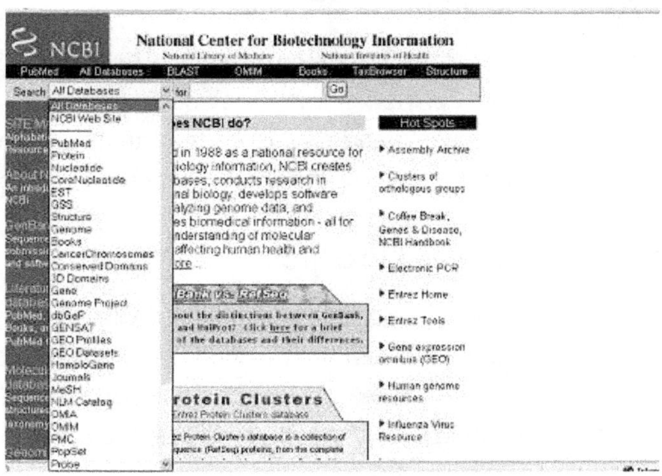

Figure 2.8 Search options in NCBI

2.9 DATABASES AND SOFTWARE

From the entire database, NCBI assumed the responsibility for the GenBank DNA sequence. The database was built from the sequences submitted by individual laboratories and through the data exchange with the international nucleotide sequence databases such as European Molecular Biology Laboratory (EMBL) and the DNA Database of Japan (DDBJ). In addition to GenBank, NCBI supports and distributes a variety of databases for the medical and scientific communities. These include the Online Mendelian Inheritance in Man (OMIM), the Molecular Modeling Database (MMDB) of 3D protein structures, the Unique Human Gene Sequence Collection (UniGene), a Gene Map of the Human Genome, the Taxonomy Browser, and the Cancer Genome Anatomy Project (CGAP), in collaboration with the National Cancer Institute. Entrez is NCBI's search and retrieval system that provides users with integrated access to sequence mapping, taxonomy, and structural data. Entrez also provides graphical views of sequences and chromosome maps. A powerful and unique feature of Entrez is its ability to retrieve related sequences, structures, and references. The journal literature is available through PubMed, a Web search interface that provides access to over 11 million journal citations in MEDLINE and contains links to full-text articles at participating publishers' websites.

BLAST is a program for sequence similarity searching developed at NCBI and is instrumental in identifying genes and genetic features. BLAST can execute sequence searches against the entire DNA database in less than 15 seconds. Additional software tools provided by NCBI include Open Reading Frame Finder (ORF Finder), Electronic PCR, and the sequence submission tools, Sequin and Bank It. All of NCBI's databases and software tools are accessible through WWW or FTP. NCBI also has email servers that provide an alternative way to access the databases for text searching or sequence similarity searching.

2.9.1 From Data and Information to Facts: GenBank and the Human Genome

GenBank is a NIH database, maintained and distributed by NCBI, that stores all known public DNA sequences. Sequence data are submitted to GenBank from individual scientists from around the world, as well as from the large centres involved in the Human Genome Project. The number of DNA sequences stored in the GenBank database, from all organisms, recently reached colossal heights and continues to grow at a rapid rate. GenBank is an international collaborative project, with partners located at the European Bioinformatics Institute in the United Kingdom and the National Institute of Genetics in Japan.

NCBI is also engaged in the essential process of annotating, or labelling, the biologically important areas of the human genome. This process includes the correct placement of known human genes into their proper genomic context as well as the prediction of previously unidentified genes from the genomic sequence. The mRNAs from the NCBI RefSeq collection are placed on the genome primarily by alignment. RefSeq mRNAs are the reference sequence standards for the human genome. These data use both computational tools and scientific judgment to determine what sequence is an appropriate representation for a gene.

2.9.2 OMIM

OMIM is a web-based catalog that contains thousands of entries for genes and genetic disorders and serves as a phenotypic companion to the Human Genome Project. The OMIM cytogenetic and morbid maps present cytogenetic locations for those genes with published locations and provide an alphabetical list of all the diseases described in OMIM.

To validate the findings generated through computer-based comparative analysis, it is essential to consider the results of wet-bench biology reported in the scientific literature. Therefore, the integration of scientific data with the literature is a necessary step for creating a unified information resource in the life sciences. To this end, individuals are provided with a direct link from OMIM to PubMed, NCBI's literature retrieval system.

PubMed provides web-based access to over 11 million citations, abstracts, and indexing terms for journal articles in the biomedical sciences. It also includes links to full-text journals. Currently, approximately 20 million searches are conducted per month, and as many as 140,000 different users seek information daily via PubMed.

2.9.3 PubMed Central (PMC)

The digital archive of life sciences journal literature is PMC. It was launched in January 2001 and offers a new model for electronic scientific communication and data retrieval. The value of PubMed Central, in addition to its role as an archive, lies in what can be done when data from diverse sources are stored in a common format in a single repository. PMC currently provides free and unrestricted access to the full text of 104 life sciences journals, with more forthcoming.

2.9.4 Discovery of Disease Genes

The integrated resources developed and disseminated by the NCBI and the Human Genome Project have led to many scientific advances like finding new genes which cause a disease. The discovery of the genes for hereditary nonpolyposis colorectal cancer (HNPCC) is one such example.

2.10 EVOLUTIONARY BIOLOGY

Homology refers to two genes sharing a common evolutionary history. A new insight into the molecular basis of a disease may come from investigating the function of homologs of a disease gene in model organisms.

Equally exciting is the potential for uncovering the evolutionary relationships and patterns between different forms of life. With the aid of nucleotide and protein sequences, it should be possible to find the ancestral ties between different organisms. By far, experience has taught us that closely related organisms have similar sequences and that more distantly related organisms have more dissimilar sequences. Proteins that show significant sequence conservation, indicating a clear evolutionary relationship, are said to be from the same protein family. By studying protein folds (distinct protein building blocks) and families, we are able to reconstruct the evolutionary relationship between two species and estimate the time of divergence between two organisms since they shared a common ancestor.

2.11 PROTEIN MODELLING

The process of evolution has resulted in the production of DNA sequences that encode proteins with specific functions. In the absence of a protein structure that has been determined by X-ray crystallography or nuclear magnetic resonance (NMR) spectroscopy, researchers can try to predict the three-dimensional structure using protein or molecular modelling. This method uses experimentally determined protein structures (templates) to predict the structure of another protein that has a similar amino acid sequence (target).

Although molecular modelling may not be as accurate at determining a protein's structure as experimental methods, it is still extremely helpful in proposing and testing various biological hypotheses. Molecular modeling also provides a starting point for researchers wishing to confirm a structure through X-ray crystallography and NMR spectroscopy. As the different genome projects are producing more sequences and since novel protein folds and families are being determined, protein modelling will become an increasingly important tool for scientists who are working to understand the normal and disease-related processes in living organisms.

2.11.1 Steps in Protein Modelling

1. Identify the proteins with known three-dimensional structures that are related to the target sequence.
2. Align the related three-dimensional structures with the target sequence and determine those structures that will be used as templates.

3. Construct a model for the target sequence based on its alignment with the template structure(s).
4. Evaluate the model against a variety of criteria to determine if it is satisfactory.

2.12 Genome Mapping

Genomic maps serve as a scaffold for orienting sequence information. If a gene, or nucleotide sequence has to localize, it must be forced to manually map the genomic region of interest, which is a time-consuming and often painstaking process. Today, with new technologies and the influx of sequence data, a number of high-quality genome-wide maps are available to the scientific community for use in their research. Computerized maps make gene hunting faster, cheaper and more practical for almost every scientist. The genetic map is used to assign a gene to a relatively small area of a chromosome. The physical map is used to closely examine the region of interest, to determine a gene's precise location. In the light of these advances, a researcher's burden has shifted from mapping a genome or genomic region of interest to navigating a vast number of websites and databases.

2.12.1 Map Viewer

NCBI's Map Viewer is a tool for visualizing whole genomes or a single chromosome that allows a user to view an organism's complete genome, integrated maps for each chromosome (when available), and/or sequence data for a genomic region of interest. When using the Map Viewer, a researcher has the option of selecting either a "Whole-Genome View" or a "Chromosome or Map View". The Genome View displays a schematic of the organism's chromosomes, whereas the Map View shows one or more detailed maps for a single chromosome. If more than one map exists for a chromosome, Map Viewer allows a simultaneous display of these maps.

The rapidly emerging field of Bioinformatics promises advances in understanding basic biological processes and, in turn, advances in the diagnosis, treatment, and prevention of many genetic diseases. Bioinformatics has transformed the discipline of biology from a purely lab-based science to an information science as well. Increasingly, biological studies begin with a scientist conducting vast number of databases and website searches to formulate specific hypotheses or to design large-scale experiments. The implications behind this change, for both science and medicine, are staggering.

2.13 PDB: THE PROTEIN DATA BANK

You will be rather surprized to know that PDB was the first ever "Bioinformatics" database built and designed to store complex three-dimensional data. It was originally developed and housed at the Brookhaven National Laboratories, but now managed and maintained by the Research Collaboratory for Structural Bioinformatics (RCSB). PDB is a collection of all publicly available three-dimensional structures of proteins, nucleic acids, carbohydrates, and a variety of other complexes experimentally determined by X-ray crystallography and NMR. It is supported by the National Science Foundation, the Department of Energy, and two units of the National Institutes of Health: the National Institute of General Medical Sciences and the National Library of Medicine.

2.14 PROTEIN MODELLING AT NCBI

NCBI's Molecular Modeling DataBase (MMDB), an integral part of the Entrez information retrieval system, is a compilation of all of the PDB three-dimensional structures of biomolecules. The difference between the two databases is that the MMDB records reorganize and validate the information stored in the database in a way that enables cross-referencing between the chemistry and the three-dimensional structure of macromolecules. By integrating chemical sequence, and structure information, MMDB is designed to serve as a resource for structure-based homology modelling and protein structure prediction.

The MMDB records have value-added information compared to the original PDB entries, including explicit chemical graph information, uniformly derived secondary structure definitions, structure domain information, literature citation matching, and molecule-based assignment of taxonomy to each biologically derived protein or nucleic acid chain.

2.15 THE ROLE OF TAXONOMY

Taxonomy provides a vivid picture of the existing organic diversity of the earth and permits a reconstruction of the phylogeny of life using the information that this branch of study provides. It reveals numerous, interesting evolutionary phenomena. Taxonomy supplies classifications that are of great explanatory value in most branches of biology.

2.16 COGs: Clusters of Orthologous Groups of Proteins

The database of Clusters of Orthologous Groups of proteins (COGs) represents an attempt at the phylogenetic classification of proteins that indicates the evolutionary relationship between organisms from complete genomes. Each COG includes proteins that are thought to be orthologous. Orthologs are genes in different species derived from a common ancestor and carried on through evolution. COGs may be used to detect similarities and differences between species for identifying protein families and predicting new protein functions and to point to potential drug targets in disease-causing species. The database is accompanied by the COGnitor program, which assigns new proteins, typically from newly sequenced genomes, to pre-existing COGs. A Web page containing additional structural and functional information is now associated with each COG. These hyperlinked information pages include systematic classification of the COG members under the different classification systems; indications of which COG members have been characterized genetically and biochemically; information on the domain architecture of the proteins constituting the COG and the three-dimensional structure of the domains if known or predictable; a succinct summary of the common structural and functional features of the COG members as well as peculiarities of individual members; and key references.

2.17 The Conserved Domain Database

The Conserved Domain Database (CDD) is a collection of sequence alignments and profiles representing protein domains conserved in molecular evolution. It includes domains from SMART and Pfam, two popular web-based tools for studying sequence domains, as well as domains contributed by NCBI. CD Search, another NCBI search service, can be used to identify conserved domains in a protein query sequence. CD-Search uses RPS-BLAST to compare a query sequence against specific matrices that have been prepared from conserved domain alignments present in CDD. Alignments are also mapped to known three-dimensional structures and can be displayed using Cn3D.

NCBI's Conserved Domain Architecture Retrieval Tool (CDART) displays the functional domains that make up a protein and lists other proteins with similar domain architectures. CDART determines the domain architecture of a query protein sequence by comparing it to the CDD, a database of conserved domain alignments, using RPS-BLAST.

The Conserved Domain Architecture Retrieval Tool then compares the protein's domain architecture to that of other proteins in NCBI's non-redundant sequence database. Related sequences are identified as those proteins that share one or more similar domains. CDART displays these sequences using a graphical summary that depicts the types and locations of domains identified within each sequence. Links to the individual sequences as well as to further information on their domain architectures are also provided. Because protein domains may be considered elementary units of molecular function and proteins related by domain architecture may play similar roles in cellular processes, CDART serves as a useful tool in comparative sequence analysis.

2.18 RPS-BLAST

RPS-BLAST is a "reverse" version of PSI-BLAST, which is described above. Both RPS-BLAST and PSI-BLAST use similar methods to derive conserved features of a protein family. However, RPS-BLAST compares a query sequence against a database of profiles prepared from readymade alignments, whereas PSI-BLAST builds alignments starting from a single protein sequence. The programs also differ in purpose: RPS-BLAST is used to identify conserved domains in a query sequence, whereas PSI-BLAST is used to identify other members of the protein family to which a query sequence belongs.

2.19 TAXONOMIC CLASSIFICATION

The binomial nomenclature for biological names of organisms is based on their taxonomic is classification, which is based on the evolutionary distances among groups of organisms. For example, species belonging to two different super kingdoms are most distantly related (their common ancestor diverged in the distant past), with progressively more exclusive groups indicated by phylum, class and so on, down to infraspecific ranks, or ranks occurring within a species. Infraspecific ranks, such as subspecies, varietas, and forma, denote the closest evolutionary relationship.

Phylogenetic systematics is the field of biology that deals with identifying and understanding the evolutionary relationships among the many different kinds of life on earth, both living (extant) and dead (extinct). Evolutionary theory states that similarity among individuals or species is attributable to common descent, or inheritance from a common ancestor. The relationships established by phylogenetic systematics are often described as species evolutionary history and, hence, its phylogeny, the historical relationships among lineages or organisms or their parts, such as their genes.

2.19.1 Phylogenetic Trees: Presenting Evolutionary Relationships

Systematics describes the pattern of relationships among taxa and is intended to understand the history of all life. In phylogenetic studies, the most convenient way of visually presenting evolutionary relationships among a group of organisms is through illustrations called phylogenetic trees.

A phylogenetic tree is composed of nodes, each representing a taxonomic unit (species, populations, individuals), and branches, which define the relationship between the taxonomic units in terms of descent and ancestry. Only one branch can connect any two adjacent nodes. The branching pattern of the tree is called the topology, and the branch length usually represents the number of changes that have occurred in the branch. This is called a scaled branch. Scaled trees are often calibrated to represent the passage of time. Such trees have a theoretical basis in the particular gene or genes under analysis. Branches can also be unscaled, which means that the branch length is not proportional to the number of changes that has occurred, although the actual number may be indicated numerically somewhere on the branch. Phylogenetic trees may also be either rooted or unrooted. In rooted trees, there is a particular node, called the root, representing a common ancestor, from which a unique path leads to any other node. An unrooted tree only specifies the relationship among species, without identifying a common ancestor, or evolutionary path.

Phylogenetic trees are a convenient way of representing evolutionary relationships among a group of organisms and can be drawn in various ways. Branches on phylogenetic trees may be scaled (top panel) representing the amount of evolutionary change, time, or both, when there is a molecular clock, or they may be unscaled (middle panel) and have no direct correspondence with either time or amount of evolutionary change. Phylogenetic trees may be rooted (top and middle panels) or unrooted (bottom panels). In the case of unrooted trees, branching relationships between taxa are specified by the way they are connected to each other, but the position of the common ancestor is not specified. For example, on an unrooted tree with five species, there are five branches (four external, one internal) on which the tree can be rooted. Rooting on each of the five branches has different implications for evolutionary relationships. The following are the components of a phylogenetic tree.

Node	Represents a taxonomic unit. This can be either an existing species or an ancestor.
Branch	Defines the relationship between the taxa in terms of descent and ancestry.
Topology	Defines the branching patterns of the tree.
Branch length	Represents the number of changes that have occurred in the branch.
Root	The common ancestor of all taxa is the root.
Distance scale	It is a measure that represents the number of differences between organisms or sequences.
Clade	A group of two or more taxa or DNA sequences that includes both their common ancestor and all of their descendents.

Terminologies in molecular phylogeny

Monophyletic Two or more DNA sequences that are derived from a single common ancestral DNA sequence.

Clade A group of monophyletic DNA sequences that make up all of the sequences included in the analysis that are descended from a particular common ancestral sequence.

Parsimony An approach that decides between different tree topologies by identifying the one that involves the shortest evolutionary pathway. This is the pathway that requires the smallest number of nucleotide changes to go from the ancestral sequence, at the root of the tree, to all of the present-day sequences that have been compared.

Molecular Clock Hypothesis This hypothesis states that nucleotide substitutions, or amino acid substitutions if proteins are being compared, occur at a constant rate, that is, the degree of difference between two sequences can be used to assign a date to the time at which their ancestral sequence diverged. The rate of molecular change differs among the groups of organisms, among genes, and even among different parts of the same gene. Furthermore, molecular clocks require calibration with fossils to determine timing of origin of clades, and thus their accuracy is crucially dependent on the fossil record, or lack thereof, for the groups under study. Fossil DNA older than about 25,000–50,000 years is virtually empty of phylogenetic signal except in rare instances, and therefore traditional morphological studies of extinct and extant organisms remain a crucial component of phylogenetic analysis.

2.20 BLAST

Let us see how the sequence analysis can be done with the help of BLAST (Basic Local Alignment Search Tool). BLAST finds regions of local similarity between sequences. The program compares nucleotide or protein sequences to sequence databases and calculates the statistical significance of matches. BLAST can be used to find the functional and evolutionary relationships between the sequences as well as help identify the members of gene families. Even though the search is based on similarity, the results are used for analysis.

2.20.1 Types of BLAST

There are five types of BLAST; they are nucleotide BLAST, protein BLAST, BLASTX, TBLASTN, and TBLASTX (Figure 2.9).

Nucleotide BLAST—Search a nucleotide database using a nucleotide query

Protein BLAST—Search a protein database using a protein query

BLASTX—Search a protein database using a translated nucleotide query

TBLASTN—Search a translated nucleotide database using a protein query

TBLASTX—Search a translated nucleotide database using a translated nucleotide query

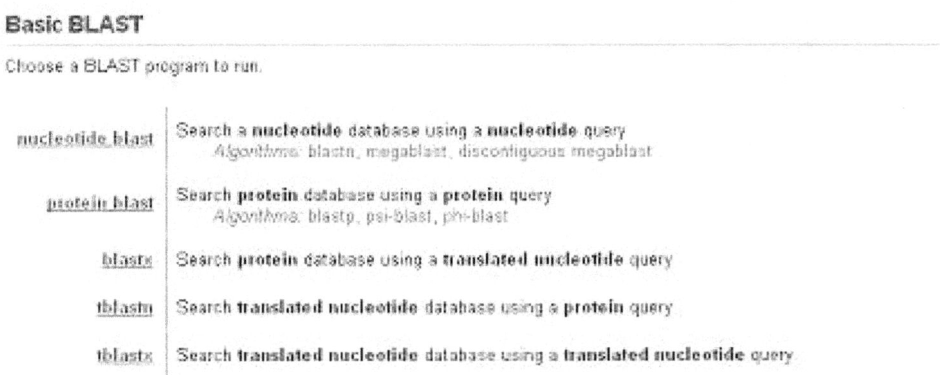

Figure 2.9 Types of BLAST

2.20.2 Specialized BLAST

The specialized BLAST (Figure 2.10) comprises the search of trace archives, conserved domains, conserved domain architecture, gene expression profile, immunoglobulins, SNPs, vector contamination.

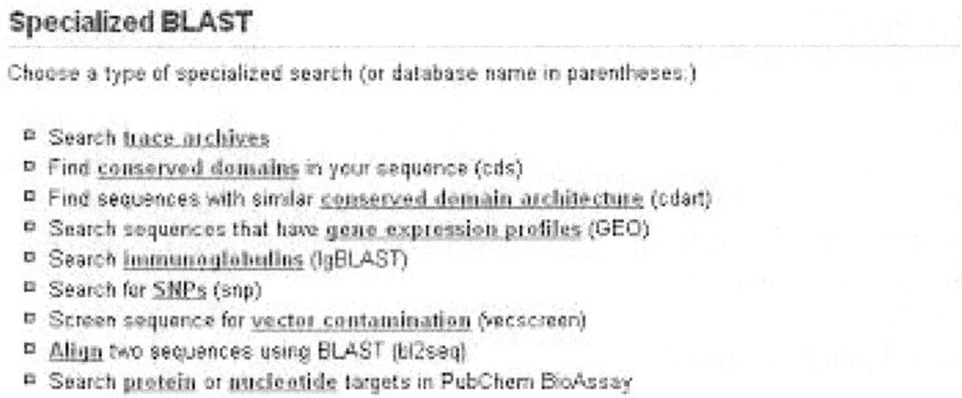

Figure 2.10 Types of specialized BLAST

2.20.3 Nucleotide BLAST

Before we go for the analysis of the result, the submission of the input in the query page is to be done. Paste your sequence in the FASTA format and then set algorithms parameter and run BLAST in the home page of nucleotide BLAST (Figure 2.11).

Figure 2.11 Home Page of nucleotide BLAST

The nucleotide sequence which has to be BLASTed for primary analysis is given as an input in the FASTA format and then in the search option the database was set as nucleotide collection following which the parameters are set and executed for the results (Figure 2.12).

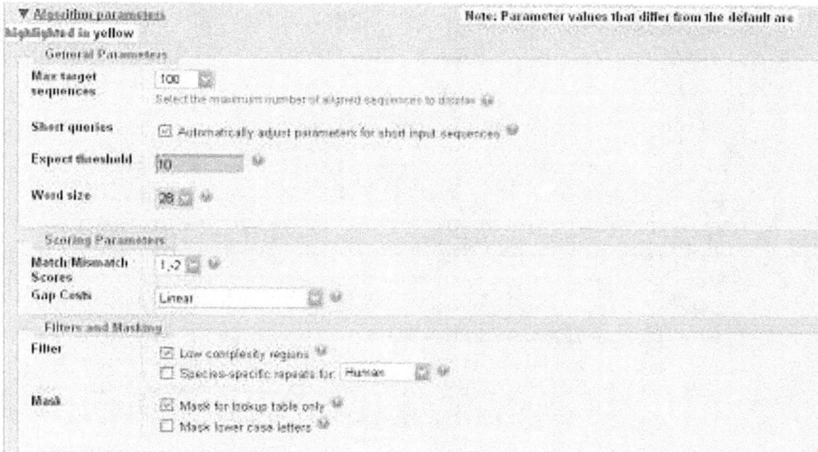

Figure 2.12 Algorithm parameters set for nucleotide BLAST

The result produced will be in the form of color key scores with unigene, GEO, Map Viewer. The alignments for each query with its subject given for interpretation.

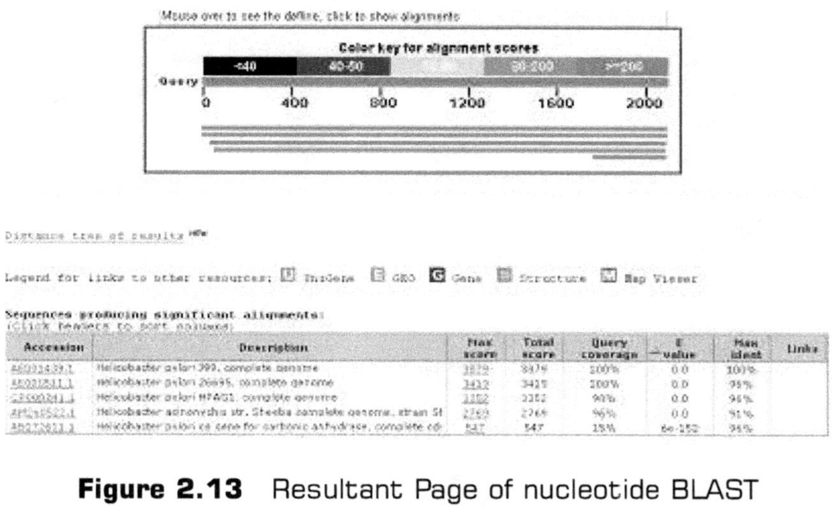

Figure 2.13 Resultant Page of nucleotide BLAST

With the help of color key score the BLAST result is analysed. The E value and query coverage plays an important role in the selection of the similar sequence from the results. Figure 2.13 shows the result of a nucleotide BLAST.

Figure 2.14 Alignment of the Query

The alignment for the query sequence is found by the alignment window (Figure 2.14). To view the tree, click the option Distance tree of results. The tree of results will be given along with the distance for interpretation You can get a tree like the one shown in Figure 2.15.

Figure 2.15 Tree analysis for the given nucleotide sequence

2.21 BioEdit

BioEdit, is a biological sequence editor that runs in Windows 95/98/NT/2000 and is intended to provide basic functions for protein and nucleic sequence editing, alignment, manipulation and analysis. Hydrophobicity and hydrophilicity profiles, interactive 2D matrix data plotting and a few more sequence manipulations were added in version 3. Version 4 added a graphical interface for drawing and annotating plasmid vectors.

BioEdit is a C++ program written in Borland's C++ Builder by a graduate student in Microbiology at North Carolina State University. This program could be much smaller and more efficient. Nevertheless, BioEdit provides an easy means for sequence alignment, output, and some analyses.

BioEdit can be used for many purposes for analysing both protein and nucleotide sequence.

The sequence to be uploaded is in the notepad where it can be open for analysis where a main window will be appearing and the open menu is given for the input of the sequence (Figure 2.16).

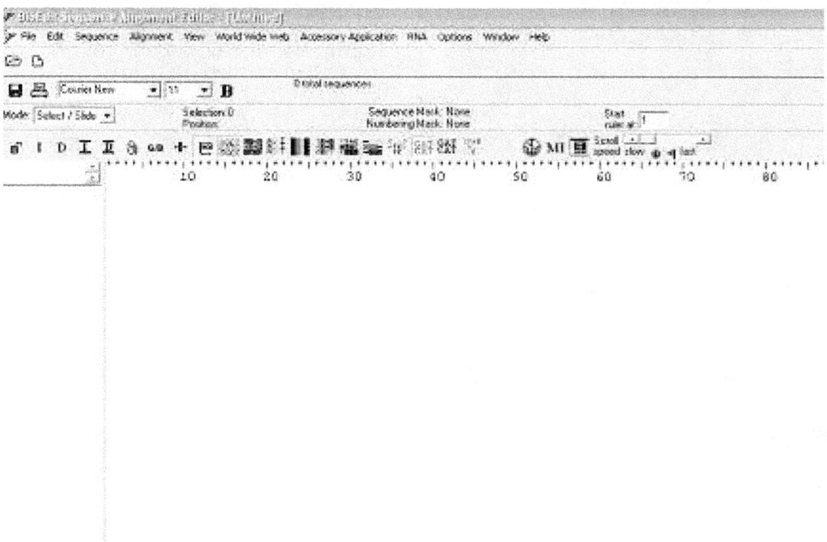

Figure 2.16 Input Window of BioEdit

2.21.1 To Analyse Nucleotides

Let us see how to analyse the nucleotides. In the menu bar the sequence option is selected and then go for nucleic acid option, a pop menu will be given with all the details like composition of nucleotides, complement, DNA to RNA, translate, restriction map, etc.

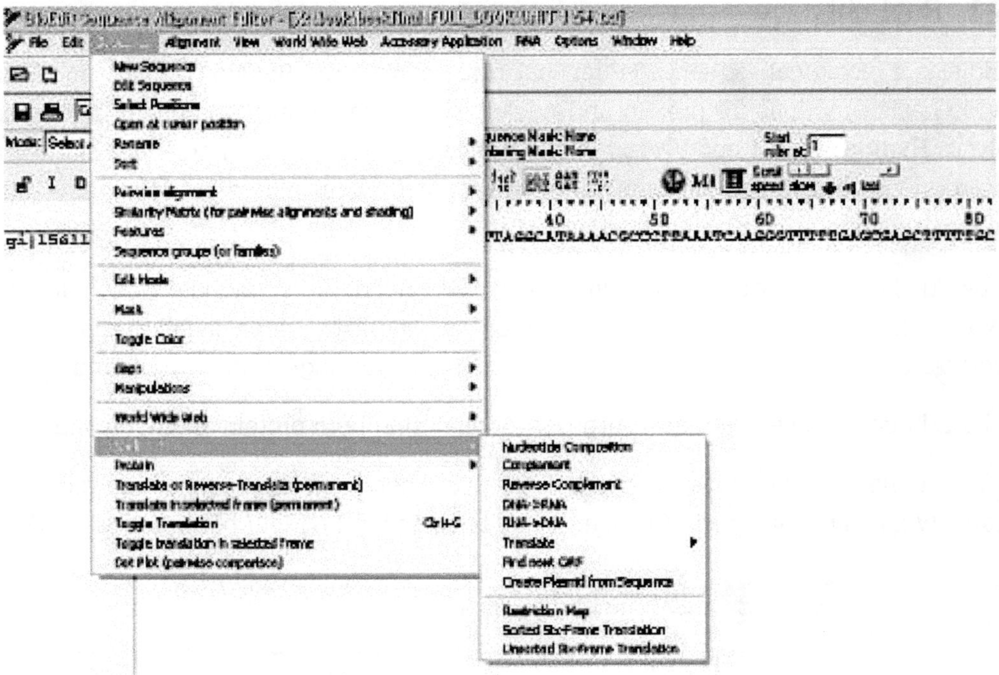

Figure 2.17 Options for analysing the sequences

Figure 2.18 Input sequence for analysis

From the Figure 2.17 and 2.18, it is clearly understood how to select the option for analysing the nucleotide. In the menu bar go for the option sequence and in the popdown menu go for nucleotide, where search options are present and those options are used for the analysis of the sequence.

2.21.2 Creation of a Plasmid

A plasmid is created from the given sequence as follows. Figure 2.19, helps us to operate the menu bar for creating the plasmid. In the Figure 2.20, a plasmid is created for *Helicobacter pylori* J99.

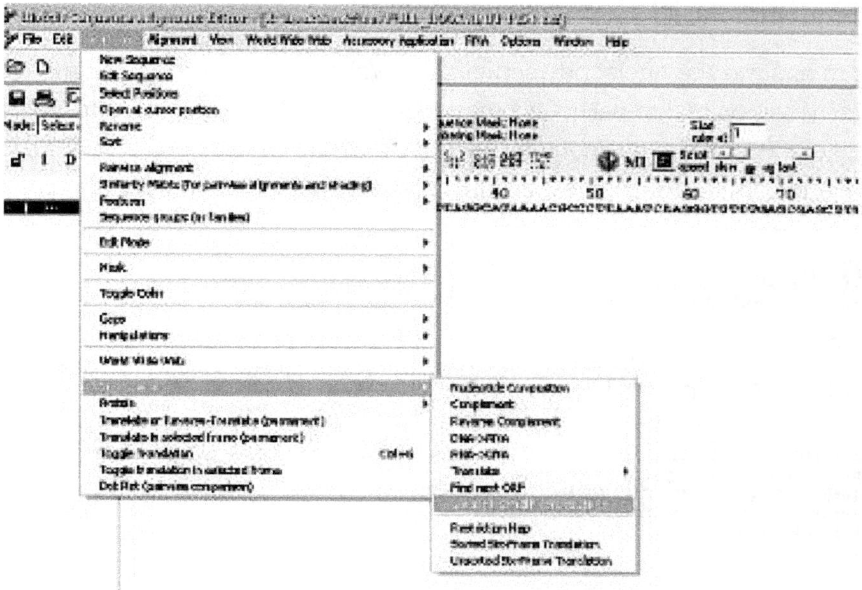

Figure 2.19 Creation of a plasmid for the input sequence

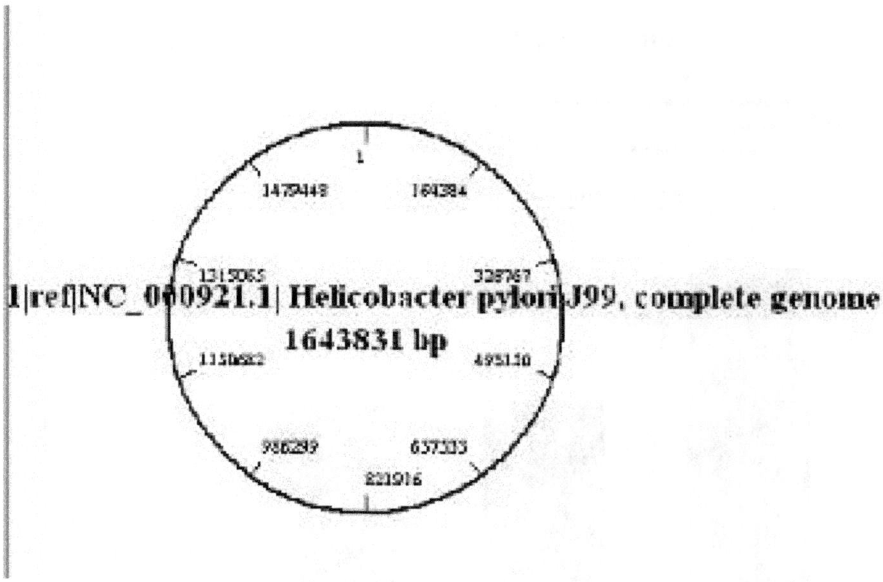

Figure 2.20 Plasmid created

I am sure you would have created a plasmid for your sequence by now, following the procedure just explained. If not, I would suggest you to try out the same.

2.21.3 Nucleic Acid Composition

If one has to analyse the nucleotide sequence, he/she has to come across the percentage of the ATGC or composition of that particular sequence. In BioEdit an option is present for analysing the nucleic acid composition of nucleotides. In Figure 2.21 the composition of A is 30.38%, C is 19.75%, G is 19.49% and T is 30.41%. Figure 2.22 illustrates the graphical representation of the composition.

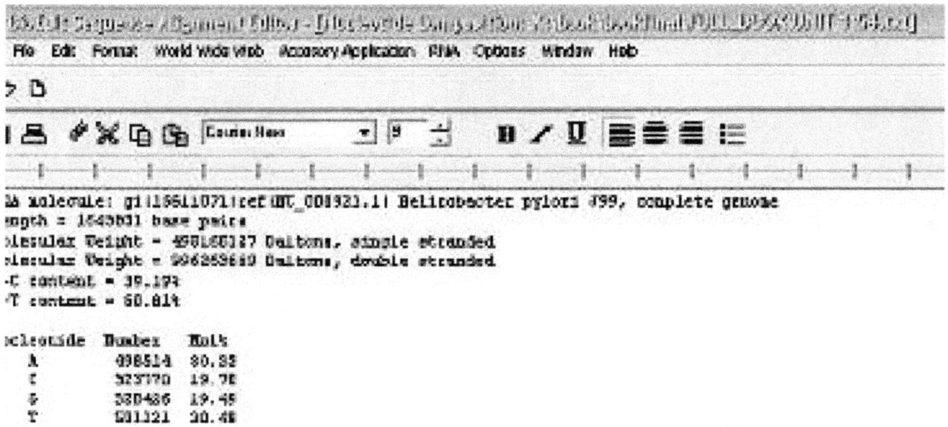

Figure 2.21 Composition of base pair

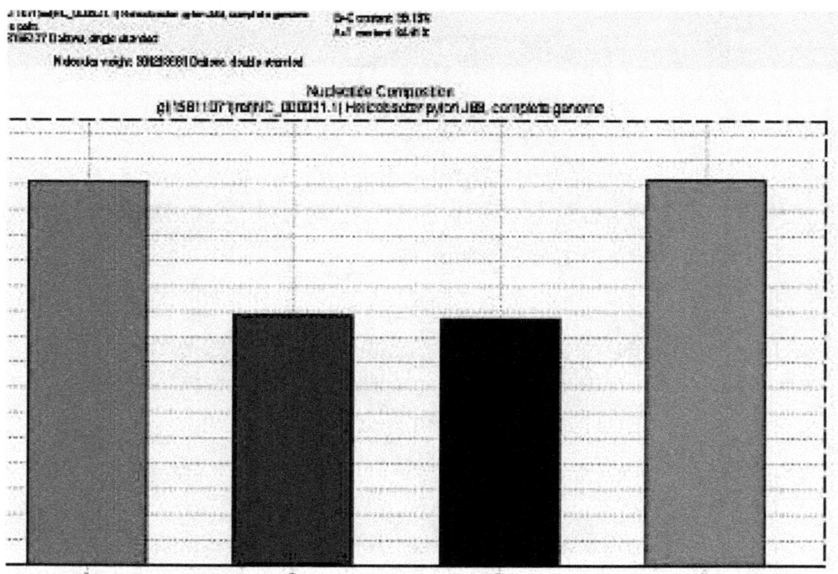

Figure 2.22 Graphical representation of the composition of base pair

2.21.4 Restriction Map

Earlier we have developed a plasmid (Figure 2.20) and now we will start creating a restriction map for the input sequence.

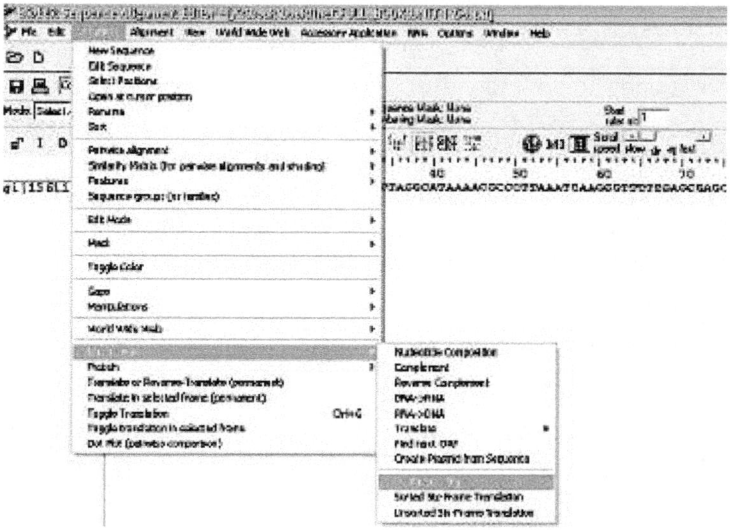

Figure 2.23 Option for Restriction Map

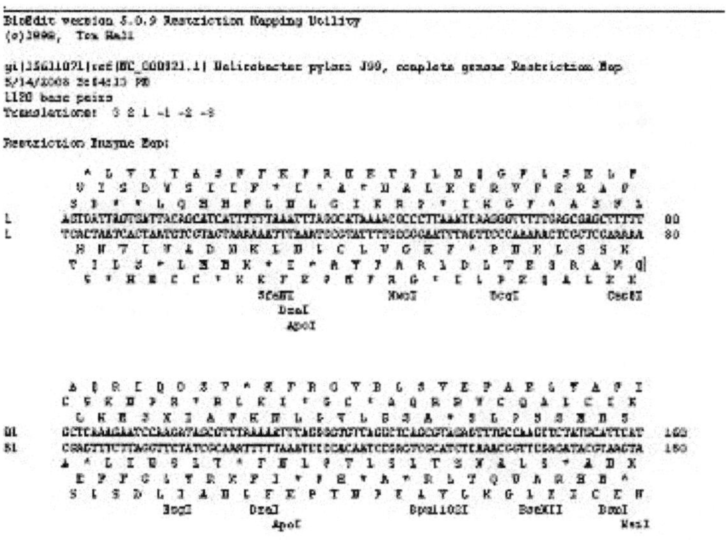

Figure 2.24 Creation of Restriction Map

The restriction map (Figure 2.23) is created for the unknown sequence by the options in the menu bar. The restriction sites with their enzymes (Figure 2.24) are easily understood from this result. I hope you are clear about nucleotides and many more options are present which can be analysed by you in the tool BioEdit.

2.21.5 Protein Analysis

To analyse a protein one should know the composition of amino acids. Based on the properties and the composition of amino acids, the type of a protein can be easily analysed.

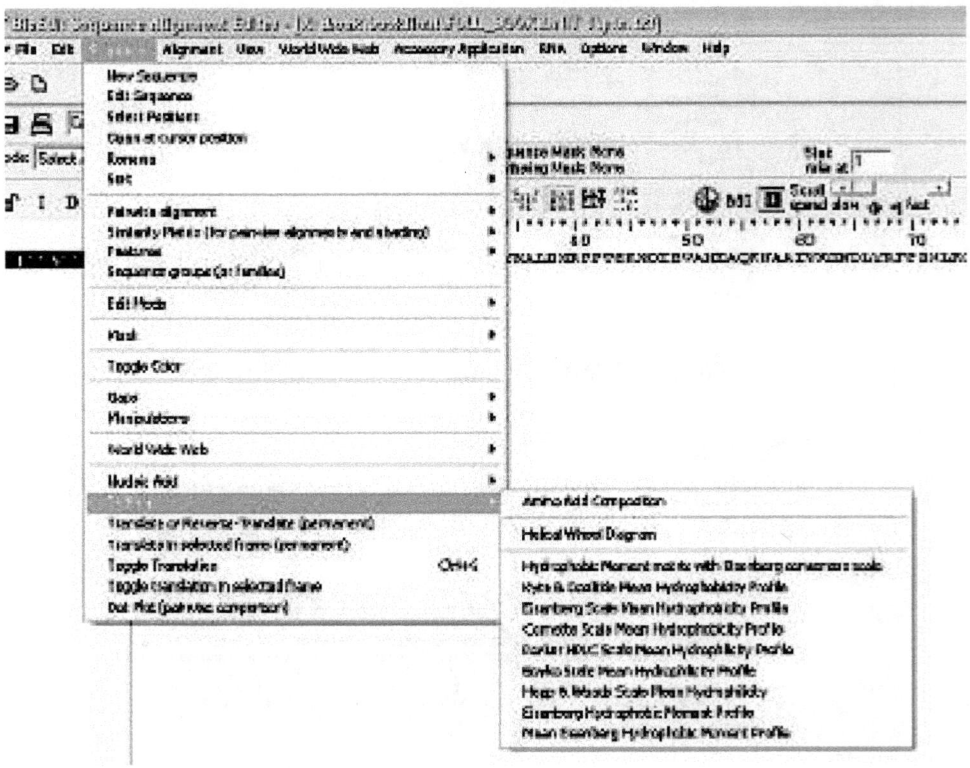

Figure 2.25 Analysing a protein sequence

Similar to analysing the nucleotide sequence in the form of A, T, G and C, protein analyses (Figure 2.25) is done based on 20 amino acids. The composition of the amino acids (Figure 2.26) are analysed and based on the percentage of individual amino acids present, one can predict the nature of the protein or the structure obtained by the protein.

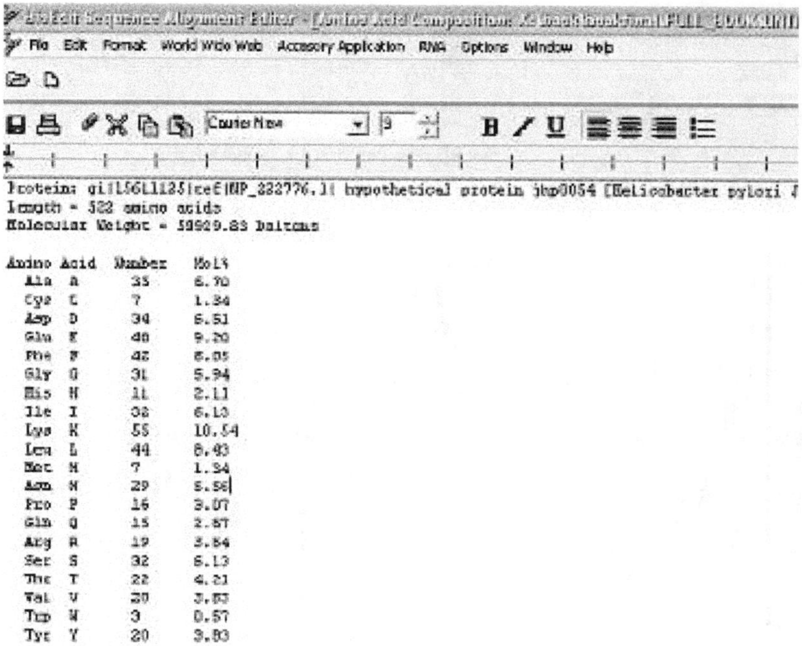

Figure 2.26 Composition of amino acids

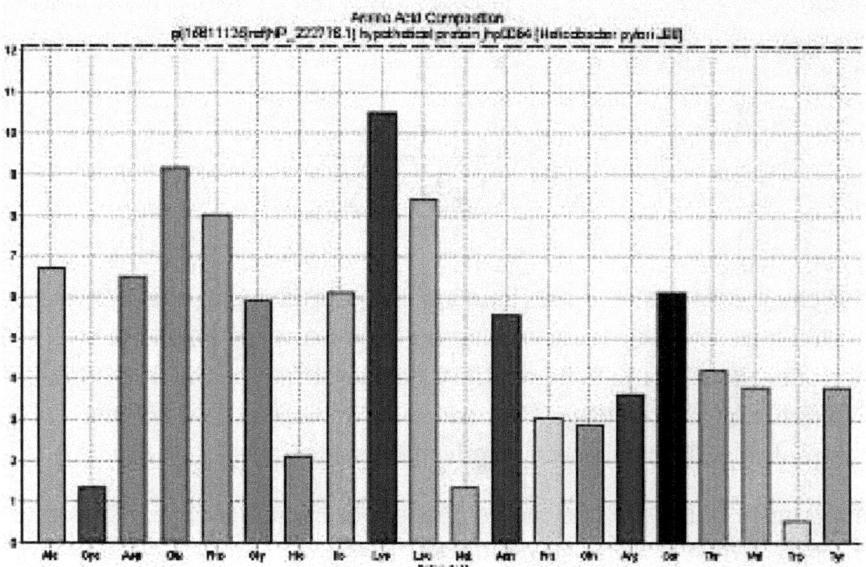

Figure 2.27 Graphical representation of amino acid composition

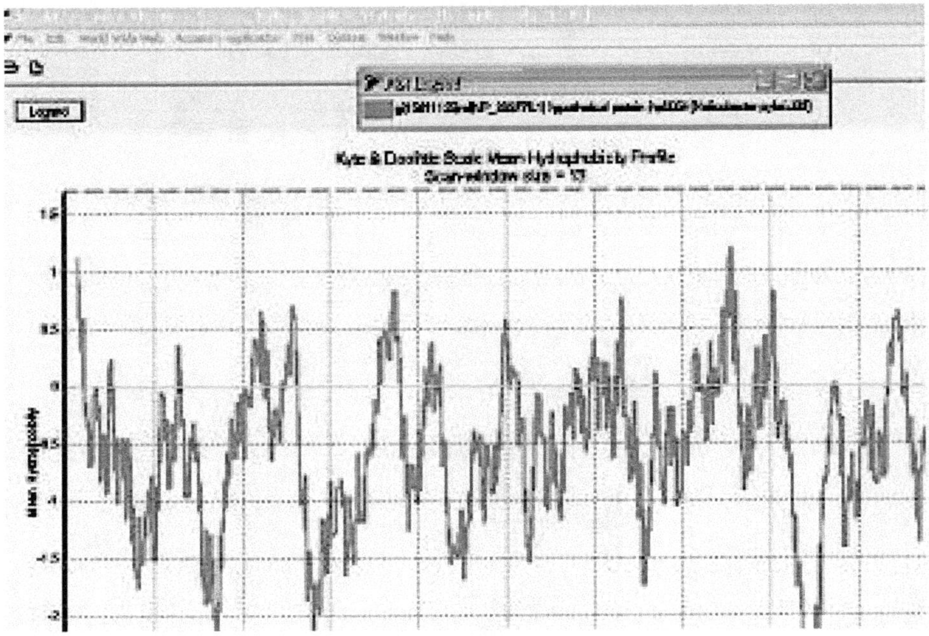

Figure 2.28 Hydrophobic plot

The graphical representation of the amino acid composition (Figure 2.27) are seen for easy understanding. Using the hydrophobic plots (Figure 2.28), the hydrophobic and hydrophilic amino acids are located.

SUMMARY

I hope in this chapter the readers would have got the basic concepts, definition, facts in Bioinformatics and also would have got a clear answer for the question what is Bioinformatics. The researchers can produce or retrieve information from the databases described in the chapter. The utility of the biological information and the use of information technology is based on the end users. The basics to access the entire NCBI databases and advanced search have been discussed.

The importance of BLAST and the component of BLAST such as nucleotide BLAST, protein BLAST, specialized BLAST were also detailed for the use of similarity search. A tool named BioEdit which is used for the sequence analysis and for creation of a plasmid and to find restriction site for the query sequence was detailed.

CRITICAL THINKING QUESTIONS

1. "Bioinformatics"—Is it a boon or bane to the research community?
2. In the integration of biology and information technology has each of these fields played an equal role in analysing the problems concerned?
3. Find out the algorithm of BLAST.

SELF-LEARNING EXERCISES

1. How will you identify the restriction sites and their restriction enzymes for your sequence of interest within a short period of time by using Bioinformatics tools/software?
2. How will you find out the origin from where your unknown sequence would have evolved?

3

GENOMICS

3.1 INTRODUCTION TO 'OMICS

The field 'Omics is defined as a broad discipline of science and engineering for analysing the interactions of biological information and objects by various 'omes. These branches include genomics, proteomics, metabolomics, expressomics and interactomics.The main focus of 'Omics is to map information on genes and proteins, to find the interaction relationships between the found/mapped genes and proteins and to create/engineer a network for the objects to understand and manipulate the regulatory mechanisms. People are working in a new biological research field to gain knowledge from the created networks of information by holistic principles and methods deviating from reductionism. To understand the 'Omics, first we should know the definition of Systems Biology in the field of Bioinformatics.

3.2 SYSTEMS BIOLOGY

Systems biology is the study of biology that focuses on complex systems in life. 'Omics is the study on large-scale information in many distinct biolayers. Bioinformatics is an information science that analyses life processes using computational tools for solving biological problems which gives an overview in biology.

You will be thinking how and when the word 'Omics would have come into existence. The original use of the suffix "ome" was in the word "Biome" and changed as "genome" which became a popular word for the complete genetic make-up of an organism. Because of the use of ome succinctly describes the new way of holistic analysis of complete genomes and proteomes, the trend of putting omics became widespread in 1995 by Bioinformatists in Cambridge, Stanford, Yale, Harvard and so on.

The words ome and omics are very convenient to describe to look at complex systems. Bioinformaticians and some molecular biologists were the first scientists to start applying the "ome" suffix widely. In the mid-1990s Bioinformatics started realizing the convenience of 'omics and termed many fields such as metabolome, textome, interactome, bacteriome, eukaryome, functome, patentome, neurome and so on.

Omics marks the realization of the importance of information processing in biology that was proposed by many biologists. It was only possible by the advent of personal computers and personal computer operating systems such as Windows, Linux and Apple.

The 'omes are now well established within biomics and genomics. Biome is the top-level -ome in biology. Biome is one of the oldest terms, but now it is changed as biomics. Biomics is the study of the whole set of biological entities such as protein, DNA, RNA and small biologically important molecules.

3.3 DEFINITION OF GENOMICS

Genomics is the study of genes and their function. Genomics aims to understand the structure of the genome, including mapping the genes and sequencing the DNA. Genomics examines the molecular mechanisms and the interplay of genetic and environmental factors in disease and is the study of all of the nucleotide sequences, including structural genes, regulatory sequences, and noncoding DNA segments in the chromosomes of an organism.

3.4 BASIC CONCEPTS OF GENOMICS

3.4.1 Nuclear and other Genomes

We all know that most of the genetic material in eukaryotes is located in the nucleus. The mitochondria and chloroplast also contain DNA, but only a small number of genes. These genes are inherited maternally. Figure 3.1 shows the cell and its components.

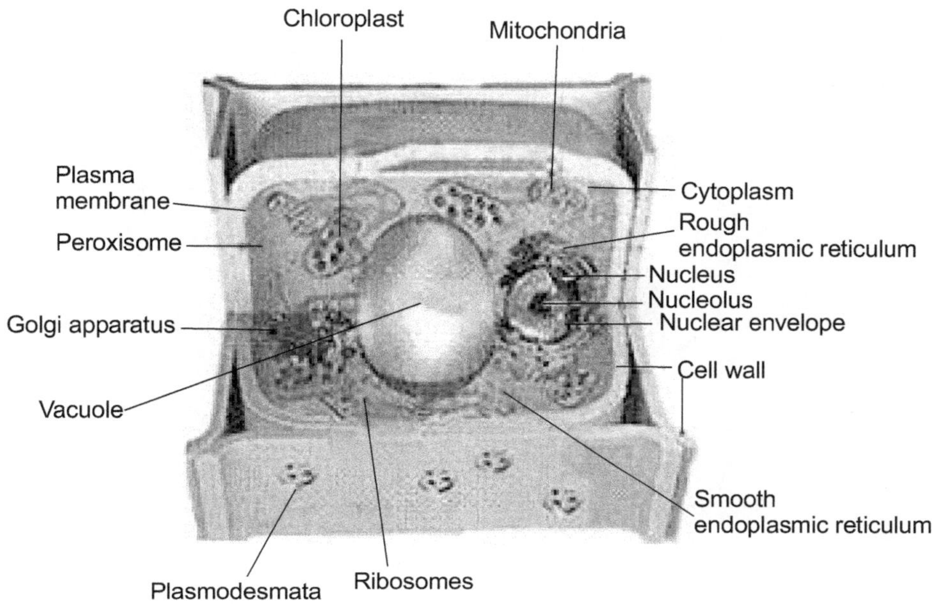

Figure 3.1 Anatomy of the plant cell

3.4.2 Transcription

In the process of transcription (Figure 3.2), the DNA sequence is copied on to molecules of messenger RNA (mRNA). Transcription is the process in which genetic information that is stored in a strand of DNA is copied into a strand of RNA. The sequence of the four bases in DNA, which are adenine (A), cytosine (C), guanine (G), and thymine (T), is preserved in the sequence of the four bases in RNA, which are A, C, G, and uracil (U). This process is similar to DNA replication in that the double helix opens so that one strand can be used as a template for the new strand, in the case of RNA.

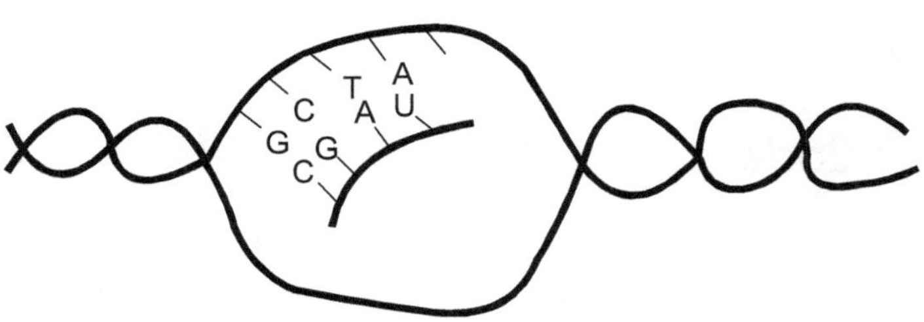

Figure 3.2 Process of transcription

3.4.3 Translation

Translation (Figure 3.3) is the process by which the genetic code carried by messenger RNA (mRNA) directs the production of proteins from amino acids.

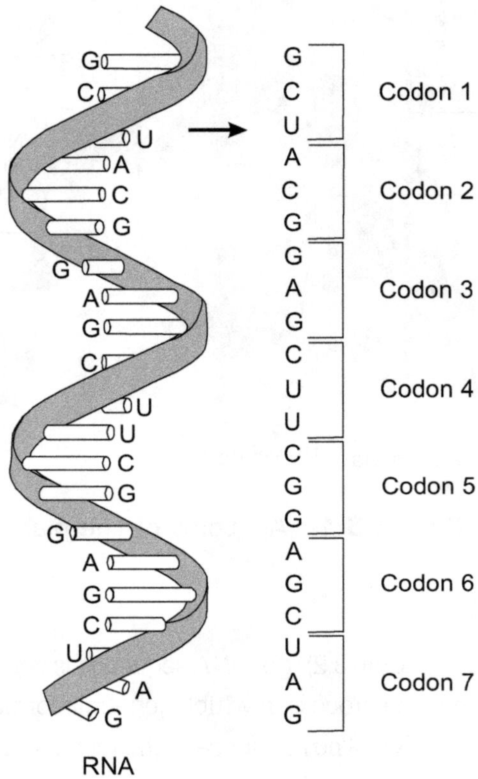

Figure 3.3 Process of translation

Sets of three nucleotides (called codons) determine which amino acid will be added next to the growing protein chain. For example, GCU, the first codon in the figure, encodes alanine. The next codon shows all the codons and their corresponding amino acids.

3.4.4 Genetic Code

There are 64 codons in the genetic code, which is made up of 20 amino acids (Figure 3.4). A, T, G and C are the "letters" of the DNA code. They stand for the adenine, thymine, guanine and cytosine, respectively that make up the nucleotide bases of DNA. Each gene's code combines four chemicals in various ways to spell out three-letter "words" that specify which amino acid is needed at every step in making a protein.

UUU phenylalanine UUC phenylalanine UUA leucine UUG leucine	UCU serine UCC serine UCA serine UCG serine	UAU tyrosine UAC tyrosine UAA stop UAG stop	UGU cystine UGC cystine UGA stop UGG tryptophan
CUU leucine CUC leucine CUA leucine CUG leucine	CCU proline CCC proline CCA proline CCG proline	CAU histidine CAC histidine CAA glutamine CAG glutamine	CGU arginine CGC arginine CGA arginine CGG arginine
AUU isoleucine AUC isoleucine AUA isoleucine AUG methionine (start)	ACU threonine ACC threonine ACA threonine ACG threonine	AAU asparagine AAC asparagine AAA lysine AAG lysine	AGU serine AGC serine AGA arginine AGG arginine
GUU valine GUC valine GUA valine GUG valine	GCU alanine GCC alanine GCA alanine GCG alanine	GAU aspartic acid GAC aspartic acid GAA glutamic acid GAG glutamic acid	GGU glycine GGC glycine GGA glycine GGG glycine

Figure 3.4 Genetic code

3.4.5 Splicing

Splicing (Figure 3.5) is a critical stage in the processing of mRNA, in which intervening sequences (introns) are removed from a primary RNA transcript (pre-RNA) and the coding exons are joined together to form the mature mRNA molecule. The mRNA is translated into proteins, which are manufactured in the ribosomes.

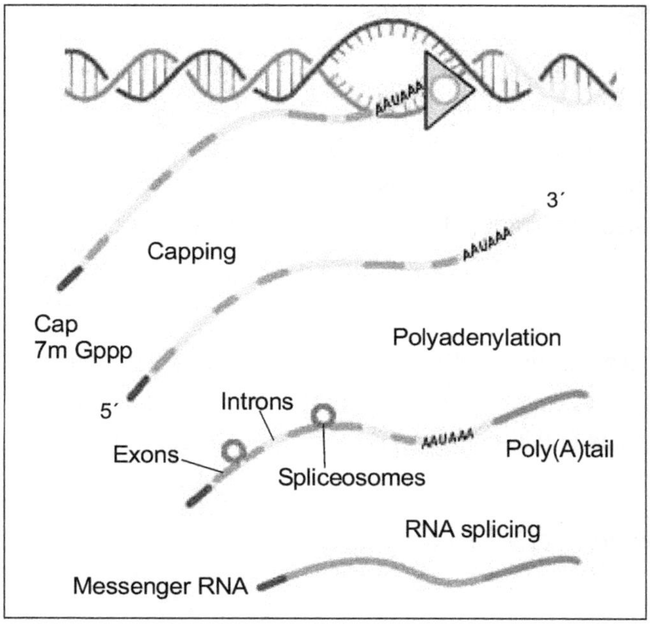

Figure 3.5 Scheme of splicing

3.5 DEFINITION OF COMPARATIVE GENOMICS

Comparative genomics helps us in the study of the basic relationships between the genomes of different species. These studies of whole genomes help researchers understand which parts of the genome in one organism are similar to those in another, how the overall structures of genes and genomes have evolved, what these findings tell us about gene expression and gene regulation, and how to interfere with these events in model organisms or humans. Comparative genomics is also a critical enabling field for functional genomics, because it gives researchers an indication of which model organisms are most useful for which studies. Some of the basic definitions in comparative genomics are as listed in the following paragraphs.

Synteny mapping The term synteny was originally defined to mean that two gene loci share the same chromosome. In the genomic context it refers to regions in which both sequence and gene order is conserved between two (closely related) species. Genes that are close together on the same chromosome will not show independent assortment during meiosis. The mapping genes does not tell which chromosomes they are on, it only identifies which genes are on the same chromosome that is called a linkage group. Linkage groups are associated with specific chromosomes in humans. Since the set of genes on specific chromosomes are said to be syntenic. The process was referred to as synteny mapping.

Paralog Paralogs are genes related by duplication within a genome. Orthologs retain the same function in the course of evolution, whereas paralogs evolve new functions, even if these are related to the original one.

SNP SNP means single nucleotide polymorphism. It is defined as DNA sequence variations that occur when a single nucleotide (A, T, C or G) in the genome sequence is altered. Each individual has many single nucleotide polymorphisms that together create a unique DNA pattern for that person. SNPs promise to significantly advance our ability to understand and treat human disease.

3.6 VISTA

VISTA (Figure 3.6) is a comprehensive suite of programs and databases for comparative analysis of genomic sequences. There are two ways of using VISTA—You can submit your own sequences and alignments for analysis (VISTA servers) or examine pre-computed whole-genome alignments of different species.

There are many types of VISTA servers.

mVISTA

rVISTA

GenomeVISTA

Phylo-VISTA

wgVISTA

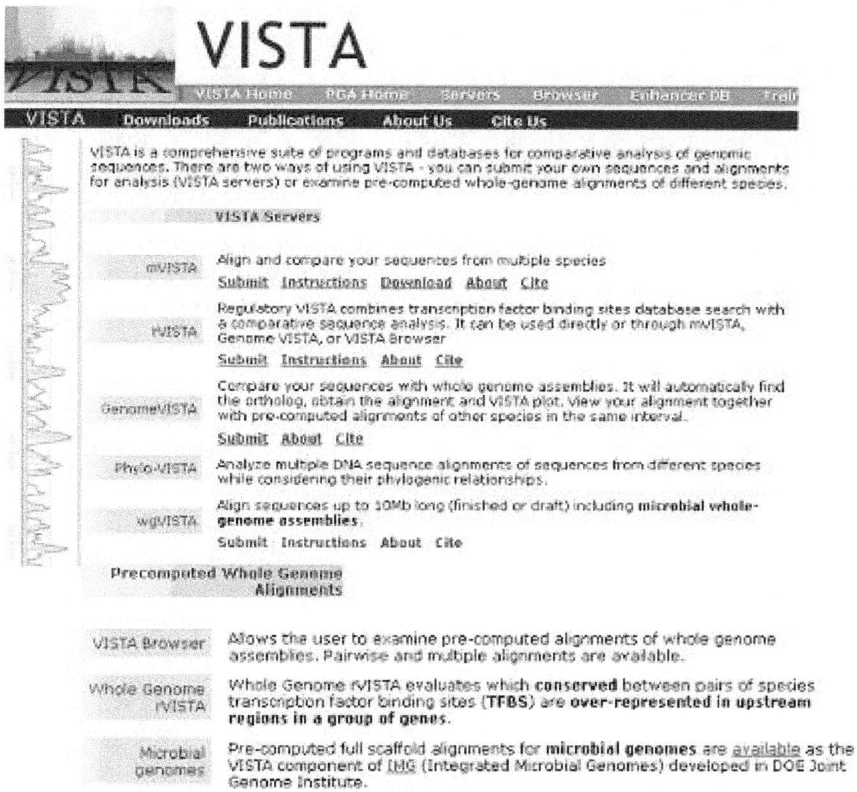

Figure 3.6 Home Page of VISTA

Figure 3.6 represents the home page of VISTA server and under this, you can find mVISTA, rVISTA, GenomeVISTA, Phylo-VISTA and wgVISTA. In mVISTA, we can align and compare sequences of multiple species. This is followed by a link related to submit instructions, downloading option, etc., The rVISTA is used to find the transcription factor binding sites. GenomeVISTA is for analysing the compared sequences with whole genome assemblies. The GenomeVISTA will find the ortholog for the species searched. Phylo-VISTA is used to analyse the phylogenetic relationships. wgVISTA is used to align sequences up to 10 megabases long. The pre-computed whole genome alignment has VISTA Browser which allows the user to examine pre-computed alignments and assemblies. Whole genome rVISTA calculates the conserved regions between the species transcription factor binding sites.

3.6.1 mVISTA

mVISTA (Figure 3.7) is a set of programs for comparing DNA sequences from two or more species up to 10 megabases long and visualize these alignments with annotation information. mVISTA has a clean output, allowing for easy identification of sequence similarities and differences, and is easily configurable, enabling the visualization of alignments of various lengths at different levels of resolution.

It is implemented as an online server that provides access to global pairwise, multiple and global (global with rearrangements) alignment tools. Stand-alone components (AVID and visualization module mVISTA) can be downloaded from the mVISTA website and used on one's own computer. Source code of the LAGAN toolkit is freely available under the GNU Public license (GPL) and available through http://lagan.stanford.edu.

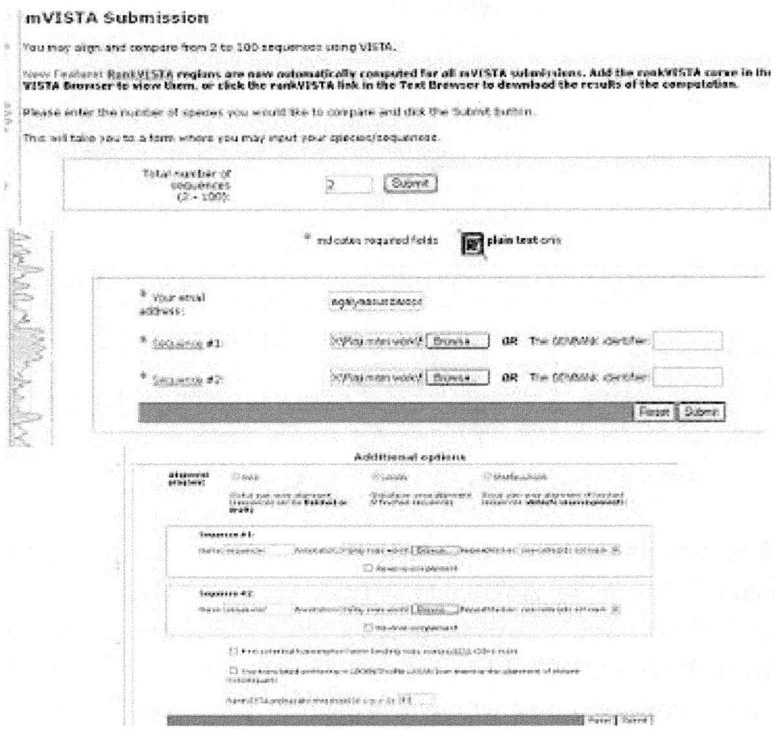

Figure 3.7 Input Page for mVISTA

Figure 3.7 shows the input page for mVISTA, where the number of sequences that can be aligned and compared is from 2 to 200 sequences. One may upload many number of sequences, but limited to 200 for analysis purpose.

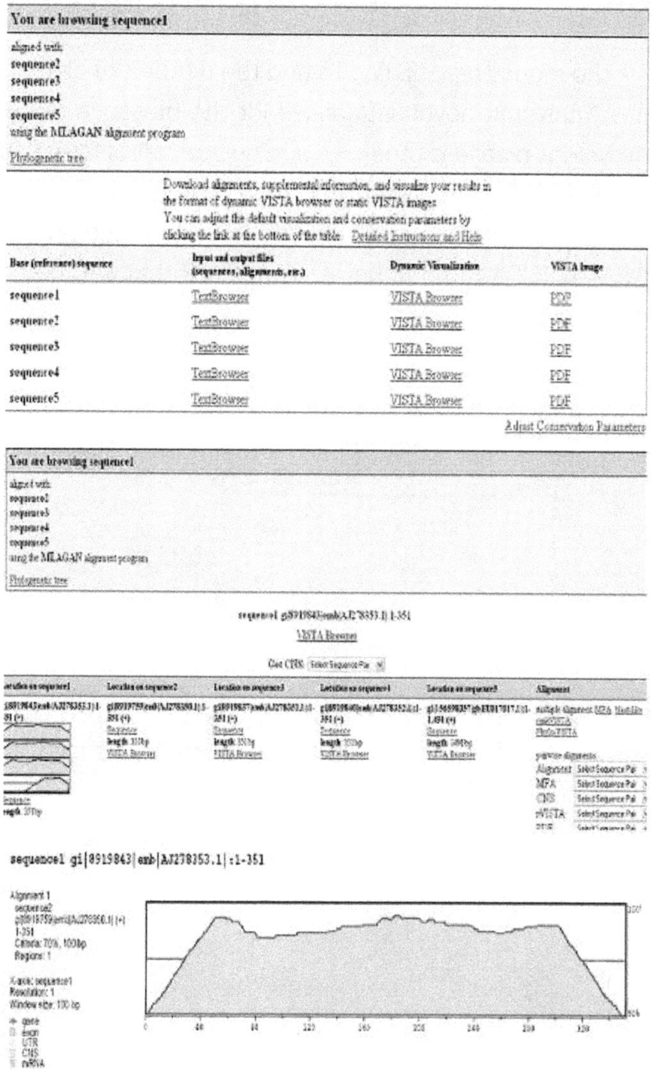

Figure 3.8 Results from mVISTA

From Figure 3.8, it is clearly understood that the alignment has been done by comparing all the sequences. Keeping the sequence 1 as our Query, the alignment is done between sequence 1 and sequence 2, sequence 1 and sequence 3, sequence 1 and sequence 4, sequence 1 and sequence 5, by using MLAGAN alingment. To the right side of the graph, many options are given as results like gene, exon, UTR, CNS, mRNA and to the left side of the graph, certain options like alignment, rVISTA and a few more options are seen. The user can click these links for further analysis. From the VISTA browser and VISTA image, the alignment is interpreted.

3.6.2 rVISTA

rVISTA (Figure 3.9) is the regulatory VISTA. To find the regulatory elements of human genome is a challenging problem. Analysing novel sequences for the presence of known transcription factor binding sites or their weight matrices produces a huge number of false positive predictions that are randomly and uniformly distributed. When database searches were combined with comparative sequence analysis, the number of predicted transcription factor binding sites was reduced by several orders of magnitude. In rVISTA the number of sequences that can be compared is from 2 to 100 sequences.

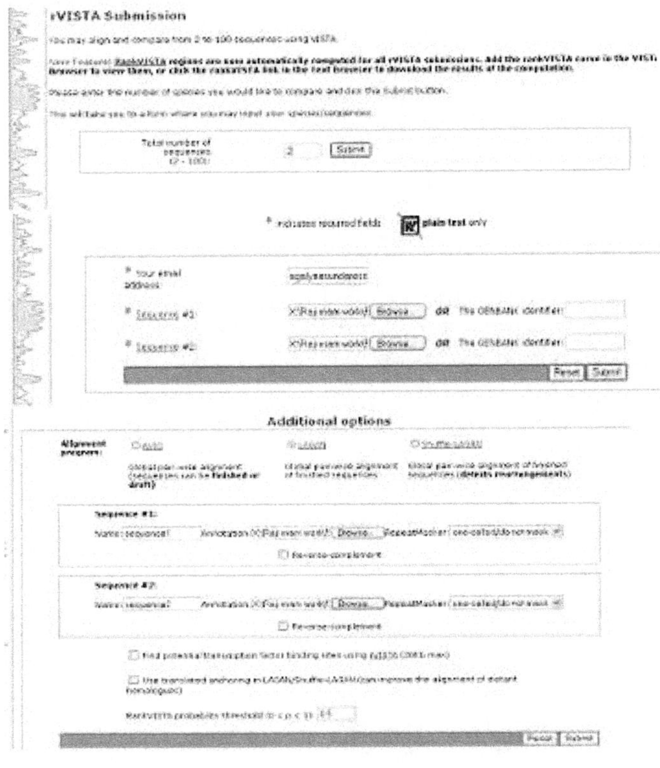

Figure 3.9 Home Page and Submission Page of rVISTA

From Figure 3.10, the regulatory regions and the transcription factor binding site can be detected with the help of the tool. The shaded regions are the regulatory sites, followed by this a graphical representation of the regulatory sites will be provided. The researcher can easily identify the intensity of the site from the graph. The regulatory site for a given set of genome can be identified and those sequences and base pairs can be extracted for further analysis.

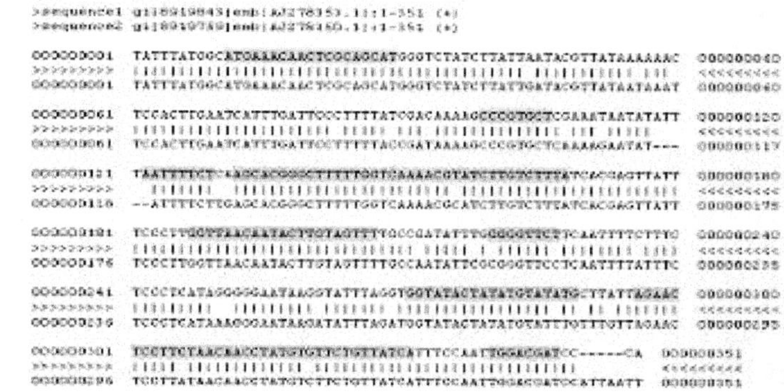

Figure 3.10 Result Page and the regulatory regions in rVISTA

3.6.3 GenomeVISTA

GenomeVISTA (Figure 3.11) is an automatic server that allows the user to find candidate orthologous regions for a draft or finished DNA sequence from any species on a base genome: human, mouse, rat and *Drosophila melanogaster.*

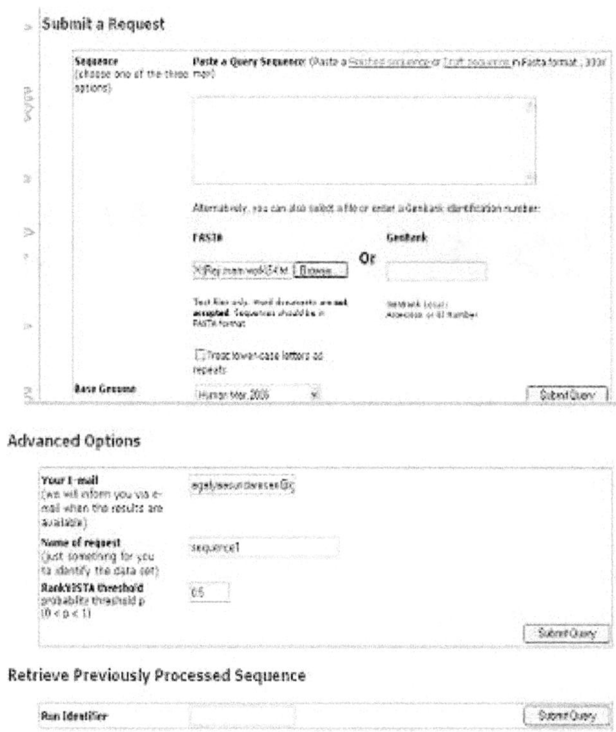

Figure 3.11 Home Page of GenomeVISTA

3.6.4 Phylo-VISTA

Phylo-VISTA (Figure 3.12) is an interactive tool for analysing multiple DNA sequence alignments by visualizing a similarity measure for DNA sequences for different species while considering their phylogenetic relationships. Features include multiple-resolution visualization for examining an alignment and easy comparison of any subtree of sequence data within the complete alignment dataset.

In Figure 3.12, the home page, the result of phylogeny and the alignment of the sequence are shown. In this server the input can be given in the FASTA format and the results can be analysed.

A phylogentic tree wil be produced as a result by comparing all the sequnces, based on the phylogeny. The alignment is also given based on the phylogeny so that we can calculate the divergence of the sequence.

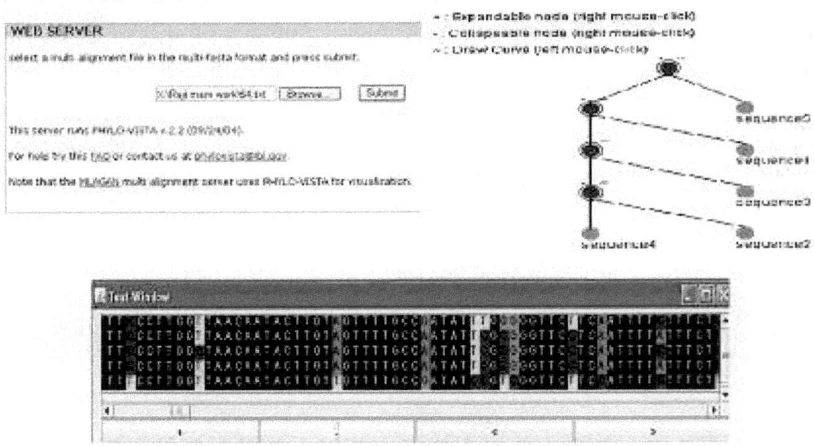

Figure 3.12 Home Page, results from Phylo-VISTA

3.6.5 wgVISTA

wgVISTA is a set of programs for comparing DNA sequences from two microbial species up to 10 megabases long and visualize these alignments with annotation information. It is implemented as an on-line server that provides access to the whole-genome alignment pipeline. The alignment program behind the pipeline is Shuffle-LAGAN, a global alignment algorithm that is able to find rearrangements (inversions, transpositions and some duplication) in a global alignment framework. The algorithm uses CHAOS local alignments to build a map of the rearrangements between the sequences, and LAGAN to align the regions of conserved synteny.

wgVISTA visualization module is the VISTA Browser, an interactive Java Applet, which is easily configurable, enabling the visualization of alignments of various lengths at different levels of resolution. The browser's clean display makes it easy to identify sequence similarities and differences.

SUMMARY

In this chapter we have learnt about genomics and its concepts, like coding, translation, etc. Comparative genomics study was also done with an online server VISTA and its components like mVISTA, rVISTA, GenomeVISTA and Phylo-VISTA and their usage in analysing the genes.

CRITICAL THINKING QUESTIONS

1. How will you detect the virulent region in influenza virus? Do comparative studies for the same?
2. How does comparative genomics help in the discovery of new genes?

SELF-LEARNING EXERCISES

1. What is the central dogma of life? Explain.
2. What will happen if there is a problem in the splicing mechanism?
3. How do mapping and SNP help in the field of genomics?

4

TRANSCRIPTOMICS

4.1 INTRODUCTION

Transcriptomics is the study of the transcription/expression level in the system.The human body is made up of molecules that fit generally into the categories of DNA, RNA, proteins, sugars, salts, fats and water which are governed by the proteins in the human body and become unbalanced if there is any protein dysfunction. Technological strategies can rapidly and almost comprehensively scan through the DNA, RNA and protein molecules of the human body in order to identify differences between individuals. These strategies are collectively known as the "-omics".

4.2 WHAT IS TRANSCRIPTOMICS

Transcriptomics refers to the comprehensive scanning of the nearly fifty thousand known genes that are transcribed into RNA molecules from the three-billion-letter human genome. Each cell utilizes different genes at different times in its development and under different physiological conditions. Transcriptomics known as global analysis of gene expression, also called genome-wide expression profiling, is one of the tools that is used to get an understanding of genes and pathways involved in biological processes. Figure 4.1 shows the steps in the process of transcriptomics. The analysis can be done by the sampling of blood or organ/tissues from which the genes are transcribed into RNA. These transcribed RNA are analysed by microarray, processed by detection, validation, transformation and standardization. Further, multivariate analysis is done for the integration of data. The idea underlying is "guilt by association", which means that genes showing similarity in expression pattern may be functionally related under the same genetic control mechanism. Common technologies for genome-wide or high-throughput analysis of gene expression are cDNA microarrays and oligo-microarrays, cDNA-AFLP and SAGE. The method of expression profiling can be done for diseased condition and implemented diagnostic tests. The power of transcriptomics is in finding the ability to sub-classify disorders that on the surface appear to be similar. This strategy has been shown elegantly for a number of disorders primarily in oncology, where the outcome and survival rates are defined and the drug response rates are based on an expression profile.

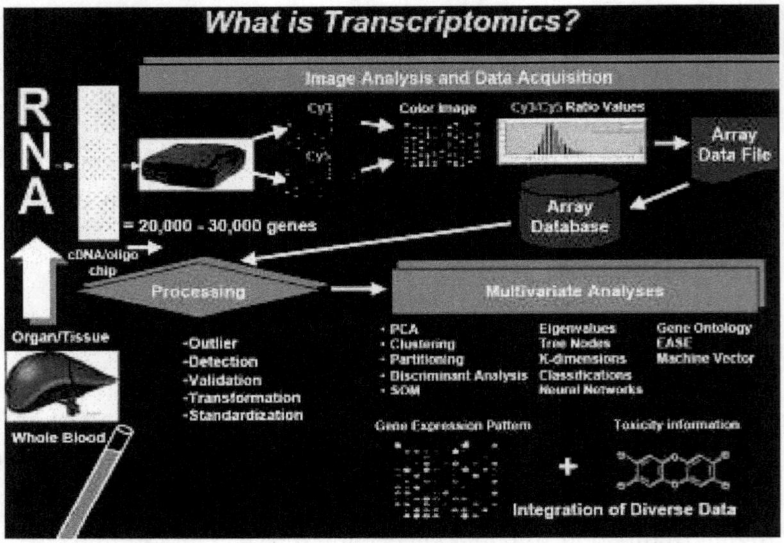

Figure 4.1 Transcriptomics

4.3 TRANSCRIPTOME

The transcriptome is the complete set of RNA products that can be produced from the genome; the transcriptome is the set of all messenger RNA (mRNA) molecules or "transcripts", produced in a population of cells. The term can be used for the total set of transcripts in a given organism, or to the specific subset of transcripts present in a particular cell type. The transcriptome is both time- and cell-specific. It encompasses all mRNA transcripts that are present in the cell at one time including modified, spliced, edited or degraded forms. The genome is static and the transcriptome is dynamic. Most of the cells contain the same genome regardless of the type of cell, stage of development or environmental conditions. Conversely, the transcriptome varies considerably in these differing circumstances due to different patterns of gene expression. Transcriptomics, the study of the transcriptome, is therefore a global way of looking at gene expression patterns.

The study of transcriptomics examines the expression level of mRNAs in a given cell population, by using high-throughput techniques based on DNA microarray technology. The mRNA is not the ultimate product of a gene, transcription is the first step in gene regulation and information about the transcript levels is needed for understanding gene regulatory networks. Thus, it is a new challenge to identify all genes, their expression patterns and their function.

4.4 TECHNIQUES TO STUDY TRANSCRIPTOMICS

There are many modern and high-throughput techniques for the analysis of gene expression and some of these are discussed below.

4.4.1 cDNA Microarray

A cDNA microarray (Figure 4.2) uses the ability of an mRNA molecule that binds specifically to, or hybridizes to, its original DNA coding sequence in the form of a cDNA template spotted on an array. cDNA microarray experiments typically involve hybridizing two mRNA samples, each of which has been converted into cDNA and labelled with its own fluorescent dye (usually a red fluorescent dye, cyanine 5 (Cy5) and a green-fluorescent dye, cyanine 3 (Cy3)), on a single glass slide that has been spotted with many number of cDNA probes. Due to the competitive binding between the two samples, the ratio of the red and green fluorescence intensities for each spot is indicative of the relative abundance of the corresponding DNA probe in the two samples. Thus the data from cDNA microarrays only provide information on the relative expression of the genes.

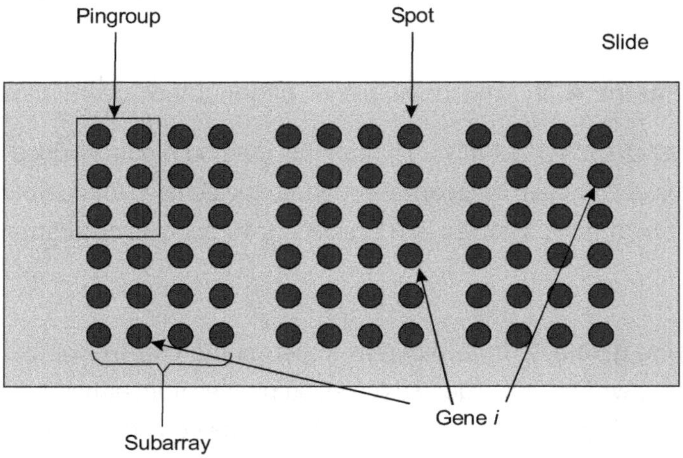

Figure 4.2 cDNA microarray slide

Probes like Gene Chip® oligonucleotide arrays from Affymetrix can be used for analysis. High-density oligonucleotide arrays provide direct information about the expression levels in an mRNA, because these microarrays usually are hybridized with only one sample. Instead of one or several cDNA clones per gene, an oligo-microarray contains two times a set of probe pairs for each gene. One set consists of "perfect match" oligonucleotides that are designed on non-conserved regions in a gene. The other set contains "mismatch oligonucleotides" that are identical to the set of "perfect match" oligonucleotides except for the nucleotide in the middle of the sequence (Figure 4.3). In the presence of a specific RNA in the hybridization solution, the perfect match probes will hybridize more strongly on average than their mismatch partners. This assumption is used to determine the presence or absence of this specific RNA. For each probeset, the value that is usually taken as representative for the expression level of the corresponding gene, is the average difference between the set of perfect match probes minus the set of mismatch probes. In addition, the mismatched oligonucleotides are used to calculate cross-hybridization and local background signals.

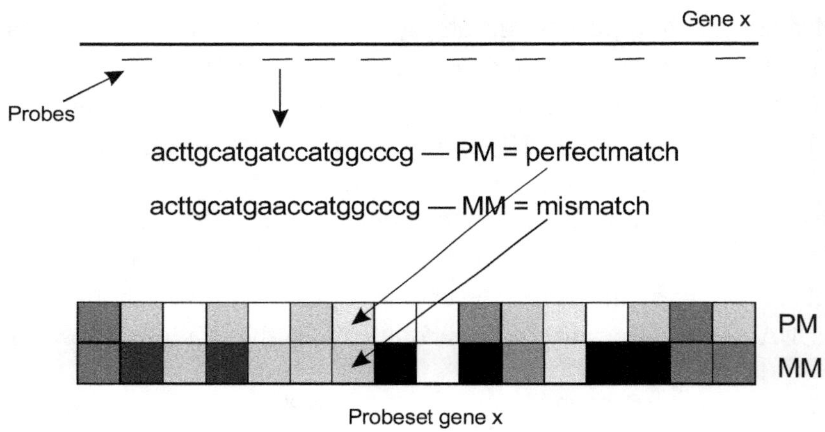

Figure 4.3 The principle of oligonucleotide microarrays

Advantages and disadvantages Though this method produces good results, there are certain advantages and disadvantages. Microarray experiments are used for comparison of gene expression profiles between two mRNA samples (e.g. treatment vs control, or treatment 1 vs treatment 2).

Advantages

- The large data sets from different experiments can be combined together in a single database, which allows gene expression profiles from either different samples or samples obtained using different treatments to be compared with each other and analysed together.
- In medical biology, microarrays are most often used for class prediction. The class prediction is based on marker genes that are specifically expressed during a certain disease or treatment.

- It is envisaged that global surveys of gene expression will identify additional marker genes that may be used to group patients into molecularly relevant categories.

Limitations

- Transcript profiling produces extremely large data sets, and even relatively simple studies can run into thousands of data points.
- It is self-evident that such data sets cannot be organized on simple spreadsheets, but instead require effective database resources for management and analysis.
- With cDNA microarrays, it is difficult to distinguish among different transcripts from genes belonging to the same gene family.
- In order to prevent cross-hybridization, long oligonucleotides may be used instead of cDNA.

Oligo-arrays

- Oligonucleotide design can avoid cross-hybridization, but it is very expensive and requires accurate gene annotation.
- Pre-made oligonucleotide arrays can be bought, which is less costly, but also less flexible than cDNA arrays, because the experimenter cannot select the probes.
- Oligo-arrays use smaller quantities of RNA.
- The hybridization signal is linearly related up to 500-fold mRNA abundance compared to 10-fold in cDNA microarrays. Oligo-arrays have a higher detection specificity (1 out of 106), but are usually only hybridized with one mRNA sample.
- Transcript profiling is done by using microarrays, which is limited to the genes that are represented on the chip.
- Only those genes for which either the DNA sequence or a cDNA clone is available can be used in microarray studies. These genes probably will not be represented on a microarray.
- When cDNA libraries are used to produce the probes, rarely expressed genes are often missing, because abundant messengers are over-represented. For this reason, many important regulatory genes will be overlooked.

4.4.2 SAGE: Serial Analysis of Gene Expression

Another technique for expression analysis is SAGE (Figure 4.4). SAGE is a method that generates sequences from cDNA fragments for the discovery of new genes and the quantification of expression levels in a certain tissue. The method is an improvement of EST sequencing; EST sequencing does not rely on previous sequence information, which made this technique very useful for the discovery of novel genes. The expression level of a gene is measured by counting the amount of ESTs that are sequenced per gene. EST sequencing is laborious and very expensive, therefore, SAGE was developed. The SAGE method is based on counting sequence tags of 10–15 bp from cDNA libraries. These tags are linked together in a chain and cloned into a vector from which they are sequenced.

In contrast to EST sequencing, SAGE is best applied to organisms whose genomic sequences are known or that have a substantial cDNA sequence database in order to identify the genes that are sequenced. Because only short pieces of sequence are sequenced per gene, the inserts with the multiple sequence tags have to be sequenced several times to reduce the amount of sequencing mistakes. In addition, a large number of tags have to be sequenced in order to quantify the expression of genes that have a low expression level.

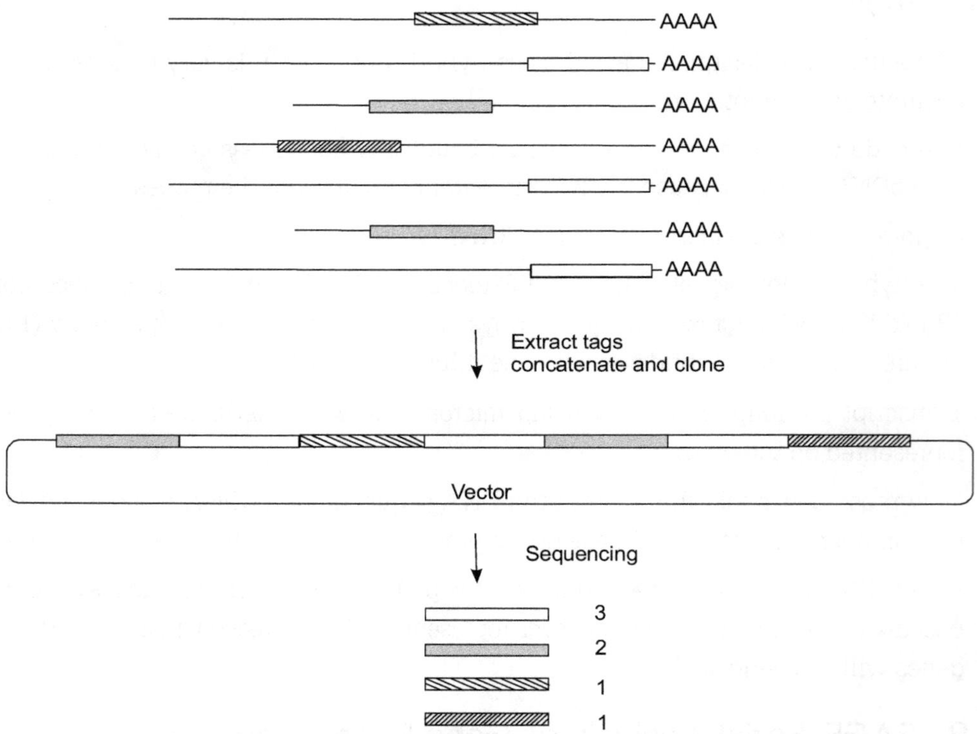

Figure 4.4 The principle of SAGE. Genes are identified based on sequence of the tags. (Picture from Yamamoto *et al.*, 2001)

Massively parallel signature sequencing (MPSS) A new technique called massively parallel signature sequencing (MPSS) is developed as an improvement to SAGE.

MPSS is a parallel sequencing method that can generate 100–1000 short sequence signatures in a single analysis, thus overcoming the principal shortcoming of SAGE. In addition, it generates longer 16–20 base pair signatures, making gene identification more accurate. However this method is very complex and not readily usable.

4.4.3 cDNA-AFLP: cDNA-amplified Fragment Length Polymorphism Analysis

The next method is the polymorphism analysis.

- cDNA-AFLP is an improvement of traditional differential display techniques. It is a PCR-based method which starts with cDNA synthesis from total or mRNA using random hexamers as primers.
- The obtained fragments are digested with two restriction enzymes; normally a 4-cutter, a 6-cutter, and adapters are ligated to the ends of the fragments.
- In the first amplification step, only those fragments are amplified that were digested by both restriction enzymes and thus have different adapters at the end.
- The complex starting mixture of cDNA is fractionated into smaller subsets by selective PCR amplification using primers on the adapters that contain one or more extra nucleotides.
- By increasing the stringency of the PCR amplification (adding more additional nucleotides to the primers), the sensitivity of the analysis can be increased. Genes with a low expression level can be detected.
- The fragments that are amplified are roughly 100–400 bp. These fragments are separated on high-resolution gels.
- The differences in the intensity of the bands that can be observed provide a good measure of the relative differences in the levels of gene expression.
- On further characterization of interesting transcripts they often require the identification of the corresponding full-length cDNA.
- cDNA-AFLP can generate a global overview of gene expression, but it involves a great amount of PCR reactions.
- In addition, separately obtained data sets cannot readily be compared, which is in contrast to SAGE and microarrays data.
- By using cDNA-AFLP, accurate gene expression profiles can be determined by quantitative analysis of band intensities.

- The sensitivity and specificity of the method allows the detection of poorly expressed genes and the determination of subtle differences in transcriptional activity.

4.5 EST (EX PRESSED SEQUENCE TAGS)

An expressed sequence tag or EST is a short sub-sequence of a transcribed spliced nucleotide sequence. They may be used to identify gene transcripts, and are instrumental in gene discovery and gene sequence determination.

4.5.1 The Procedure

- Isolate the mRNA from a particular tissue.
- Treat it with reverse transcriptase.
- Reverse transcriptase is a DNA polymerase that uses RNA as its template and is able to make genetic information flow in the reverse (RNA→DNA) of its normal direction (DNA→RNA). This produces complementary DNA (cDNA).
- Sequence 200–500 nucleotides at both the 5′ and 3′ ends of each cDNA.
- Examine the database of the organism's genome to find a matching sequence.
- ESTs can be mapped to specific chromosome locations using physical mapping techniques, like radiation hybrid mapping or FISH. Alternatively, if the genome of the organism that originated the EST has been sequenced, one can align the EST sequence to that genome.
- In this respect, ESTs become a tool to refine the predicted transcripts for those genes, which leads to prediction of their protein products, and eventually of their function.
- The ESTs are obtained from tissue, organ, disease state—e.g. cancer gives information on the conditions in which the corresponding gene is acting. ESTs contain enough information to permit the design of precise probes for DNA microarrays that then can be used to determine the gene expression.

dbEST dbEST is a division of GenBank established in 1992. As for GenBank, data in dbEST is directly submitted by laboratories worldwide and is not curated.

EST contigs Many distinct expressed sequence tags are often partial sequences that correspond to the same mRNA of an organism. In an effort to reduce the number of expressed sequence tags for downstream gene discovery analyses, several groups assembled expressed sequence tags into EST contigs. Example of resources that provide EST contigs include TIGR gene indices and Unigene.

STACK Constructing EST contigs is not trivial and may yield artefacts (contigs that contain two distinct gene products). When the complete genome sequence of an organism is available and transcripts are annotated, it is possible to bypass contig assembly and directly match transcripts with ESTs. This is used in the TissueInfo system and makes it easy to link annotations in the genomic database to tissue information provided by EST data.

TissueInfo High-throughput analyses of ESTs often encounter similar data management challenges. A first challenge is that tissue provenance of EST libraries is described in plain English in dbEST.

This makes it difficult to write programs that can non-ambiguously determine that two EST libraries were sequenced from the same tissue. Similarly, disease conditions for the tissue are not annotated in a computationally friendly manner.

For example, cancer origin of a library is often mixed with the tissue name (e.g. the tissue name "glioblastoma" indicates that the EST library was sequenced from brain tissue and the disease condition is cancer).

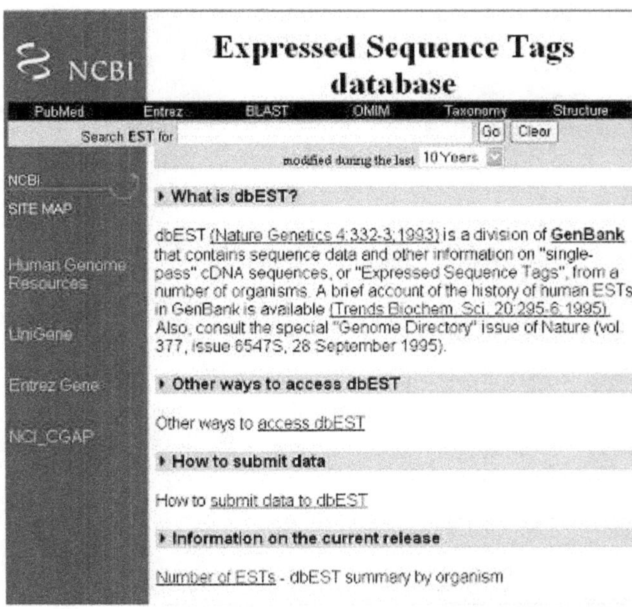

Figure 4.5 Home Page of dbEST

With the notable exception of cancer, the disease condition is often not recorded in dbEST entries. The TissueInfo project was started in 2000 to help with these challenges. The project provides curated data (updated daily) to disambiguate tissue origin and disease state (cancer/non-cancer), offers a tissue ontology that links tissues and organs by "is part of" relationships (i.e., formalizes knowledge that hypothalamus is part of brain, and that brain is part of the central nervous system) and distributes open-source software for linking transcript annotations from sequenced genomes to tissue expression profiles calculated with data in dbEST.

From Figure 4.5, the home page of dbEST is accessed from NCBI. These databases have additional links of Human Genome Resource, UniGene, Entrez Gene, etc. The search for EST is done. This EST is specific for a particular search.

Figure 4.6 shows the search done for lung cancer and the results in the summary format was retained for analysis.

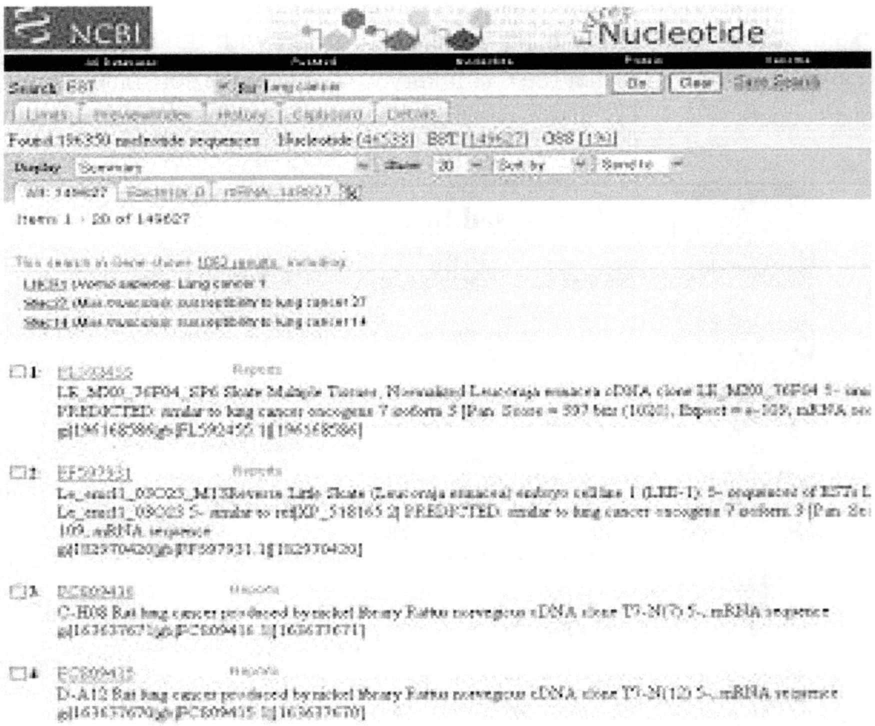

Figure 4.6 EST for lung cancer

Figure 4.7 shows the result page for the EST searched for lung cancer. The result page has headers like Identifiers with their dbEST Id, EST Name, GenBank Accession no., clone information, the type of clone, primers used for sequencing, quality and the predicted results. The sequence is also shown so that the retrieval of the sequence can be easily taken for further analysis.

Figure 4.7 Result Page for lung cancer

4.6 STANFORD MICROARRAY DATABASE

The Stanford Microarray Database (SMD) stores raw and normalized data from microarray experiments, and provides web interfaces for researchers to retrieve, analyse and visualize their data. The two immediate goals for SMD are to serve as a storage site for microarray data from ongoing research at Stanford University, and to facilitate the public dissemination of that data once published, or released by the researcher. Of paramount importance is the connection of microarray data with the biological data that pertains to the DNA deposited on the microarray (genes, clones, etc.). SMD makes use of many public resources to connect expression information to the relevant biology, including SGD.

Figure 4.8 shows the home page of SMD. The access to the database is possible by registered or public data login. On the Home Page the recent publications, source, caryoscope are all given as inks. The login can be done in public login for accessing the data. The basic search and advanced search are the two types of search in SMD, the advanced search option is chosen for analysis.

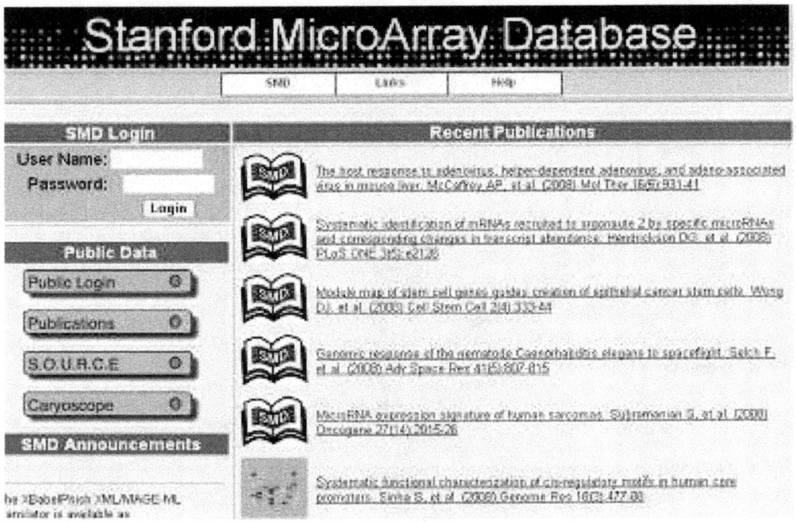

Figure 4.8 Home Page of SMD

In Figure 4.9, the login has been done on public login and the advanced search option is chosen and the selection of experimenter category, subcategory and organism is done. The options and/or are present and based on our query, the result will be produced.

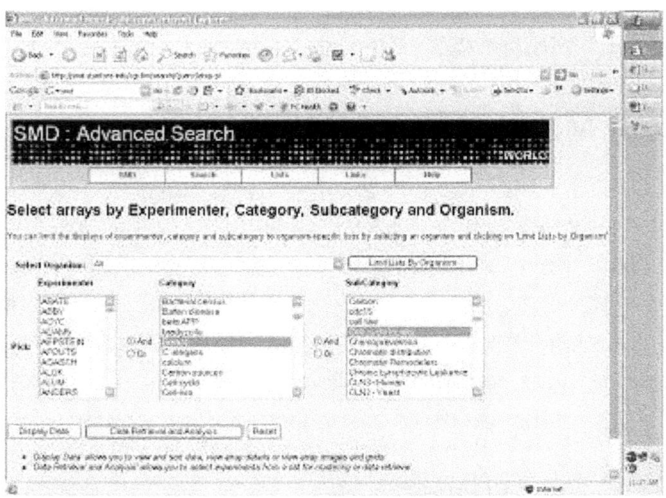

Figure 4.9 Home Page of SMD advanced search with selection

Figure 4.10 describes the Result Page of SMD search, which gives the result for all the categories chosen.

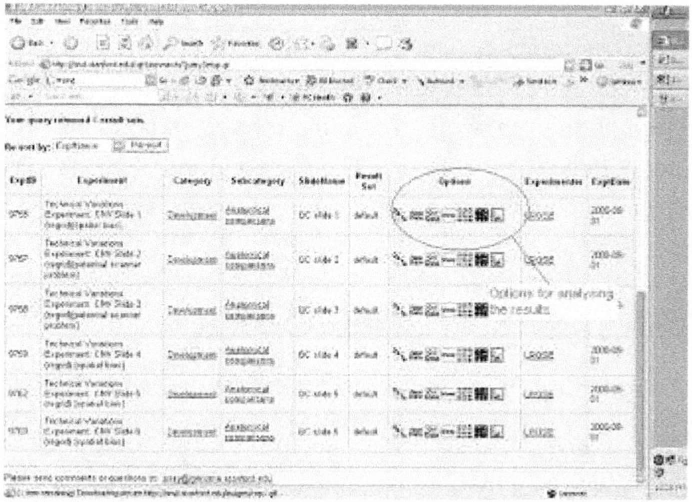

Figure 4.10 Result Page of SMD

Figure 4.11 shows the entire result set where the data can be analysed, for sorting the data. The category or subcategory can be retrieved from the options as view and sort, Download Raw data, Array details, Array image, Plot array, align to chromosomes.

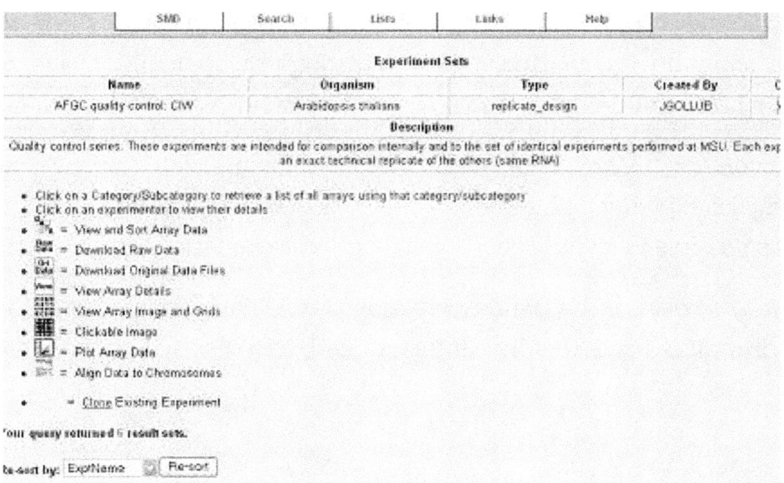

Figure 4.11 Result Set from SMD

4.7 mRNA REGULATION

Microarray studies reveal the relative amounts of different mRNAs in the cell, levels of mRNA that are not directly proportional to the expression level of the proteins they code for. The number of protein molecules synthesized using a given mRNA molecule as a template is highly dependent on translation-initiation features of the mRNA sequence; The complete protein complement of a cell or organism is known as the proteome. The human genome contains the complete set of genes required to build a functional human being. The transcription of genes to produce RNA is the first stage of gene expression. The transcriptome is the complete set of RNA transcripts produced by the genome at any one time.

4.8 THE INTEREST IN TRANSCRIPTOMICS

There are many different ways in which the large-scale analyses of gene expression patterns are useful which include:

To find an answer to how gene expression occurs We can find which genes are activated by a particular transcription factor. This is done by comparing gene expression patterns in tissues, in which the particular transcription factor is either active or inactive.

The general discovery experiments There is no particular hypothesis but can be used to identify interesting genes, like the genes which are highly expressed in brain tumours and not in healthy brain tissue, and whether this can be used for drug targets or diagnostic markers?

Disease classification The single markers are not sufficient to distinguish between two similar diseases, as is often the case in cancer. Testing the expression profiles of a larger number of genes can provide accurate diagnosis.

Functional annotation Many DNA sequences have been isolated to have no known function. If they show similar expression patterns to a characterized gene, it is likely their functions are similar. It is sometimes possible to identify conserved regulatory elements in such genes.

To identify drug target If the gene expression profile caused by a mutation is similar to that caused by a drug, it is likely the drug interacts with and inactivates the protein affected by the mutation.

SUMMARY

Since transcriptomics deals with the transcription level of the central dogma of life,one can study the expression level by microarrays, a technique which helps scientists to find the normal versus diseased genes. SAGE is a method that generates sequences from cDNA fragments for the discovery of new genes and the quantification of expression levels in a certain tissue. cDNA-AFLP is an improvement of display techniques. It is a PCR-based method which starts with cDNA synthesis from total or mRNA using random hexamers as primers. An expressed sequence tag is a sequence of a transcribed spliced nucleotide sequence. It is used to identify gene transcripts, gene discovery and gene sequence determination.

CRITICAL THINKING QUESTIONS

1. What is the role of transcriptomics in research?
2. Sketch the process of transcriptomics.
3. Explain SMD.

SELF-LEARNING EXERCISES

1. With the help of ESTs, refine any transcript of interest.
2. Analyse the basic results for any cancer in SMD.

METABOLOMICS

5.1 INTRODUCTION

Metabolomics is the systematic study of the unique chemical interactions specific for cellular processes, specifically the study of small molecules and their metabolic profiles produced by active, living cells under different conditions and times in their life cycles.

5.2 METABOLOME

Metabolome refers to the complete set of small-molecule metabolites such as metabolic intermediates, hormones and other signalling molecules, and secondary metabolites which are found in the biological sample. In January 2007, scientists at the University of Alberta and the University of Calgar completed the first draft of the human metabolome.

They have catalogued and characterized 2500 metabolites, 1200 drugs and 3500 food components that can be found in the human body. Metabolomics research carries a huge responsibility for providing a detailed description of human pathways and their working.

One of the challenges of systems biology is to integrate proteomic, transcriptomic, and metabolomic information to give a more complete picture of living organisms.

5.3 METABOLITES

Metabolites are the intermediary products of metabolism. The term metabolite usually refers to a small molecule. A primary metabolite is directly involved in the normal growth, development, and reproduction. A secondary metabolite is not directly involved in normal growth, development, and reproduction, but usually has important ecological function, e.g. antibiotics and pigments. The metabolome forms a large network of metabolic reactions, where the metabolites from one enzymatic chemical reaction will be the initial compound for another chemical reaction and is called as intermediary metabolism.The history of metabolomics began when the chromatographic separation techniques were initially used in the detection of metabolites during the late 1960s, which marks the technical origin of the field. The bioanalytical methods used for metabolomics analysis have been adapted from existing biochemical techniques. Two characteristics common to metabolomic research are:

- Metabolites are profiled without bias towards a specific metabolite or group of metabolites.
- Relationships between the metabolites are characterized, currently mostly by multivariate methods.

5.4 ENZYME CLASSIFICATION

Enzymes are necessary for metabolism. Based on their catalysing reactions, the Nomenclature Committee of the **International Union of Biochemistry and Molecular Biology** (IUBMB) recommended the following classification. To understand metabolism, one should know the enzyme classification.

5.4.1 Class 1 Oxidoreductases

Oxidoreductases catalyse a variety of oxidation–reduction reactions. Common names include dehydrogenase, oxidase, reductase and catalase. The substrate that is oxidized is regarded as the hydrogen donor. The systematic name is based on **donor:acceptor oxidoreductase** . The common name will be **dehydrogenase,** wherever this is possible; as an alternative, **reductase** can be used. **Oxidase** is only used in cases where O_2 is the acceptor.

5.4.2 Class 2 Transferases

Transferases are enzymes that transfer a group, e.g. a methyl group or a glycosyl group, from one compound (generally regarded as donor) to another compound (generally regarded as acceptor). The systematic names are formed according to the scheme **donor:acceptor grouptransferase.** The common names are normally formed according to **acceptor group transferase** or **donor** grouptransferase. The donor is a cofactor (coenzyme) charged with the group to be transferred.Common names include acetyltransferase, methylase, protein kinase and polymerase. The first three subclasses play major roles in the regulation of cellular processes. Polymerase is essential for the synthesis of DNA and RNA.

5.4.3 Class 3 Hydrolases

Hydrolases catalyse hydrolysis reactions where a molecule is split into two or more smaller molecules by the addition of water. Common examples are given below.

Proteases split protein molecules, e.g. HIV protease and caspase. HIV protease is essential for HIV replication. Caspase plays a major role in apoptosis.

Nucleases splits nucleic acids (DNA and RNA). Based on the substrate type, they are divided into **RNase** and **DNase**. RNase catalyses the hydrolysis of RNA, and DNase acts on DNA. They may also be divided into **exonuclease** and **endonuclease**. The exonuclease progressively splits off single nucleotides from one end of DNA or RNA. The endonuclease splits DNA or RNA at internal sites.

Phosphatase catalyses dephosphorylation (removal of phosphate groups). Example: calcineurin. The immunosuppressive drugs FK506 and Cyclosporin A are the inhibitors of calcineurin.

5.4.4 Class 4 Lyases

Lyases catalyse the cleavage of C–C, C–O, C–S and C–N bonds by means other than hydrolysis or oxidation. The systematic name is formed according to the pattern **substrate group-lyase**. The hyphen is an important part of the name, and to avoid confusion should not be omitted, e.g. hydro-lyase not "hydrolyase". In the common names, expressions like decarboxylase, aldolase, dehydratase (in case of elimination of CO_2, aldehyde, or water) are used. In cases where the reverse reaction is much more important, or the only one demonstrated, synthase (not synthetase) may be used in the name. Common names include decarboxylase and aldolase.

5.4.5 Class 5 Isomerases

Isomerases catalyse geometric or structural changes within one molecule. According to the type of isomerism, they may be called **racemases, epimerases,** *cis–trans*-**isomerases, isomerases, tautomerases, mutases or cycloisomerases**. Examples include rotamase, protein disulphide isomerase (PDI), epimerase and racemase.

5.4.6 Class 6 Ligases

Ligases are enzymes joining together two molecules coupled with the hydrolysis of a diphosphate bond in ATP or a similar triphosphate. The systematic names are formed on the system X:Y ligase (ADP-forming). In earlier editions of the list the term synthetase has been used for the common names. Examples include peptide synthase, aminoacyl-tRNA synthetase, DNA ligase and RNA ligase.

5.5 PATHWAY DATABASES

To study about metabolomics we should know the basic concepts of metabolism and metabolic databases. Many of us are familiar with metabolism, so we will learn the concepts of databases, i.e., metabolic databases, which relate the pathways of reactions and their enzymes.

KEGG is a collection of metabolic pathways, which include graphical pathway of maps for all known metabolic pathways from various organisms. Ortholog group tables containing conserved, functional units in a molecular pathway or assembly as well as comparative lists of genes for a given functional unit in different organisms are also available.

EcoCyc, the database for *E.coli*, describes the genome and the biochemical machinery of *E. coli*, which provides the molecular and functional aspects of *E. coli* to understand to facilitate system-level understanding of cell. It also supports computational studies of metabolism, such as pathway design, evolutionary studies, and simulations.

The *E.coli* pathways retrieved from EcoCyc database is used to interface and visualize the layout of genes present in the individual pathway of complete biochemical reactions.

BRENDA, the enzyme database, has comprehensive information on enzymes and enzymatic reactions. It is one of several databases nested within the metabolic pathway database set of the SRS5 sequence retrieval system at EBI.

The WIT metabolic reconstruction project produces metabolic reconstructions for sequenced, or partially sequenced, genomes. It currently provides a set of over 25 such reconstructions in varying states of completion. Over 2900 pathway diagrams are available, associated with functional roles and linked to ORFs.

Boehringer Mannheim biochemical pathways is a searchable database of metabolic pathways, enzymes, substrates and products. Based on a given search, it produces a graphic representation of the relevant pathway(s) within the context of an enormous metabolic map. Neighbouring metabolic reactions can then be viewed through links to adjacent maps.

BioSilico is a web-based database system that facilitates the search and analysis of metabolic pathways. Heterogeneous metabolic databases including LIGAND, ENZYME, EcoCyc and MetaCyc are integrated in a systematic way, thereby allowing users to efficiently retrieve the relevant information on enzymes, biochemical compounds and reactions. In addition, it provides well-designed view pages for more detailed summary information. BioSilico is developed as an extensible system with a robust systematic architecture.

5.6 KEGG: KYOTO ENCYCLOPEDIA OF GENES AND GENOMES

Kyoto Encyclopedia of Genes and Genomes (KEGG) is a bioinformatic metabolic database containing information on genes, proteins, reactions, and pathways. Kyoto Encyclopedia of Genes and Genomes (KEGG) is a knowledge base for systematic analysis of gene functions in terms of the networks of genes and molecules. The major component of KEGG (Figure 5.1) is the PATHWAY database that consists of graphical diagrams of biochemical pathways including most of the known metabolic pathways and some of the known regulatory pathways. KEGG maintains the GENES database for the gene catalogs of all organisms with complete genomes and selected organisms with partial genomes, which are continuously re-annotated, as well as the LIGAND database for chemical compounds and enzymes. Each gene catalog is associated with the graphical genome map for chromosomal locations that is represented by Java Applet. KEGG develops and provides various computational tools, for reconstructing biochemical pathways from the complete genome sequence and for predicting gene regulatory networks from the gene expression profiles. The main reason for constructing the database is that there is an increasing amount of genomic information for understanding the principles of high-order biological systems, such as the cell, the organism, and the biosphere, are formed, as well as for medical, industrial, and other practical applications. Since 1995 the development of knowledge-based methods for uncovering higher-order systemic behaviours of the cell and the organism from genomic information stored in KEGG, and associated Bioinformatics technologies are developed both for basic research and practical applications. The "KEGG Organisms" section, which is divided into eukaryotes and prokaryotes, encompasses many organisms for which gene and DNA information can be searched by typing in the enzyme of choice. KEGG is a part of the research projects of the Kanehisa Laboratories in the Bioinformatics Center of Kyoto University and the Human Genome Center of the University of Tokyo. KEGG has lots of entry points like KEGG2, KEGG pathway, KEGG Brite, KEGG Genes, KEGG ligand and KEGG FTP site, KEGG Drug, KEGG Glycan, KEGG reaction and KAAS.

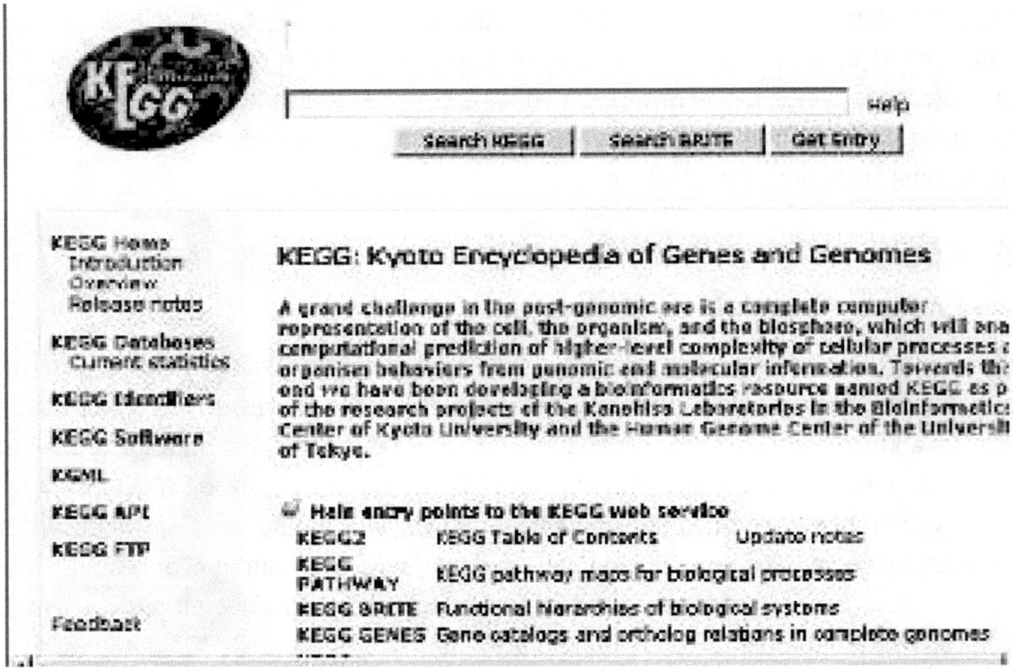

Figure 5.1 Home Page of KEGG

KEGG2 (Figure 5.2) has the table of contents of all the links and entries in the database. They are categorized as:

1. ***System information*** KEGG PATHWAY is the collection of network of pathways for all the organisms which are found in the database. In this the pathway new maps can be updated or can be retrieved for research.

KEGG BRITE is a collection of hierarchical classifications of representing the knowledge-based result on various aspects of biological systems. In contrast to KEGG pathway, which is limited to molecular interactions and reactions, KEGG BRITE incorporates many different types of relationships. The mapping of genomic and molecular data to KEGG BRITE supplements the KEGG pathway mapping for inferring higher-order functions.

2. ***Genomic information*** KEGG GENES is a collection of gene catalogs for all complete genomes and some partial genomes generated from publicly available resources, mostly NCBI RefSeq. They are subject to SSDB computation and manual KO assignment (gene annotation). DGENES for draft genomes (eukaryotes only) and EGENES for EST consensus contigs (mostly

plants) are meant to supplement the repertoire of KEGG organisms, and they are given automatic KO assignment with GENES used as a reference data set. This site covers all genes.

3. ***Chemical information*** In the chemical information site (Figure 5.2) the details of the KEGG LIGAND contains chemical substances and reactions that are relevant to life. It is a composite database consisting of COMPOUND, DRUG, GLYCAN, REACTION, RPAIR, and ENZYME databases.

4. ***KEGG organism*** Specific for organisms.

5. ***KEGG disease*** Human disease and infectious disease.

6. ***KEGG drug*** KEGG DRUG (Figure 5.3) is a chemical structure-based information resource for all approved drugs in Japan and USA. Each chemical structure is identified by the D number, and is associated with generic names, trade names, efficacy, target information, etc. KEGG DRUG is maintained in the KEGG LIGAND relational database.

7. ***KEGG glycan: study on glycobiology*** The glycan structure map is another representation of glycan biosynthesis. The composite structure map (CSM) is computationally generated to represent all possible structures in the KEGG GLYCAN structure database. CSM can be used to examine the structural repertoire inferred from the genomic or transcriptomic repertoire. The KEGG GLYCAN database is a collection of experimentally determined glycan structures. It contains all unique structures taken from CarbBank, structures entered from recent publications, and structures present in KEGG pathways. The G number, such as G00197 for CD65, identifies each structure. KEGG GLYCAN is maintained in the KEGG LIGAND relational database.

8. ***KEGG compound*** Specific for lipid metabolism.

9. ***KEGG reaction*** Mainly for xenobiotics metabolism. KEGG ENZYME is an implementation of the Enzyme Nomenclature (EC number) of the IUBMB/IUPAC biochemical nomenclature committee. KEGG ENZYME is based on the ExplorEnz database at Trinity College, Dublin, and is maintained in the KEGG LIGAND relational database with additional annotation of reaction and organism information.

KAAS (KEGG Automatic Annotation Server) provides functional annotation of genes by BLAST comparisons against the manually curated KEGG GENES database. The result contains KO (KEGG Orthology) assignments and automatically generated KEGG pathways (Figure 5.4).

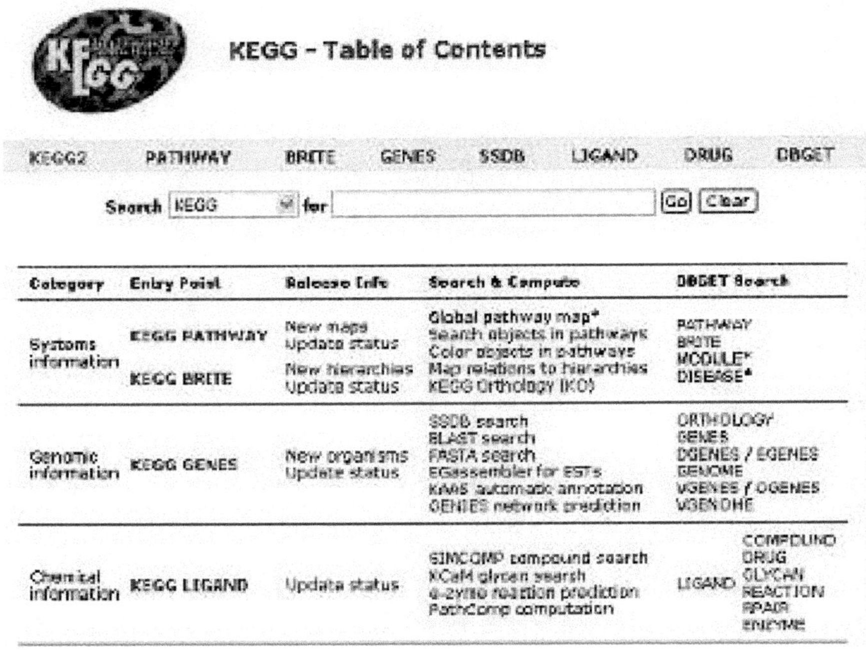

Figure 5.2 Table of Contents

Figure 5.3 Entries in KEGG

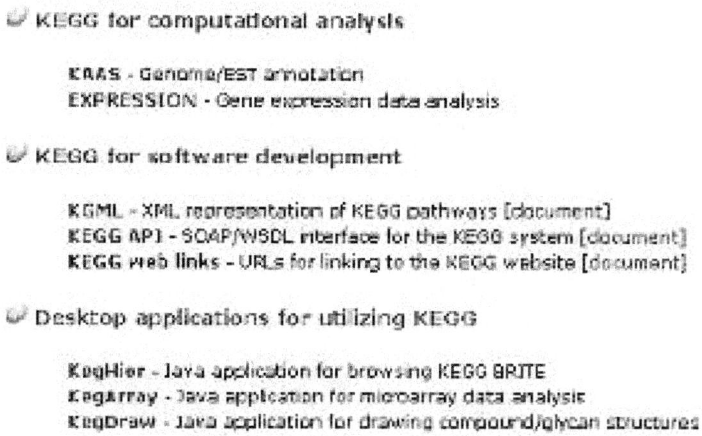

KEGG for computational analysis

KAAS - Genome/EST annotation
EXPRESSION - Gene expression data analysis

KEGG for software development

KGML - XML representation of KEGG pathways [document]
KEGG API - SOAP/WSDL interface for the KEGG system [document]
KEGG web links - URLs for linking to the KEGG website [document]

Desktop applications for utilizing KEGG

KegHier - Java application for browsing KEGG BRITE
KegArray - Java application for microarray data analysis
KegDraw - Java application for drawing compound/glycan structures

Figure 5.4 Entries in KEGG

A systems biological approach can be done based on drug targets, lead compounds modelling and simulation.

Figure 5.5 shows the route map of KEGG and how these pathways are interconnected for the systems biological approach.

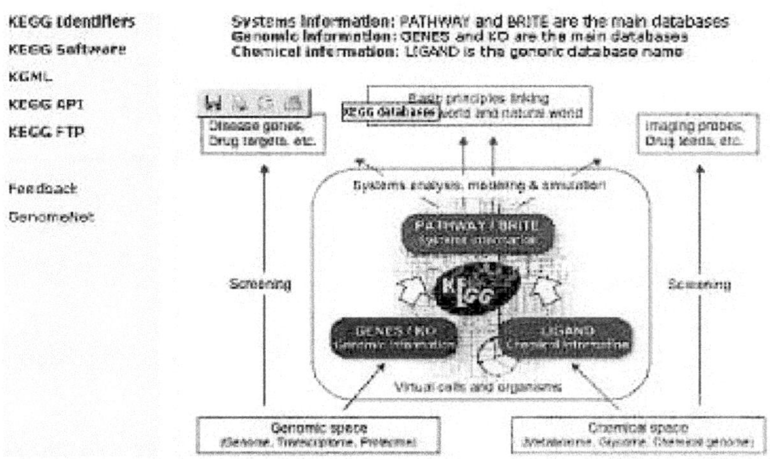

Figure 5.5 Route Map of KEGG

The database comprises (Figure 5.6) of many modules to find the maps for pathways, disease-related orthologs, EST, metabolites, drugs, sugar derivatives, enzymes and reactions. The auxillary databases cover viral genes and genomes, etc.

Database	Content
KEGG PATHWAY	Pathway maps
KEGG BRITE	Functional hierarchies
KEGG MODULE	Pathway modules (to be released)
KEGG DISEASE	Diseases (to be released)
KEGG ORTHOLOGY	KEGG orthology (KO) groups
KEGG GENES	Genes in high-quality genomes
KEGG DGENES	Genes in draft genomes
KEGG EGENES	Genes as EST contigs
KEGG GENOME	Organisms with complete genomes
KEGG SSDB	Sequence similarities and best hit relations
KEGG COMPOUND	Metabolites
KEGG DRUG	Drugs
KEGG GLYCAN	Glycans
KEGG ENZYME	Enzymes
KEGG REACTION	Enzymatic reactions
KEGG RPAIR	Reactant pairs and chemical transformations

In addition, the following auxiliary databases are maintained without any annotation at KEGG.

Database	Content
KEGG VGENOME	Viral genomes
KEGG VGENES	Viral genes
KEGG OGENES	Organelle genes
KEGG EXPRESSION	Gene expression profiles

Figure 5.6 Database in KEGG

5.6.1 KEGG Identifiers

KEGG object identifiers KEGG objects (Figure 5.7) are biological entities from molecular to higher levels. Each object is identified by a five-digit number prefixed by an uppercase alphabet prefixed by a two- or three-letter code .

KEGG Identifiers

KEGG Object Identifier

KEGG objects are biological entities from molecular to higher levels. Each object (except for GENES) is identified by the KEGG object identifer, consisting of a five-digit number prefixed by an upper-case alphabet, such as K05032 and D00336, c prefixed by a 2-4 letter code for PATHWAY and BRITE, such as map00010 and br08301.

Prefix	Content	Database
K	Gene/protein ortholog group	ORTHOLOGY
C	Chemical compound	COMPOUND
D	Drug	DRUG
G	Glycan	GLYCAN
R	Reaction	REACTION
A	Reactant pair alignment	RPAIR
(code)	Pathway map	PATHWAY
(code)	Hierarchical text file	BRITE
M	Pathway module	
H	Human disease	
T	Organism	GENOME

Enter KEGG object identifiers to retrieve corresponding database entries:

(Example) R07326 R00623 R00754 R01036 R04805 R04880 R06917

Figure 5.7 KEGG identifiers

Figure 5.8 shows the KEGG entries, i.e., how the entries will be present in the pathway as prefixes.

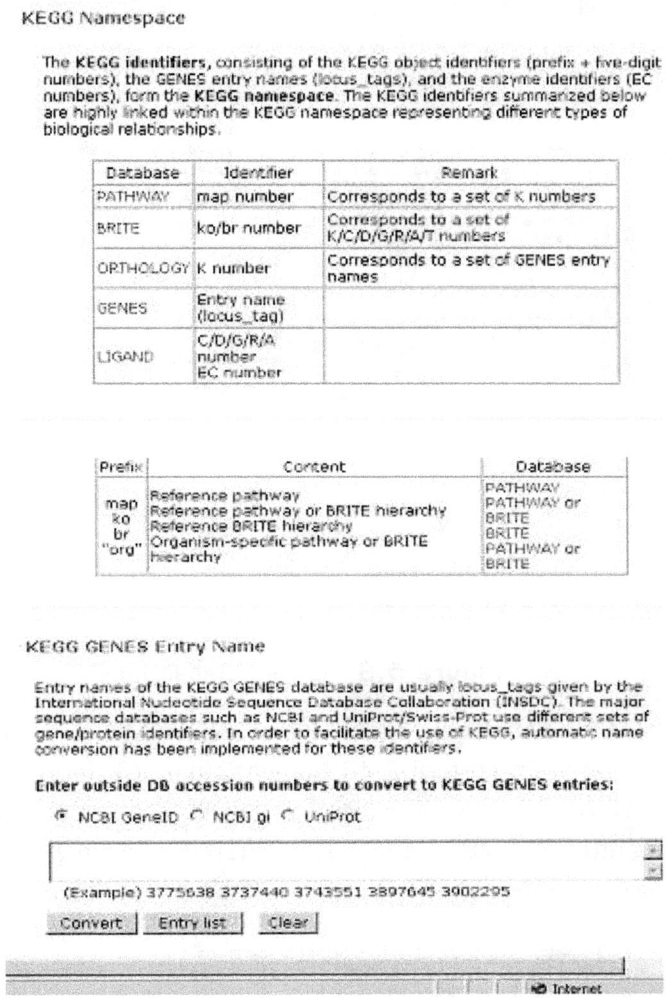

KEGG Namespace

The **KEGG identifiers,** consisting of the KEGG object identifiers (prefix + five-digit numbers), the GENES entry names (locus_tags), and the enzyme identifiers (EC numbers), form the **KEGG namespace**. The KEGG identifiers summarized below are highly linked within the KEGG namespace representing different types of biological relationships.

Database	Identifier	Remark
PATHWAY	map number	Corresponds to a set of K numbers
BRITE	ko/br number	Corresponds to a set of K/C/D/G/R/A/T numbers
ORTHOLOGY	K number	Corresponds to a set of GENES entry names
GENES	Entry name (locus_tag)	
LIGAND	C/D/G/R/A number EC number	

Prefix	Content	Database
map	Reference pathway	PATHWAY
ko	Reference pathway or BRITE hierarchy	PATHWAY or BRITE
br	Reference BRITE hierarchy	BRITE
"org"	Organism-specific pathway or BRITE hierarchy	PATHWAY or BRITE

KEGG GENES Entry Name

Entry names of the KEGG GENES database are usually locus_tags given by the International Nucleotide Sequence Database Collaboration (INSDC). The major sequence databases such as NCBI and UniProt/Swiss-Prot use different sets of gene/protein identifiers. In order to facilitate the use of KEGG, automatic name conversion has been implemented for these identifiers.

Enter outside DB accession numbers to convert to KEGG GENES entries:

NCBI GeneID NCBI gi UniProt

(Example) 3775638 3737440 3743551 3897645 3902295

Convert Entry list Clear

Internet

Figure 5.8 KEGG Entries

5.6.2 Links to Outside Database

The links are found from NCBI, TIGR, Sanger, Expasy, PubChem, CarBank and many more databases.

The links to outside the database are shown in Figure 5.9. The GENES database is linked to NCBI, Entry Gene, OMIM, UniProt/Swiss-Prot, TIGR-CMR, Sanger GeneDB.

The GENOME database is linked to NCBI Taxonomy, RefSeq, JGI, TIGR, Sanger. The ENZYME database is linked to IUBMB, Explor Enz, BRENDA, ExPASy UM-BBD, ERGO. The COMPOUND database is connected to PubChem, 3DMET, ChEBI. The DRUG database has the external link to PubChem. GLYCAN database provides the other accession to CarbBank. The ORTHOLOGY database has external links to gene ontology, TCDB, COG, CAZy.

Links to Outside Databases

In addition to the links within the KEGG namespace, the KEGG identifiers are alsc linked to outside database entries. The links usually represent the same biological meaning, such as the same gene or protein in the same organism, enabling the user to integrate information from different sources. In this respect, KEGG is a portal site to more specialized databases in various fields of biomedica sciences.

Database	Identifier	Outside links	
GENES	Entry name (locus_tag)	NCBI gi number Entrez Gene OMIM UniProt/Swiss-Prot	TIGR CMR Sanger GeneDB Other genome DBs
GENOME	Organism code (T number)	NCBI Taxonomy RefSeq JGI	TIGR Sanger Other genome DBs
ENZYME	EC number	IUBMB ExplorEnz BRENDA	ExPASy UM-BBD ERGO
COMPOUND	C number	PubChem ChEBI	3DMET
DRUG	D number	PubChem	
GLYCAN	G number	CarbBank	
ORTHOLOGY	K number	Gene Ontology COG	TCDB CAZy

Figure 5.9 Other Databases

Figure 5.10 has the KEGG database as BRITE and PATHWAY. The BRITE has ko number and br number. The ko number is linked to IUBMB Nomenclature, IUPHAR Nomenclature, Swiss-Prot, TRANSFAC, TCDB, GPCR, NURSA, MEROPS, and PROW. The br number has the external links with IUPAC Nomenclature and LIPID MAPS. The PATHWAY has map number and is linked with UM-BBD.

KEGG pathways and BRITE functional hierarchies are higher-level objects, wh are developed based on published literature and selected databases. Links t outside databases are given as a form of acknowledgement in this developm

Database	Identifier	Outside links	
BRITE	ko number	IUBMB Nomenclature IUPHAR Nomenclature SWISS-PROT Documents TRANSFAC	TCDB GPCR NURSA MEROPS PROW
	br number	IUPAC Nomenclature LIPID MAPS	
PATHWAY	map number	UM-BBD	

Figure 5.10 KEGG Database

5.6.3 KEGG Organism Code

KEGG organism code can be used to retrieve the code for organisms with different prefixes. The KEGG organism code (Figure 5.11) entry has database entry and organism entry.

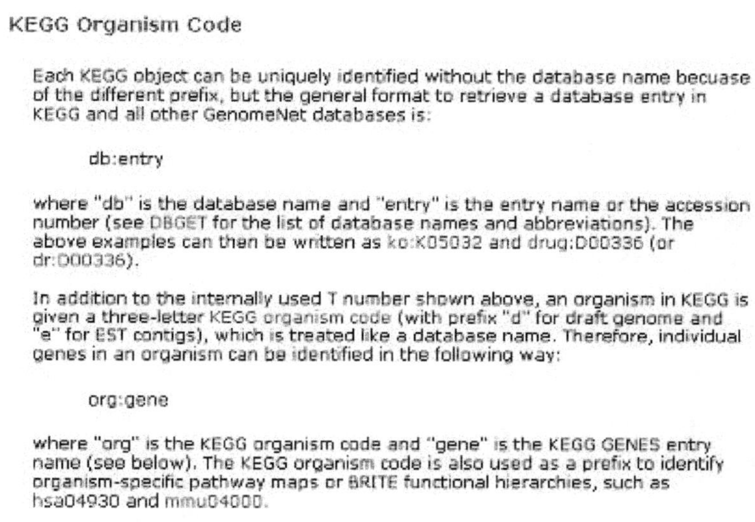

KEGG Organism Code

Each KEGG object can be uniquely identified without the database name becuase of the different prefix, but the general format to retrieve a database entry in KEGG and all other GenomeNet databases is:

db:entry

where "db" is the database name and "entry" is the entry name or the accession number (see DBGET for the list of database names and abbreviations). The above examples can then be written as ko:K05032 and drug:D00336 (or dr:D00336).

In addition to the internally used T number shown above, an organism in KEGG is given a three-letter KEGG organism code (with prefix "d" for draft genome and "e" for EST contigs), which is treated like a database name. Therefore, individual genes in an organism can be identified in the following way:

org:gene

where "org" is the KEGG organism code and "gene" is the KEGG GENES entry name (see below). The KEGG organism code is also used as a prefix to identify organism-specific pathway maps or BRITE functional hierarchies, such as hsa04930 and mmu04000.

Figure 5.11 KEGG organism code

1. *Metabolic pathways in KEGG* There are many pathways seen with respect to specific metabolism: They are carbohydrate, energy, lipid, nucleotide, amino acid, glycan biosynthesis, biosynthesis of polyketides and nonribosomal peptides, cofactors and vitamins, secondary metabolites, xenobiotics.

2. *Genetic information processing* The necessary information related to genetics like transcription, translation, folding, sorting and degradation can be retrieved from genetic information processing.

3. *Environmental processing* To understand the metabolic activities in the environment like membrane transport, signal transduction, signalling molecules and interaction.

4. *Cellular process* The concept of cell motility, cell growth and death, cell communication, endocrine system, immune systen, nervous system, sensory system, development, behaviour can be studied in cellular process.

5. *Human diseases* Cancer, immune disorders, neurodegenerative disease, metabolic and infectious diseases.

6. ***Drug development*** Antibiotics, antineoplastics, nervous system agents, drugs, target based classification, skeleton based classification.

The KEGG classifiers (Figure 5.12) have many subdivisions like metabolism, Genetic information processing, environmental information processing, cellular process, cancers, drug development and more. These main classes are again subdivided for ease of the user.

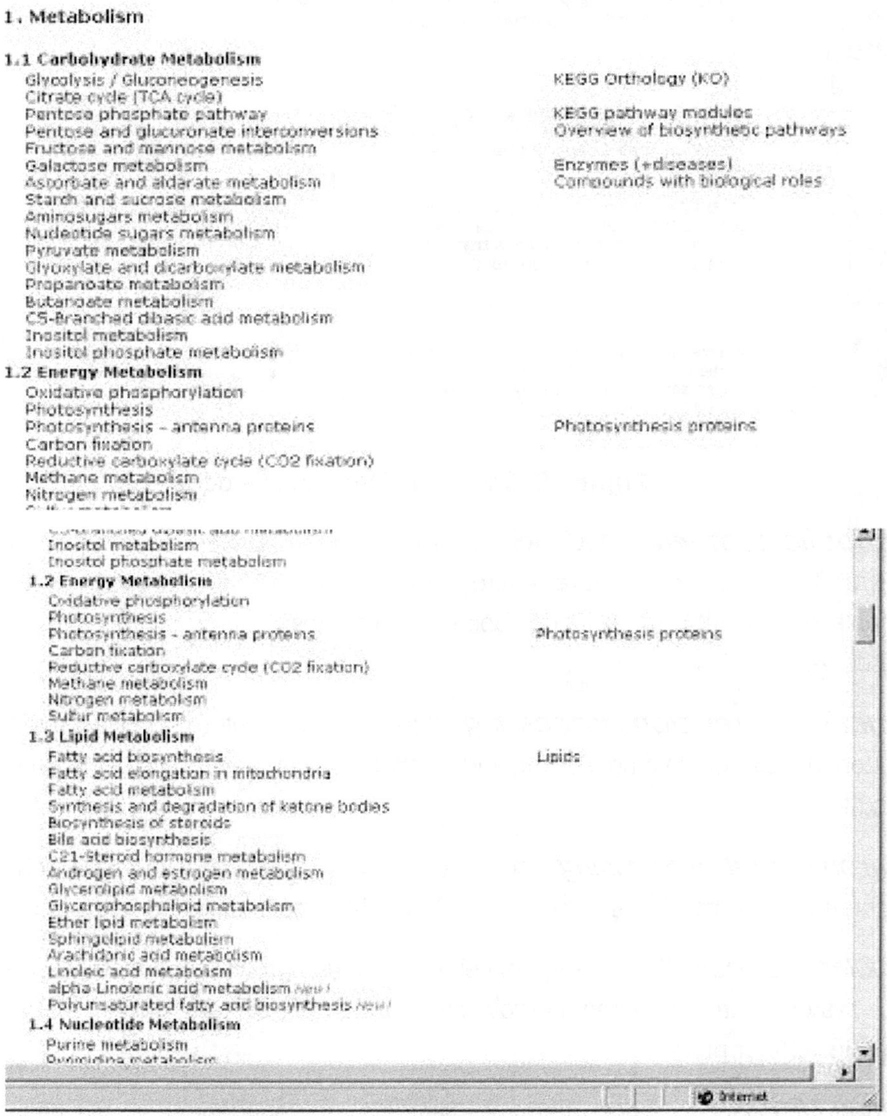

Figure 5.12 KEGG Classifiers *(Continues)*

Lysine degradation
Arginine and proline metabolism
Histidine metabolism
Tyrosine metabolism
Phenylalanine metabolism
Tryptophan metabolism
Phenylalanine, tyrosine and tryptophan biosynthesis
Urea cycle and metabolism of amino groups

1.6 Metabolism of Other Amino Acids

beta-Alanine metabolism
Taurine and hypotaurine metabolism
Aminophosphonate metabolism
Selenoamino acid metabolism
Cyanoamino acid metabolism
D-Glutamine and D-glutamate metabolism
D-Arginine and D-ornithine metabolism
D-Alanine metabolism
Glutathione metabolism

1.7 Glycan Biosynthesis and Metabolism

N-Glycan biosynthesis — KEGG GLYCAN
High-mannose type N-glycan biosynthesis
N-Glycan degradation — Composite structure map
O-Glycan biosynthesis — Monosaccharide codes
Chondroitin sulfate biosynthesis
Heparan sulfate biosynthesis — Glycosyltransferases
Keratan sulfate biosynthesis — Proteoglycans
Glycosaminoglycan degradation
Lipopolysaccharide biosynthesis
Peptidoglycan biosynthesis
Glycosylphosphatidylinositol(GPI)-anchor biosynthesis
Glycosphingolipid biosynthesis - lactoseries

Biosynthesis of ansamycins
Biosynthesis of type II polyketide backbone
Biosynthesis of type II polyketide products
Polyketide sugar unit biosynthesis
Nonribosomal peptide structures
Biosynthesis of siderophore group nonribosomal peptides
Biosynthesis of vancomycin group antibiotics

1.9 Metabolism of Cofactors and Vitamins

Thiamine metabolism
Riboflavin metabolism
Vitamin B6 metabolism
Nicotinate and nicotinamide metabolism [UMBBD link]
Pantothenate and CoA biosynthesis
Biotin metabolism
Lipoic acid metabolism
Folate biosynthesis
One carbon pool by folate
Retinol metabolism
Porphyrin and chlorophyll metabolism
Ubiquinone biosynthesis

1.10 Biosynthesis of Secondary Metabolites

Terpenoid biosynthesis — Phytochemical compounds
Diterpenoid biosynthesis
Monoterpenoid biosynthesis
Limonene and pinene degradation
Indole and ipecac alkaloid biosynthesis
Carotenoid biosynthesis
Brassinosteroid biosynthesis
Phenylpropanoid biosynthesis
Flavonoid biosynthesis
Anthocyanin biosynthesis *New!*

Figure 5.12 KEGG Classifiers *(Continues)*

2. Genetic Information Processing

2.1 Transcription

RNA biosynthesis (illustration)	Transcription factors
RNA polymerase	
Basal transcription factors	

2.2 Translation

Protein biosynthesis (illustration)	Translation factors
Ribosome	Ribosome
Aminoacyl-tRNA biosynthesis	

2.3 Folding, Sorting and Degradation

Protein export	
Type II secretion system	
Type III secretion system	
Type IV secretion system	
SNARE interactions in vesicular transport	SNAREs
Regulation of autophagy	
Ubiquitin mediated proteolysis	
Proteasome	

2.4 Replication and Repair

DNA polymerase

3. Environmental Information Processing

3.1 Membrane Transport

ABC transporters	Transporters (+diseases)
Phosphotransferase system (PTS)	

3. Environmental Information Processing

3.1 Membrane Transport

ABC transporters Transporters (+diseases)

Phosphotransferase system (PTS)

3.2 Signal Transduction

Two-component system	Two-component system
MAPK signaling pathway	
ErbB signaling pathway	
Wnt signaling pathway	
Notch signaling pathway	
Hedgehog signaling pathway	
TGF-beta signaling pathway	
VEGF signaling pathway	
Jak-STAT signaling pathway	
Calcium signaling pathway	
Phosphatidylinositol signaling system	
mTOR signaling pathway	

3.3 Signaling Molecules and Interaction

Neuroactive ligand-receptor interaction	Receptors and channels (+diseases)
Cytokine-cytokine receptor interaction	Cytokines
ECM-receptor interaction	Cell adhesion molecules (CAMs)
Cell adhesion molecules (CAMs)	CAM ligands
	CD molecules
	GTP-binding proteins

Figure 5.12 KEGG Classifiers *(Continues)*

4. Cellular Processes

4.1 Cell Motility
Bacterial chemotaxis — Bacterial motility proteins
Flagellar assembly
Regulation of actin cytoskeleton

4.2 Cell Growth and Death
Cell cycle
Cell cycle - yeast
Apoptosis
p53 signaling pathway

4.3 Cell Communication
Focal adhesion
Adherens junction
Tight junction
Gap junction

4.4 Endocrine System
Insulin signaling pathway
Adipocytokine signaling pathway
PPAR signaling pathway
GnRH signaling pathway
Progesterone-mediated oocyte maturation
Melanogenesis
Renin-angiotensin system

4.5 Immune System
Hematopoietic cell lineage
Complement and coagulation cascades
Toll-like receptor signaling pathway
Natural killer cell mediated cytotoxicity

5.1 Cancers
Colorectal cancer
Pancreatic cancer
Glioma
Thyroid cancer
Acute myeloid leukemia
Chronic myeloid leukemia
Basal cell carcinoma
Melanoma
Renal cell carcinoma
Bladder cancer
Prostate cancer
Endometrial cancer
Small cell lung cancer
Non-small cell lung cancer

5.2 Neurodegenerative Diseases
Alzheimer's disease
Parkinson's disease
Amyotrophic lateral sclerosis (ALS)
Huntington's disease
Dentatorubropallidoluysian atrophy (DRPLA)
Prion disease

5.3 Metabolic Disorders
Type I diabetes mellitus
Type II diabetes mellitus
Maturity onset diabetes of the young

5.4 Infectious Diseases
Cholera — Infectious diseases
Epithelial cell signaling in Helicobacter pylori infection

Figure 5.12 KEGG Classifiers *(Continues)*

6. Drug Development

6.1 Chronology: Antibiotics
Penicillins
Cephalosporins - parenteral agents
Cephalosporins - oral agents
Aminoglycosides - Streptomyces
Aminoglycosides - Micromonospora
Tetracyclines
Macrolides and ketolides
Quinolones
Rifamycins

6.2 Chronology: Nervous System Agents
Hypnotics
Anxiolytics
Anticonvulsants
Local analgesics
Opioid analgesics
Antipsychotics
Antipsychotics - phenothiazines
Antipsychotics - butyrophenones
Antidepressants

6.3 Chronology: Other Drugs
Azoles
Quinolines
Statins
Sulfonamide derivatives - sulfa drugs
Sulfonamide derivatives - diuretics
Sulfonamide derivatives - hypoglycemic agents

KEGG DRUG

Therapeutic category of drugs
+ Target
+ Map number

Figure 5.12 KEGG Classifiers

In KEGG database (Figure 5.13) the metabolic pathways for any organism can be retrieved which are specified in the databases. The pathway can be accessed directly from the home page or type "glycolysis" and go.

KEGG PATHWAY Database

Wiring diagrams of molecular interactions, reactions, and relations

KEGG2 PATHWAY BRITE GENES SSDB LIGAND DRUG DBGET

Pathway Maps

KEGG PATHWAY is a collection of manually drawn pathway maps representing our knowledge on the molecular interaction and reaction networks for:

1. **Metabolism**
   Carbohydrate Energy Lipid Nucleotide Amino acid Other amino acid
   Glycan PK/NRP Cofactor/vitamin Secondary metabolite Xenobiotics
2. **Genetic Information Processing**
3. **Environmental Information Processing**
4. **Cellular Processes**
5. **Human Diseases**

and also on the structure relationships (KEGG drug structure maps) in:

6. **Drug Development**

Search PATHWAY for glycolysis Go Clear

bfind mode bget mode

Figure 5.13 Home Page of KEGG Database

hdu00010
Glycolysis / Gluconeogenesis - Haemophilus ducreyi
hha00010
Glycolysis / Gluconeogenesis - Halorhodospira halophila
hhe00010
Glycolysis / Gluconeogenesis - Helicobacter hepaticus
hin00010
Glycolysis / Gluconeogenesis - Haemophilus influenzae (serotype d)
hip00010
Glycolysis / Gluconeogenesis - Haemophilus influenzae PittEE
hiq00010
Glycolysis / Gluconeogenesis - Haemophilus influenzae PittGG
hit00010
Glycolysis / Gluconeogenesis - Haemophilus influenzae 86-028NP (nontypeable)
hma00010
Glycolysis / Gluconeogenesis - Haloarcula marismortui
hne00010
Glycolysis / Gluconeogenesis - Hyphomonas neptunium
hpa00010
Glycolysis / Gluconeogenesis - Helicobacter pylori HPAG1
hpj00010
Glycolysis / Gluconeogenesis - Helicobacter pylori J99
hpy00010
Glycolysis / Gluconeogenesis - Helicobacter pylori 26695
hsa00010
Glycolysis / Gluconeogenesis - Homo sapiens (human)
hso00010
Glycolysis / Gluconeogenesis - Haemophilus somnus
hwa00010
Glycolysis / Gluconeogenesis - Haloquadratum walsbyi
iho00010
Glycolysis / Gluconeogenesis - Ignicoccus hospitalis
ilo00010
Glycolysis / Gluconeogenesis - Idiomarina loihiensis
jan00010
Glycolysis / Gluconeogenesis - Jannaschia sp. CCS1
ko00010
Glycolysis / Gluconeogenesis - Reference pathway (KO)

Figure 5.14 Result for Pathways

The result (Figure 5.14) obtained from result page can be analysed and the required pathway can be taken for the study. In this the organism name, the KEGG Entry for the organism and the number of the pathway are specified. For example

hsa	00010
Homo sapiens	Number for glycolysis

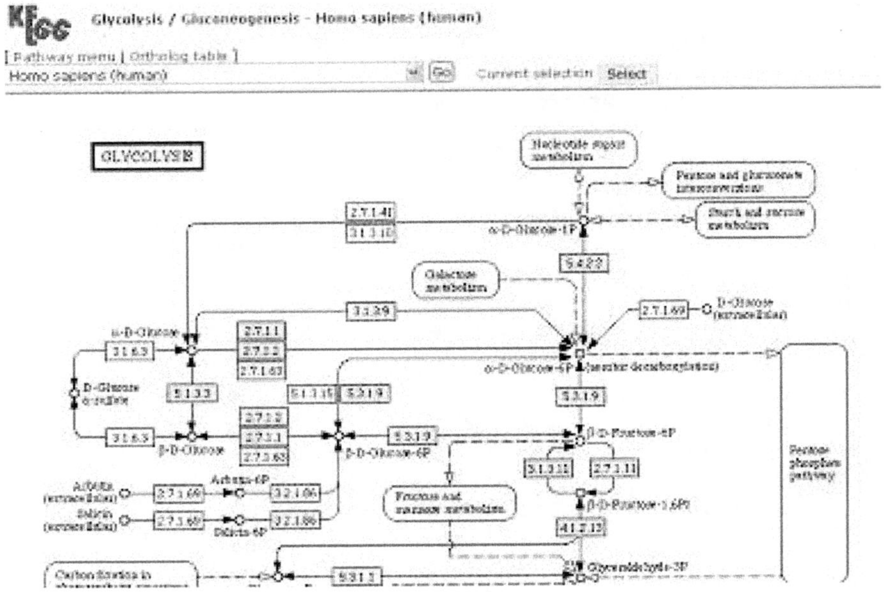

Figure 5.15 Pathway for Glycolysis

Figure 5.15 interprets the pathway for glycolysis for humans. The shaded box determines the expression of the enzyme in the particular organism.

For the single pathway (Figure 5.16) the reference pathways are given with respect to organisms.

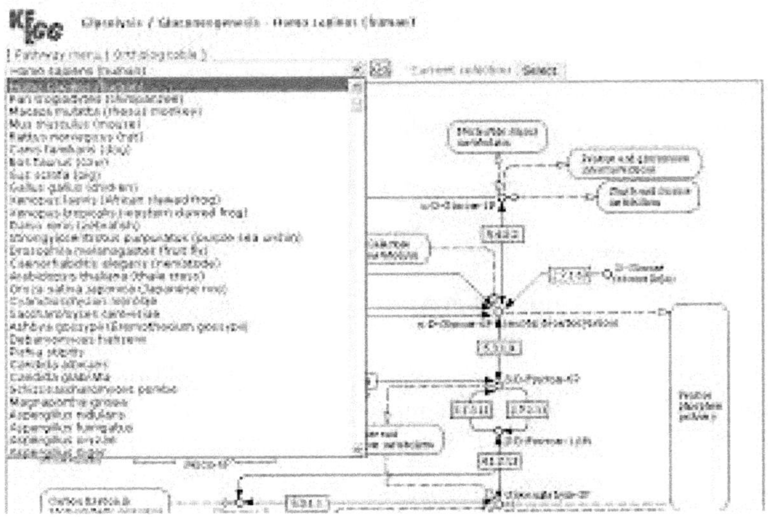

Figure 5.16 Reference Menu for Pathways

Figure 5.17, is used to study the KEGG Pathways, the maps provided are used. The arrow keys, symbols and colours are used to interpret the results. The Ortholog table in Figure 5.18 shows the details of glycolysis pathway.

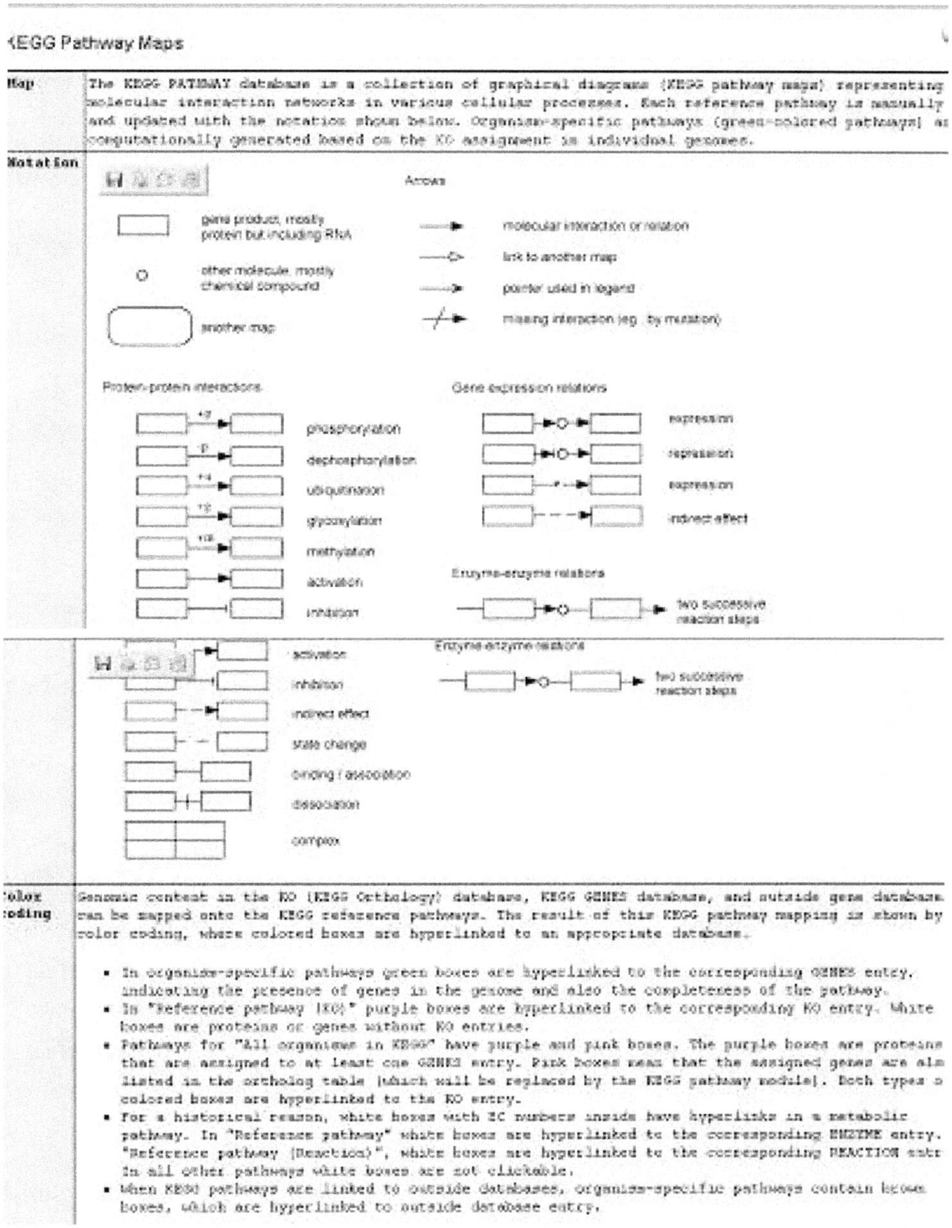

Figure 5.17 KEGG Pathway Maps

5.6.4 Ortholog Table

Glycolysis / Gluconeogenesis

See also: Pyruvate/oxoglutarate oxidoreductases

Current selection Select

Organism	E2.7.1.11	E4.1.2.13		E5.3.1.1	E1.2.1.12	E2.7.2.3	E5.4.2.1	E4.2.1.11
	Phospho-fructokinase	Aldolase		Triose-phosphate isomerase	Glyceraldehyde-3P dehydrogenase	Phosphoglycerate kinase	Phosphoglycerate mutase	Enolase
		class II	class I					
eco [P | G | T]	b3916(pfkA) b1723(pfkB)	b2925(fba)	b2097 (dhnA)	b3919(tpiA)	b2927(epd) b1779(gapA) b1417(gapC_1) b1416(gapC_2)	b2926(pgk)	b0755(gpmA) b4395(gpmB) b3612(yibO)	b2779(eno)
sty [P | G | T]	STY1785 STY3809	STY3226	STY2370	STY3789	STY3228(1.2.1.-) STY1825	STY3227	STY0004 STY4093 STY4932	STY3081
ype [P | G | T]	YPO0078	YPO0928 YPO3960		YPO0085	YPO0922 (1.2.1.-) YPO2157	YPO0921	YPO0064 YPO0455 YPO1533	YPO3376
sfl [P | G | T]	SF3994 SF1507	SF2910	SF2159	SF3997	SF2912(1.2.1.-) SF1444 SF1795 SF1796	SF2911	SF0549 SF3651 SF4427	SF2794
buc [P | G | T]	BU305	BU451		BU307	BU298	BU450	BU304	BU417
wbr [P | G | T]		WGLp306		WGLp274	WGLp108	WGLp305	WGLp218	WGLp353
hin [P | G | T]	HI0982	HI0524		HI0678	HI0001	HI0525	HI0757	HI0988
pmu [P | G | T]	PM0069	PM1861 PM1373		PM1311 PM1640	PM0224	PM1860	PM1506 PM0634	PM1871
xfa [P | G | T]	XF0274		XF0826	XF0303	XF0457	XF0823	XF1893	XF1291

Figure 5.18 KEGG Ortholog Table

5.7 MALAYSIAN GENOMIC RESOURCE CENTRE (MGRC)

MGRC (Figure 5.19) has provided an easy access to ultra-fast sequence analysis tools for researchers and scientists who require high-throughput applications and new database platforms. By empowering life scientists in this aspect, the critical bottlenecks in research areas such as drug discovery, comparative genomics, microarray chip design and personalized medicine can be alleviated or overcome. This is provided in conjunction with Malaysian Biotechnology Corporation an Agency under the Ministry of Science, Technology and Innovation and is wholly-owned by the Ministry of Finance Incorporated. BiotechCorp is governed by the Biotechnology Implementation Council and advised by the Biotechnology International Advisory Panel. The services commenced on 1 July 2005. The online Bioinformatics services platform is managed and operated by Malaysian Genomics Resource Centre, which was incorporated on 18 May 2004. The centre for worldwide operations is currently based in Kuala Lumpur, Malaysia. MGRC centre is based upon a cluster of HP Itanium servers.

MGRC uses SynaSuite applications created by Synamatix, a pioneer bioinformatics technology developer. These applications have been specially adapted for online use in the form of SynaSITE, Synamatix's new ASP product-offering. The company was chosen as MGRC's key software partner because of the extensibility of SynaSuite and the analytical performance and storage efficiencies of the SynaBASE database platform. The integrated suite of applications available at MGRC is built on the SynaBASE database platform, a novel structured network database, which efficiently stores genetic data based on patterns and their interrelationships. Through utilizing this novel pattern-based approach in storing data, limitations in analysis speeds and scalability commonly associated with flat files and relational database platforms are overcome. MGRC's current set of analysis tools enables scientists to conduct a wide array of research activities ranging from high-throughput genomics searches and mining, identification of novel data hits and motifs through to more exhaustive whole-genome comparisons and microarray chip design.

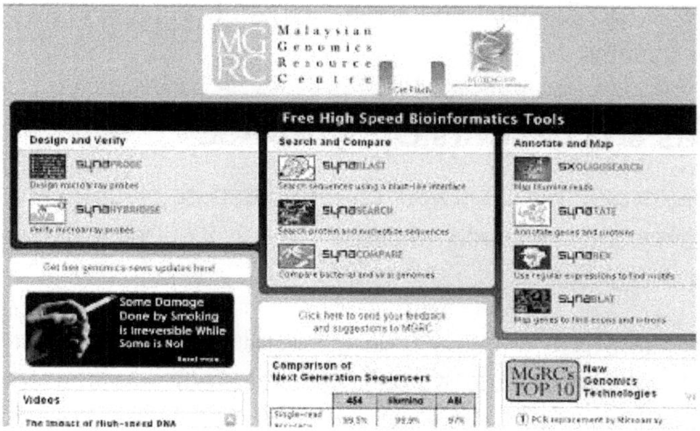

Figure 5.19 MGRC Home Page

5.7.1 Applications and Descriptions

SynaBLAST SynaBLAST is a sequence similarity search tool specifically designed to search sequences stored in SynaBASE.

SynaSearch SynaSearch is a powerful search tool that matches query sequences against all existing data stored in SynaBASE and is capable of providing results in milliseconds.

SynaMine or SynaTate SynaMine is an application for identification of potential functional domains and other regions of interest. SynaTate enables ultra-fast analysis of patterns within sequences to quantify frequency and significance, as well as mapping of key words and annotations to patterns.

SynaCompare SynaCompare is a high-speed and exhaustive genome-to-genome comparison tool. SynaCompare enables in-depth comparisons between a query and target sequence, chromosome or whole genome stored in SynaBASE. SynaCompare is capable of conducting analyses in the order of 100 Mb vs 100 Mb within minutes.

SynaRex This utilises the power of regular expressions to rapidly identify sequences that share a common biological motif or pattern.

SynaProbe This identifies highly specific and sensitive probes suitable for the rapid design of whole genome microarrays. SynaProbe is capable of designing whole human genome arrays in under six hours.

SynaHybridise SynaHybridise is a sequence verification tool that searches microarray probes to confirm specificity and uniqueness to validate the probes.

SynaBlat or SynaMap This application is designed for accurately mapping transcript data to genome sequences. It provides fast and accurate identification of exon positions and splice sites within a particular genome.

5.7.2 Features and Benefits of SynaBASE

SynaBASE is a unique structured network pattern database platform enabling ultra-fast genomic analysis and can accelerate research analyses 100- or 1000-fold compared to applications built upon conventional databases.

SynaBASE database platform eliminates data redundancy as each unique pattern is stored only once, the more data that is stored into SynaBASE the more efficient it becomes. Incremental database size begins to flatten out as more data is added. One can easily analyse 10s or 100s of genomes on a single integrated platform.

Scalable performance Access speed scales at Log N Base A, where N is the database size and A is the alphabet size, which analyse very large datasets.

Significance A unique, mathematical approach to annotate genome-based data upon predictability of patterns in SynaBASE.Use of a single unbiased score to identify a wide variety of functional domains, SNPs, active sites, promoters and genes. Quickly annotates poorly annotated genomes or confirm existing annotations.

Novel data structure and algorithms to perform exhaustive and accurate pattern identification Each position from 1 to n, for each unique pattern construct is stored in SynaBASE. One can identify novel domains and matches due to the 100% exhaustive nature of SynaBASE.

SynaBASE is an extensible platform for a wide array of life-science applications. Users have the choice of using ready-made high-throughput SynaSUITE tools, integrating existing applications to SynaBASE platform or simply building completely new applications to suit the specific research requirements.

Hardware independence and future proof of performance scalability Acceleration in performance analysis and enhanced scalability are not dependent on hardware. One can manage and analyse progressively more complex volumes of biological data without investing in costly and high-maintenance server farms and clusters.

Hosted services without cost implications We can store custom or proprietary data in a secure environment without the commonly associated implications of hardware and personnel costs.

Bespoke application development services We can match individual researcher's needs and specifications. Custom application development or customization services are available to enable rapid, low-cost optimization of existing tools or development of completely new modules specific to the individual research activity.

Integrated applications to conduct exhaustive and extensive research efficiently The user can easily monitor workflows across all applications, save time and easily transfer data outputs from one application to the other. Moving from comparative genomics to microarray probe design, the evaluation and hybridization of the result, enables us to identify the functional motifs.

5.7.3 SynaBLAST

SynaBLAST is a sequence similarity search tool specifically designed to search sequences stored in SynaBASE. SynaBLAST is the same as that of NCBI BLAST. SynaBLAST allows protein–protein, nucleotide–protein and nucleotide–nucleotide searches. The Figures 5.20 through 5.23 depict the series of steps involved in using SynaBLAST.

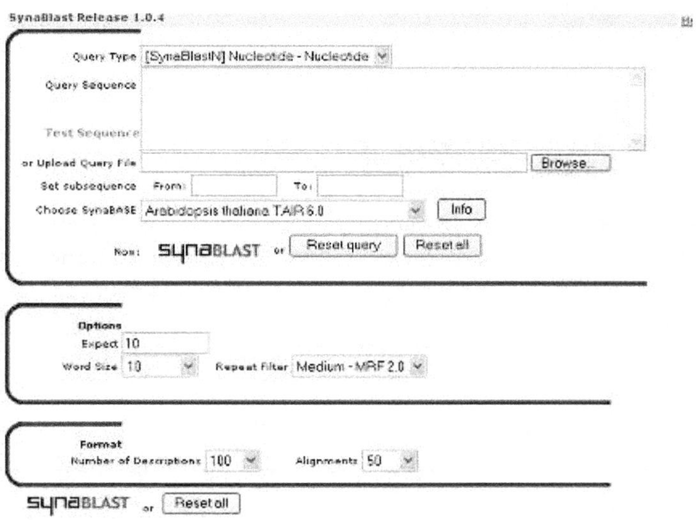

Figure 5.20 Home Page of SynaBLAST

Step 1 Specify a query type and submit a sequence

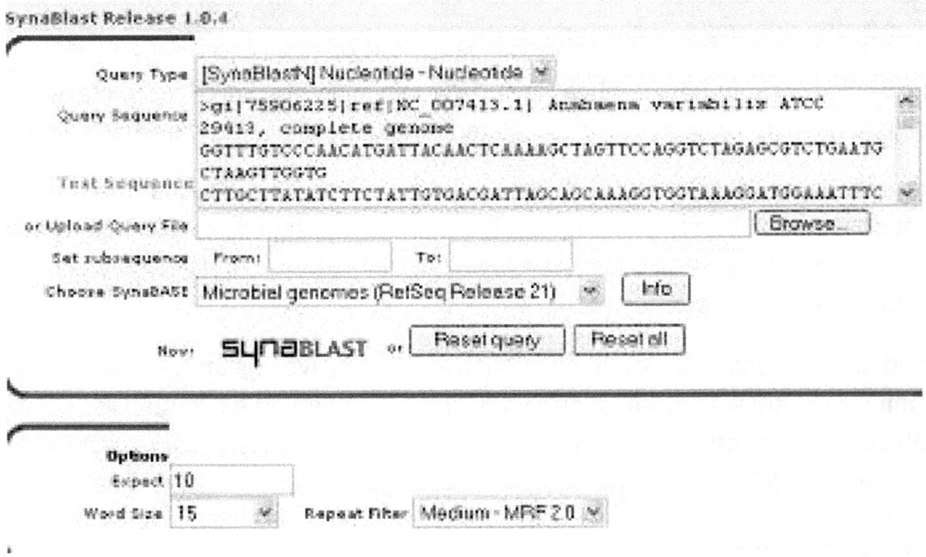

Figure 5.21 Query Submission Page in MGRC

Step 2 Specify a target sequence

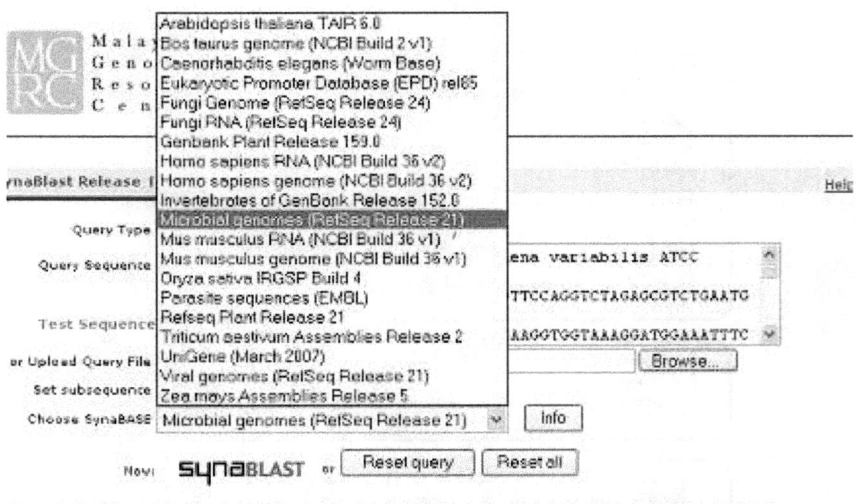

Figure 5.22 Choosing the Database in SynaBLAST

Step 3 Standard query option

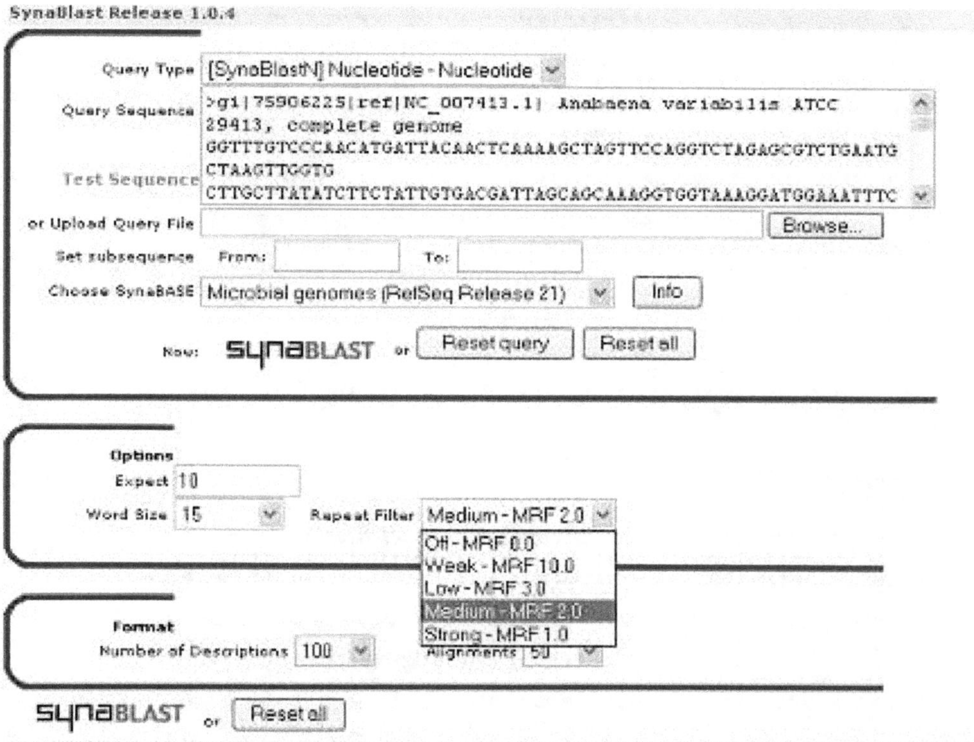

Figure 5.23 Choosing of a Filter

Step 4 Result and interpretation.

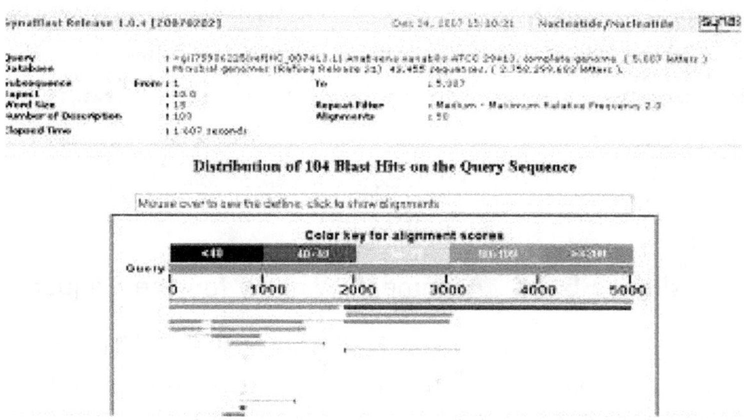

Figure 5.24 Alignment Window for the Sequence *(Continues)*

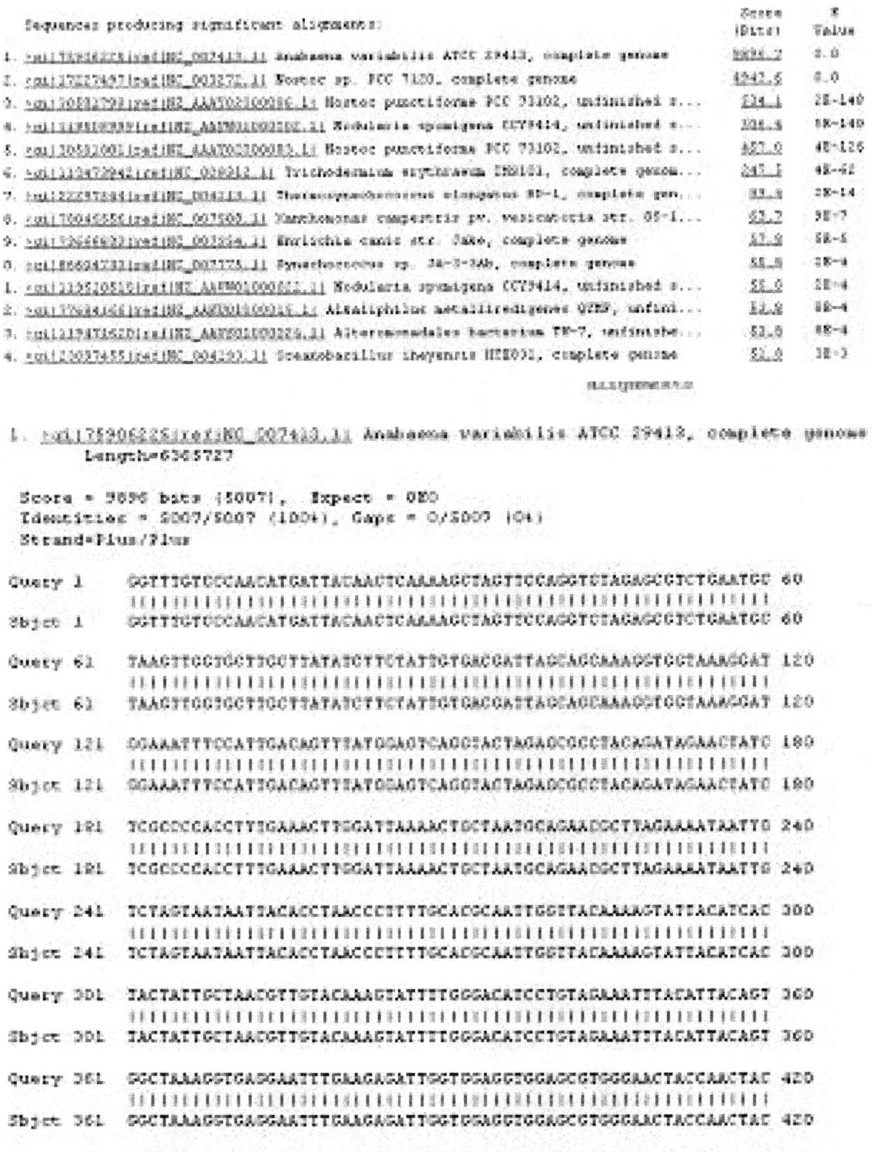

Figure 5.24 Alignment Window for the Sequence

The results and interpretation (Figure 5.24) shows the BLAST hit of the Query sequence based on the color key score. The score and the e-value are provided on the alignment. The alignment of the query and the sequence is given to find how much they match and how much they mismatch.

5.7.4 SynaSearch

SynaSearch is a powerful search tool that matches query sequences against all existing data stored in SynaBASE and is capable of providing results in milliseconds. The Figures 5.25 through 5.28 represent the methodology of using SynaSearch.

Step 1 Query screen

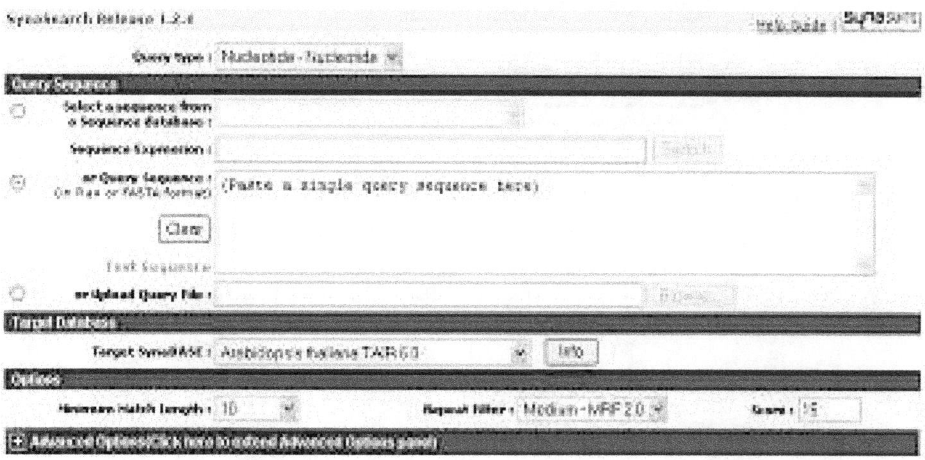

Figure 5.25 Home Page of SynaSearch

Step 2 Submit and specify target sequence

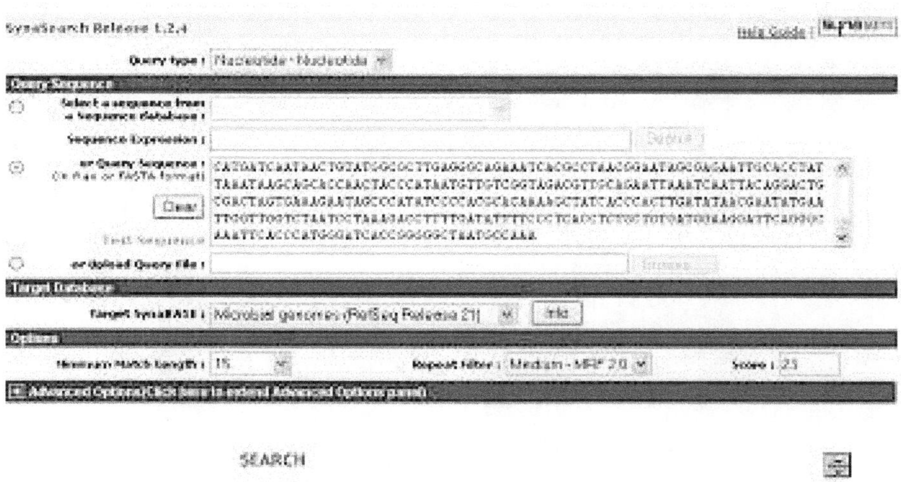

Figure 5.26 Input Page for SynaSearch

Step 3 Result and interpretation

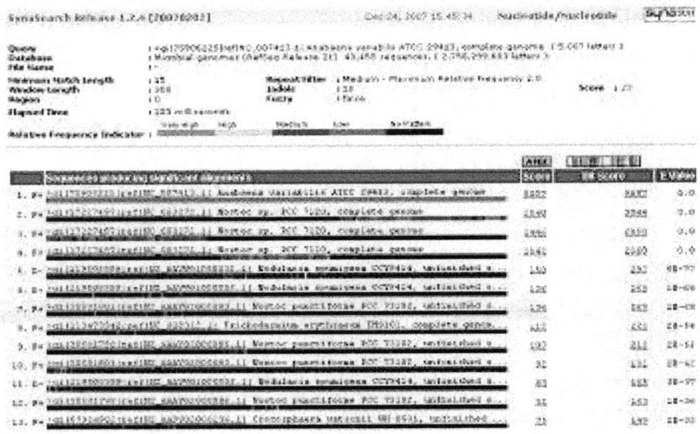

Figure 5.27 Result for SynaSearch

To view the alignment click score and bit score link for graphical view.

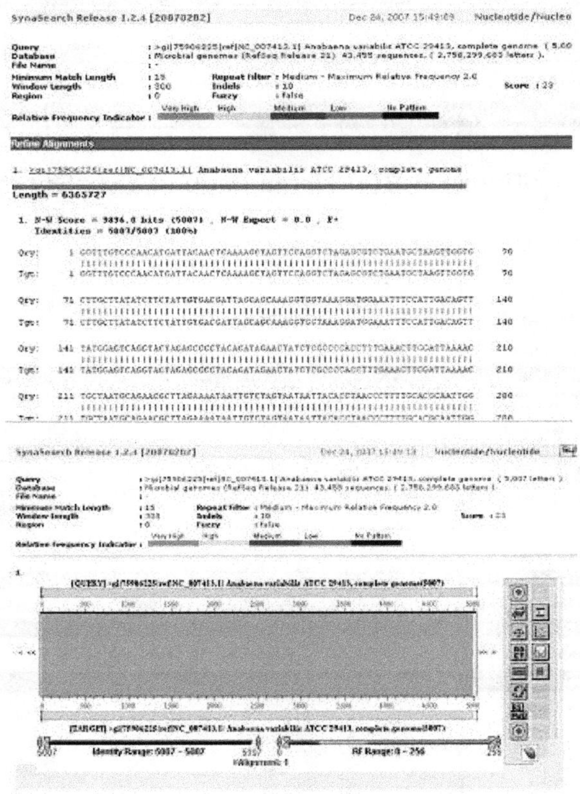

Figure 5.28 Result Analysis

The result of SynaSearch (Figure 5.27 and Figure 5.28) shows the exact matter of the query in the existing data stored in SynaBASE.

5.7.5 SynaMine

SynaMine is an application for identification of potential functional domains and other regions of interest. SynaTate enables ultra-fast analysis of patterns within sequences to quantify frequency and significance, as well as mapping of key words and annotations to patterns.

Step 1 In the Query page of SynaTate (Figure 5.29), the sequences are submitted as nucleotides or proteins. In SynaTate there are two options—TATE and SEARCH. Initially we go for the TATE option. The graphical representation (Figures 5.30 and 5.31) of the results are produced for the analysis.

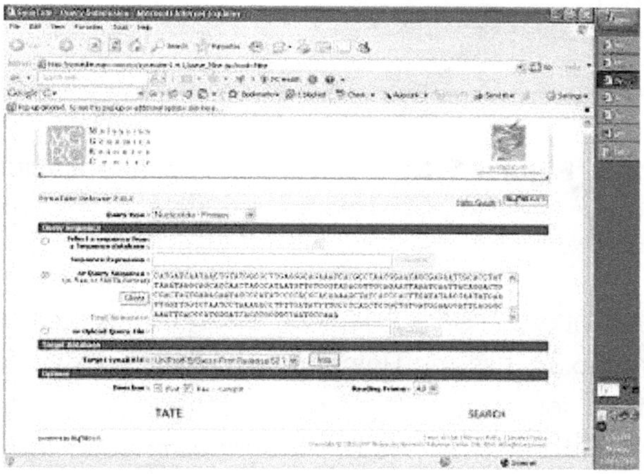

Figure 5.29 Home Page of SynaTate

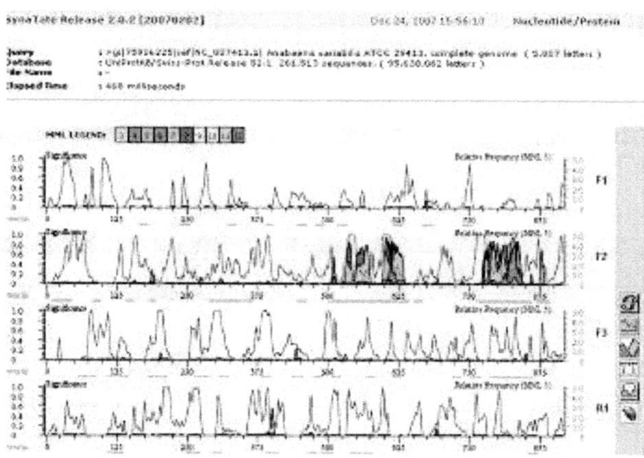

Figure 5.30 Result Page for SynaTate

Result The first output interface is a dynamic graph that displays the conserved sequence motifs along the sequence length from the analysis of patterns. It shows the plot in terms of SIGNIFICANCE along a Query sequence length.

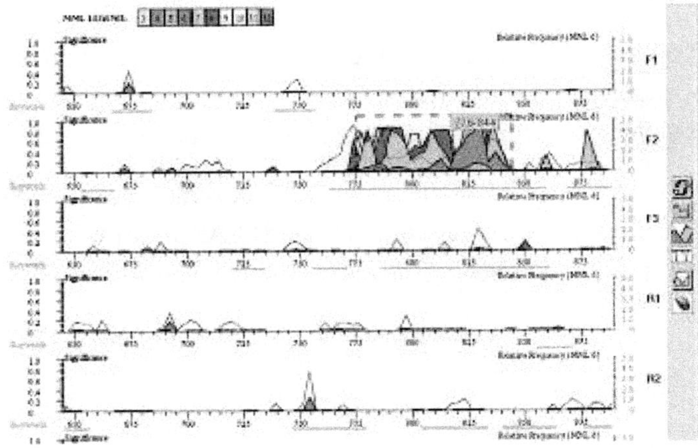

Figure 5.31 Graphical representation

5.7.6 SynaRex

Utilizes the power of regular expressions to rapidly identify sequences that share a common biological motif or pattern. Figure 5.32, has the Home Page of SynaRex, where the nucleotide or protein sequence are submitted for analysis. The result from these Query sequences will be as in Figure 5.33 which shows the area of the regular expression along with the expression table.

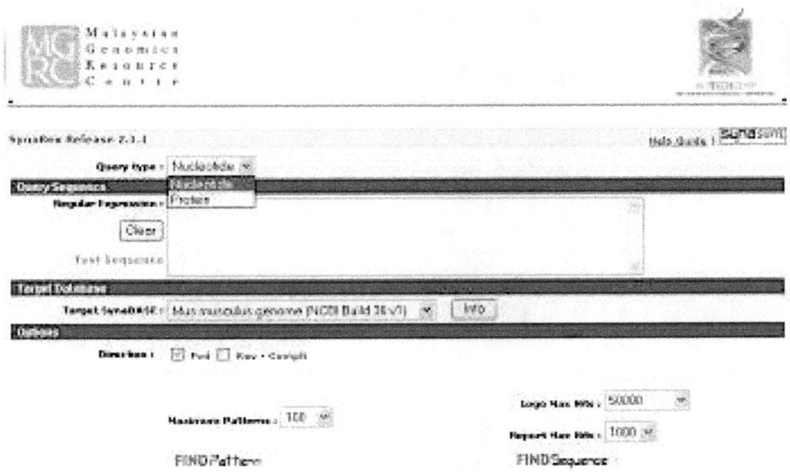

Figure 5.32 Home Page of SynaRex

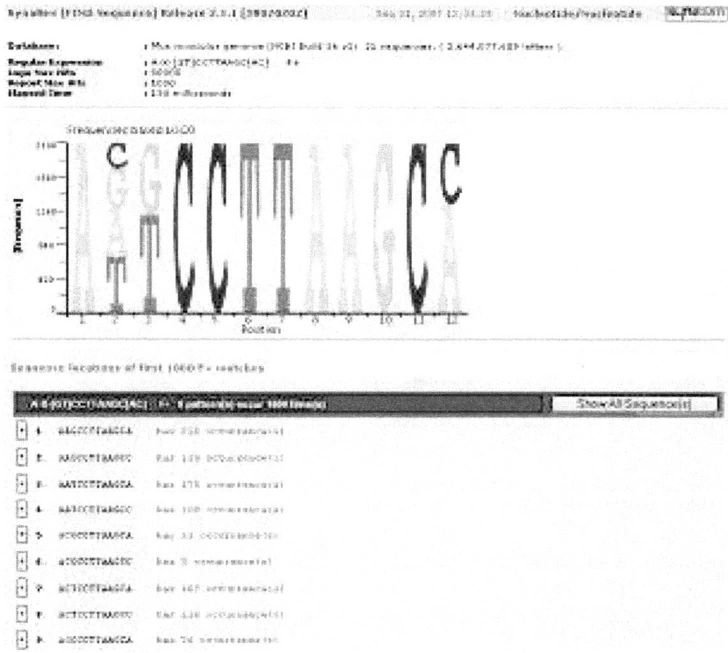

Figure 5.33 Result Page of SynaRex

5.7.7 SynaProbe

Identifies highly specific and sensitive probes suitable for the rapid design of whole genome microarrays. SynaProbe is capable of designing whole human genome arrays in under six hours.

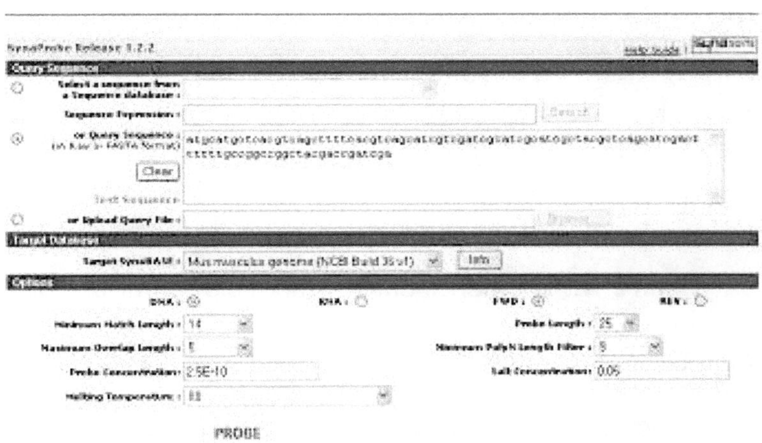

Figure 5.34 Home Page of SynaProbe

The SynaProbe is used to identify the probe region in the sequence. The Query sequence is input in the Home Page (Figure 5.34) and then the submit button is pressed. The probes will be designed for the Query sequence by the help of the graphical data (Figure 5.35) and the formation of the bar in the base line with a no. denoting the number of probes present. The Figure 5.36 shows the sequence present in the probes along with its score.

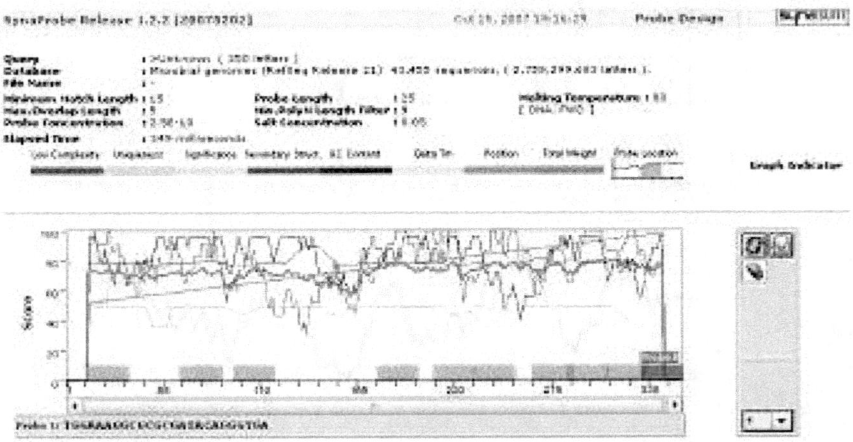

Figure 5.35 Result Page and identification of probe

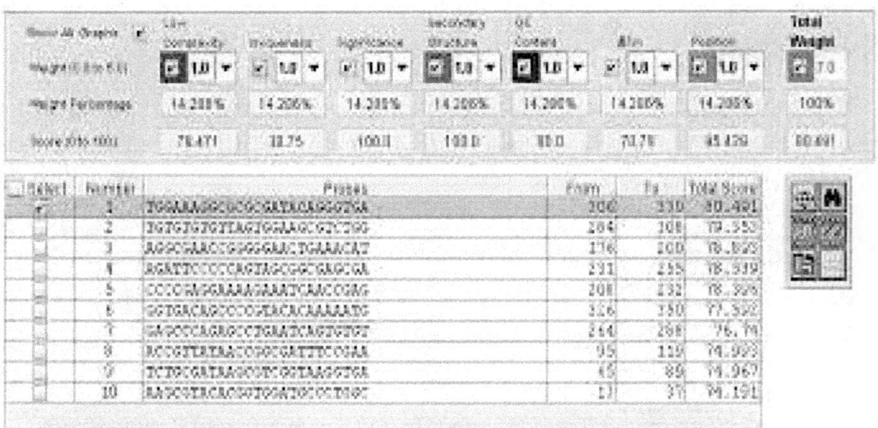

Figure 5.36 Identification and isolation of a probe

5.7.8 SynaBLAT

SynaBlat application designed for accurately mapping transcript data to genome sequences. It provides fast and accurate identification of exon positions and splice sites within a particular genome.

Figures 5.37 and 5.38, clearly denote the mapping transcript region of a genome with the start and stop end.

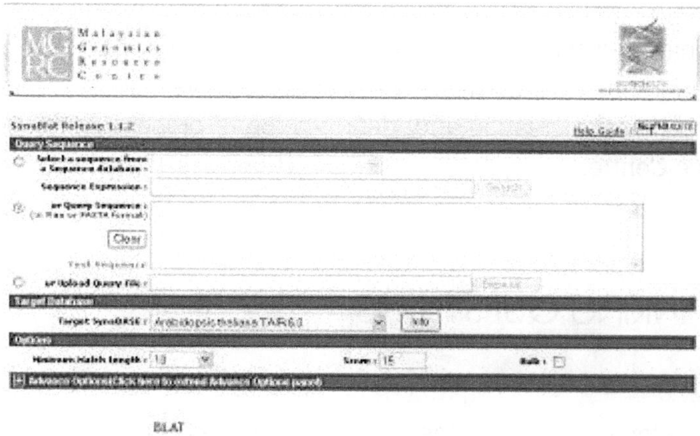

Figure 5.37 Home Page for SynaBlat

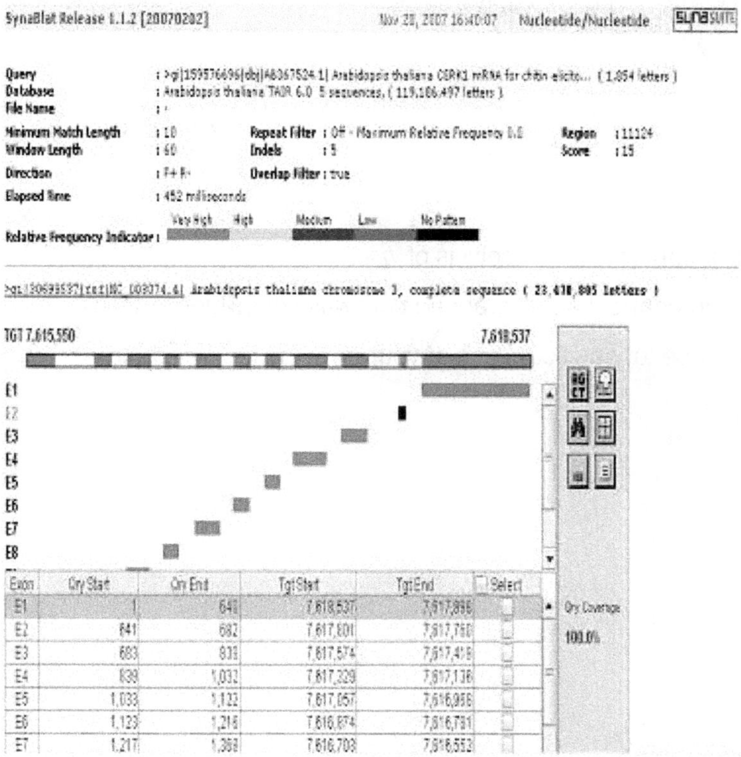

Figure 5.38 Result Page for SynaBlat

SUMMARY

In this chapter we have discussed in detail about the concepts of Metabolomics, a field related to metabolism and its regulation. Many pathway databases were studied and KEGG was studied in detail. Along with this the creation and identification of probes, motifs were done in Malaysian Genomic Resource Centre.

CRITICAL THINKING QUESTIONS

1. Define metabolomics.What are the possible outcomes in metabolomics?
2. How many pathway databases are present and how many are specific databases.
3. What is a probe and where is it used?

SELF-LEARNING EXERCISES

1. Retrieve the pathway for glycolysis of *hsa*.
2. Compare and contrast the metabolic pathways of influenza virus.
3. Prepare a probe for any sequence of your interest.

GLOSSARY

Bioinformatics A field of science in which biology, computer science, and information technology merge into a single discipline to analyse biological information using computers and statistical techniques.

BLAST A set of programs used to perform fast similarity searches. Nucleotide sequences can be compared with nucleotide sequences in a database using BLASTN, for example. Complex statistics are applied to judge the significance of each match. Reported sequences may be homologous to, or related to the query sequence. The BLASTP program is used to search a protein database for a match against a query protein sequence. There are several other flavours of BLAST.

Cell The basic unit of all living organisms.

Codon A sequence of three nucleotides (in a DNA or mRNA), that encodes a specific amino acid to be incorporated into a protein.

Comparative genomics Study of the relationship of genome structure and function across different biological species.

Complementary DNA DNA copies of the mRNA expressed in a specified tissue. cDNA sequencing has the advantage of only representing expressed genes. Since only ~3% of the vast quantity of DNA in the human genome are coding sequences, cDNA sequencing is particularly useful in certain situations.

DNA sequencing The experimental process of determining the nucleotide sequence of a region of DNA. This is done by labelling each nucleotide (A, C, G or T) with either a radioactive or fluorescent marker which identifies it. There are several methods of applying this technology, each with their advantages and disadvantages. High-throughput laboratories frequently use automated sequencers, which are capable of rapidly reading large numbers of templates. Sometimes, the sequences may be generated more quickly than they can be characterized.

Dynamic programming A large class of programming algorithms that are based on breaking a large problem down to incremental steps so that, at any given stage, optimal solutions are always known.

Eukaryote An organism whose genetic information is, in contrast to prokaryotes (such as bacteria), contained in a separate cellular compartment: the nucleus. Besides algae, fungi and protozoa, all multicellular, cell-differentiating organisms including plants and animals are eukaryotes. In addition to their nuclear genome, all eukaryotic cells contain small additional, extranuclear genomes, which are contained in mitochondria (in all eukaryotes) and in plastids (only in eukaryotes that can perform photosynthesis).

FASTA A computer program based on the method of W. Pearson and D. Lipman to search for similarities between one sequence (the query) and any group of sequences (the database).

Genbank A large database and data repository of nucleic acid and protein sequences at the National Library of Medicine (USA). http://www.ncbi.nlm.nih.gov/

Genes The functional and physical unit of heredity passed from parent to offspring. Genes are pieces of DNA, and most genes contain the information for making a specific protein.

Genetic material Loss of genetic diversity between and within populations of the same species over time; or reduction of the genetic basis of a species due to human intervention, environmental changes, etc.

Genomics The study of genes and their function.

Metabolomics The study and identification of metabolic products of cells that may be used to distinguish a disease state from a healthy state.

NCBI National Center for Biotechnology Information (USA); created by the United States Congress in 1988, this organization was set up to develop information systems to support the biological research community.

Nucleotide The basic structural unit of nucleic acids (DNA or RNA).

'Omics A broad discipline of science and engineering for analysing the interactions of biological information.

Organelles A similar set of little organs, within cells.

PDB Brookhaven Protein Data Bank; a database and format of files, which describe the 3D structure of a protein or nucleic acid, as determined by X-ray crystallography or nuclear magnetic resonance (NMR) imaging. The molecules described by the files are usually viewed locally by dedicated software, but can sometimes be visualized on the World Wide Web.

PIR A database of translated GenBank nucleotide sequences. PIR is a redundant (*See* Redundancy) protein sequence database. The database is divided into four categories:

1. PIR1 — Classified and annotated.
2. PIR2 — Annotated.
3. PIR3 — Unverified.
4. PIR4 — Unencoded or untranslated.

Prokaryotes A unicellular organism having cells lacking membrane-bound nuclei; bacteria are the prime example but also included are blue-green algae and actinomycetes and mycoplasma.

Protein A large group of nitrogenous organic compounds that are essential constituents of living cells and consist of polymers of amino acids. They are essential in the diet of animals for growth and for repair of tissues and can be obtained from meat, eggs, milk and legumes.

Redundancy The presence of more than one identical item represents redundancy. In Bioinformatics, the term is used with reference to the sequences in a sequence database. If a database is described as being redundant, more than one identical (redundant) sequence may be found. If the database is said to be non-redundant (nr), the database managers have attempted to reduce the redundancy. The term is ambiguous with reference to genetics, and as such, the degree of non-redundancy varies according to the database manager's interpretation of the term. One can argue whether or not two alleles of a locus define the limit of redundancy, or whether the same locus in different, closely related organisms constitutes redundancy. Non-redundant databases are, in some ways, superior, but are less complete. These factors should be taken into consideration when selecting a database to search.

Ribosomes Minute round particles composed of RNA and protein that are found in the cytoplasm of living cells and serve as the sites of assembly for polypeptides encoded by messenger RNA. They attach to mRNA and move down it, one codon at a time, and then stop until tRNA brings the required amino acid; when a ribosome reaches a stop codon, it falls apart and releases the completed protein molecule for use by the cell "the ribosome is the site of protein synthesis".

Swiss-Prot A non-redundant (*See* Redundancy) protein sequence database, thoroughly annotated and cross-referenced. A subdivision is TrEMBL.

Systems Biology The study of biology that focuses on complex systems in life.

Transcription The process by which DNA is used as a template for the synthesis of an RNA molecule.

Transcriptome A collection of all the gene transcripts present in a cell.

Transcriptomics A global way of looking at gene-expression patterns. This can involve measurements of thousands of genes simultaneously with microarrays or measurements of small numbers of genes that are facilitated by global sequence information from EST or genome-sequencing projects.

Translation The process of protein synthesis from an mRNA template, occurring at the ribosome.

UniGene Database of unique human genes, at NCBI. Entries are selected by near identical presence in GenBank and dbEST databases. The clusters of sequences produced are considered to represent a single gene.

WOL (Web Ontology Language) An XML format that adds meaning to the content of web files so that computer programs can read/use web-based resources.

REFERENCES

Arthur M. Lesk. (2002). *Introduction to Bioinformatics.* Oxford University Press, London.

Bastiaan J.H. Jansen and Joost Schalkwijk. (2003). "Transcriptomics and proteomics of human skin." *Brief. Funct. Genomics and Proteomics.* 1(4): 326–341.

Cynthia Gibas and Per Janbeck. (2001). *Developing Bioinformatics Computer Skills.* O'Reilly Media, Inc., USA.

Donald Voet and Judith G. Voet. (1995). *Biochemistry,* 2nd edn. John Wiley, USA.

Julian O.D. Coleman, Richard P. Haslem and Andrew L. Downie. (2005). "Transcriptomics and Proteomics: Tools for Optimising Phytoremediation Activities." *Z. Naturforsch.* 60: 544–8.

Kanetkar, Y. (1991). *Let Us C,* 4th edn. BPB Publications, New Delhi.

Kanetkar, Y. (2001). *Understanding Pointers in C,* 3rd edn. BPB Publications, New Delhi.

Lehninger Albert, L., David L. Nelson and Michael M. Cox. (1993). *Principles of Biochemistry,* 2nd edn. Worth Publishers, New York.

Oliver Fiehn. (2001). "Combining genomics, metabolome analysis and biochemical modeling to understand metabolic networks." *Comparative and Functional Genomics.* 2(3): 155–168.

Venugopal, K.R. and Prasad, S.R. (2007). *Mastering C.* Tata McGraw-Hill Publishing Company Limited, New Delhi.

Web References:

http://www.lexicon-biology.com/biology/definition_24.html

http://blast.ncbi.nlm.nih.gov/Blast.cgi

http://www.mbio.ncsu.edu/BioEdit/BioEdit.html

http://www.systemsbiology.nl

http://web-books.com/MoBio/Free/Ch2E1.htm

http://www.chem.qmul.ac.uk/iubmb/enzyme/

http://www.ncbi.nlm.nih.gov/pubmed/

http://medline.cos.com/

http://www.ncbi.nlm.nih.gov/sites/entrez?db=nucest&cmd=search&term

http://genome-www5.stanford.edu/

http://www.genome.jp/kegg/

http://www.mgrc.com.my/

http://www.web-books.com/MoBio/Free/Ch2A2.htm

INDEX

C

V

W

X

Z

www.ingramcontent.com/pod-product-compliance
Lightning Source LLC
Chambersburg PA
CBHW061151240926
51846CB00006B/980
* 9 7 8 8 1 8 0 9 4 0 6 3 7 *